A History of Mathematics

Contents

Chapter 1

History of mathematics

The area of study known as the **history of mathematics** is primarily an investigation into the origin of discoveries in mathematics and, to a lesser extent, an investigation into the mathematical methods and notation of the past.

Before the modern age and the worldwide spread of knowledge, written examples of new mathematical developments have come to light only in a few locales. The most ancient mathematical texts available are *Plimpton 322* (Babylonian mathematics c. 1900 BC),[2] the *Rhind Mathematical Papyrus* (Egyptian mathematics c. 2000-1800 BC)[3] and the *Moscow Mathematical Papyrus* (Egyptian mathematics c. 1890 BC). All of these texts concern the so-called Pythagorean theorem, which seems to be the most ancient and widespread mathematical development after basic arithmetic and geometry.

The study of mathematics as a subject in its own right begins in the 6th century BC with the Pythagoreans, who coined the term "mathematics" from the ancient Greek $\mu\acute{\alpha}\theta\eta\mu\alpha$ (*mathema*), meaning "subject of instruction".[4] Greek mathematics greatly refined the methods (especially through the introduction of deductive reasoning and mathematical rigor in proofs) and expanded the subject matter of mathematics.[5] Chinese mathematics made early contributions, including a place value system.[6][7] The Hindu-Arabic numeral system and the rules for the use of its operations, in use throughout the world today, likely evolved over the course of the first millennium AD in India and were transmitted to the west via Islamic mathematics through the work of Muḥammad ibn Mūsā al-Khwārizmī.[8][9] Islamic mathematics, in turn, developed and expanded the mathematics known to these civilizations.[10] Many Greek and Arabic texts on mathematics were then translated into Latin, which led to further development of mathematics in medieval Europe.

From ancient times through the Middle Ages, bursts of mathematical creativity were often followed by centuries of stagnation. Beginning in Renaissance Italy in the 16th century, new mathematical developments, interacting with new scientific discoveries, were made at an increasing pace that continues through the present day.

1.1 Prehistoric mathematics

The origins of mathematical thought lie in the concepts of number, magnitude, and form.[11] Modern studies of animal cognition have shown that these concepts are not unique to humans. Such concepts would have been part of everyday life in hunter-gatherer societies. The idea of the "number" concept evolving gradually over time is supported by the existence of languages which preserve the distinction between "one", "two", and "many", but not of numbers larger than two.[11]

Prehistoric artifacts discovered in Africa, dated 20,000 years old or more suggest early attempts to quantify time.[12]

Most evidence is against the Lebombo bone (ca. 43,500 yr BC) being a mathematical object, but the Ishango bone, found near the headwaters of the Nile river (northeastern Congo), may be more than 20,000 years old and consists of a series of tally marks carved in three columns running the length of the bone. Common interpretations are that the Ishango bone shows either the earliest known demonstration of sequences of prime numbers[13] or a six-month lunar calendar.[14] In the book *How Mathematics Happened: The First 50,000 Years*, Peter Rudman argues that the development of the concept of prime numbers could only have come about after the concept of division, which he dates to after 10,000 BC, with prime numbers probably not being understood until about 500 BC. He also writes that "no attempt has been made to explain

Ἐπὶ τῆς δοθείσης εὐθείας πεπερασμένης τρίγωνον ἰσόπλευρον συστήσασθαι.

Ἔστω ἡ δοθεῖσα εὐθεῖα πεπερασμένη ἡ ΑΒ.

Δεῖ δὴ ἐπὶ τῆς ΑΒ εὐθείας τρίγωνον ἰσόπλευρον συστήσασθαι.

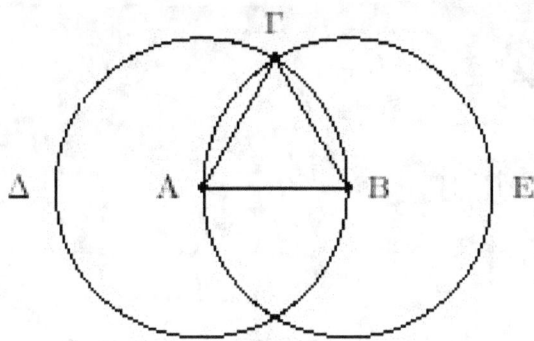

Κέντρῳ μὲν τῷ Α διαστήματι δὲ τῷ ΑΒ κύκλος γεγράφθω ὁ ΒΓΔ, καὶ πάλιν κέντρῳ μὲν τῷ Β διαστήματι δὲ τῷ ΒΑ κύκλος γεγράφθω ὁ ΑΓΕ, καὶ ἀπὸ τοῦ Γ σημείου, καθ᾽ ὃ τέμνουσιν ἀλλήλους οἱ κύκλοι, ἐπὶ τὰ Α, Β σημεῖα ἐπεζεύχθωσαν εὐθεῖαι αἱ ΓΑ, ΓΒ.

Καὶ ἐπεὶ τὸ Α σημεῖον κέντρον ἐστὶ τοῦ ΓΔΒ κύκλου, ἴση ἐστὶν ἡ ΑΓ τῇ ΑΒ· πάλιν, ἐπεὶ τὸ Β σημεῖον κέντρον ἐστὶ τοῦ ΓΑΕ κύκλου, ἴση ἐστὶν ἡ ΒΓ τῇ ΒΑ. ἐδείχθη δὲ καὶ ἡ ΓΑ τῇ ΑΒ ἴση· ἑκατέρα ἄρα τῶν ΓΑ, ΓΒ τῇ ΑΒ ἐστὶν ἴση. τὰ δὲ τῷ αὐτῷ ἴσα καὶ ἀλλήλοις ἐστὶν ἴσα· καὶ ἡ ΓΑ ἄρα τῇ ΓΒ ἐστὶν ἴση· αἱ τρεῖς ἄρα αἱ ΓΑ, ΑΒ, ΒΓ ἴσαι ἀλλήλαις εἰσίν.

Ἰσόπλευρον ἄρα ἐστὶ τὸ ΑΒΓ τρίγωνον, καὶ συνέσταται ἐπὶ τῆς δοθείσης εὐθείας πεπερασμένης τῆς ΑΒ.

[Ἐπὶ τῆς δοθείσης ἄρα εὐθείας πεπερασμένης τρίγωνον ἰσόπλευρον συνέσταται]· ὅπερ ἔδει ποιῆσαι.

A proof from Euclid's Elements, *widely considered the most influential textbook of all time.*[1]

why a tally of something should exhibit multiples of two, prime numbers between 10 and 20, and some numbers that are almost multiples of 10."[15] The Ishango bone, according to scholar Alexander Marshack, may have influenced the later development of mathematics in Egypt as, like some entries on the Ishango bone, Egyptian arithmetic also made use of multiplication by 2; this, however, is disputed.[16]

Predynastic Egyptians of the 5th millennium BC pictorially represented geometric designs. It has been claimed that megalithic monuments in England and Scotland, dating from the 3rd millennium BC, incorporate geometric ideas such as circles, ellipses, and Pythagorean triples in their design.[17]

All of the above are disputed however, and the currently oldest undisputed mathematical usage is in Babylonian and dynastic Egyptian sources.

1.2 Babylonian mathematics

Main article: Babylonian mathematics
See also: Plimpton 322

Babylonian mathematics refers to any mathematics of the peoples of Mesopotamia (modern Iraq) from the days of the

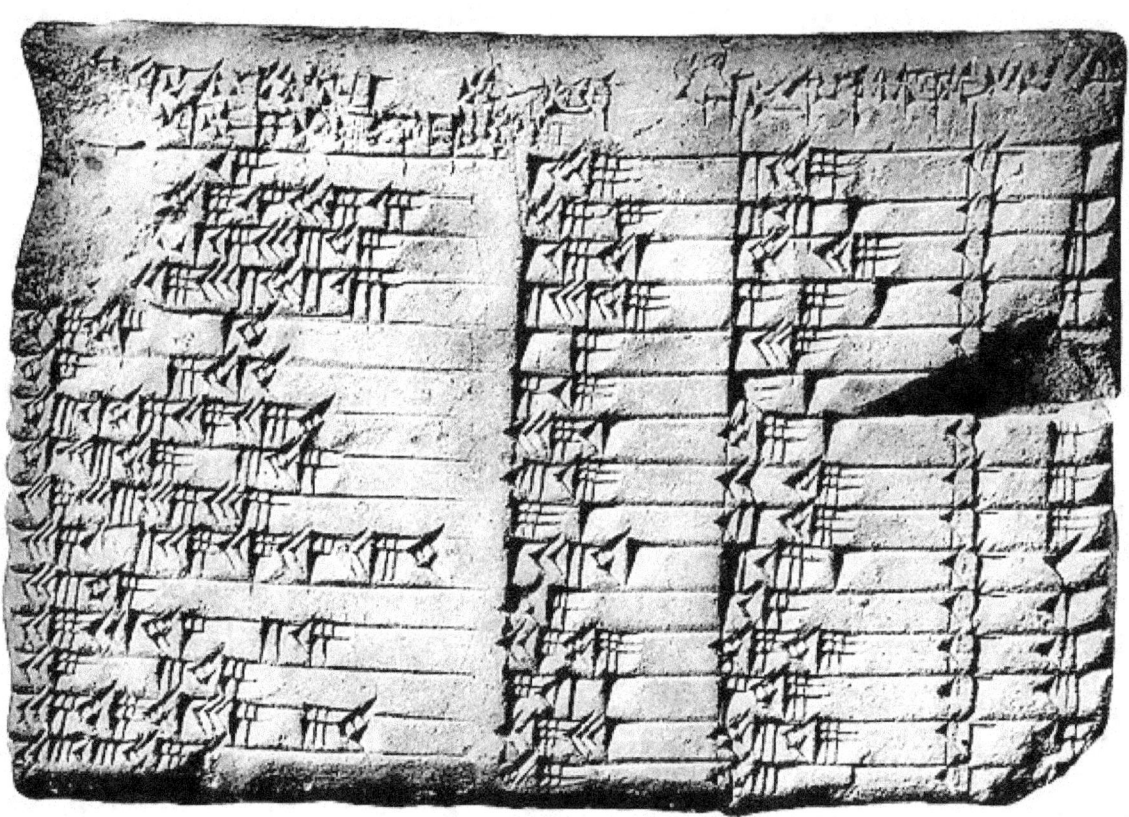

The Babylonian mathematical tablet Plimpton 322, dated to 1800 BC.

early Sumerians through the Hellenistic period almost to the dawn of Christianity.[18] It is named Babylonian mathematics due to the central role of Babylon as a place of study. Later under the Arab Empire, Mesopotamia, especially Baghdad, once again became an important center of study for Islamic mathematics.

In contrast to the sparsity of sources in Egyptian mathematics, our knowledge of Babylonian mathematics is derived from more than 400 clay tablets unearthed since the 1850s.[19] Written in Cuneiform script, tablets were inscribed whilst the clay was moist, and baked hard in an oven or by the heat of the sun. Some of these appear to be graded homework.

The earliest evidence of written mathematics dates back to the ancient Sumerians, who built the earliest civilization in Mesopotamia. They developed a complex system of metrology from 3000 BC. From around 2500 BC onwards, the Sumerians wrote multiplication tables on clay tablets and dealt with geometrical exercises and division problems. The earliest traces of the Babylonian numerals also date back to this period.[20]

The majority of recovered clay tablets date from 1800 to 1600 BC, and cover topics which include fractions, algebra, quadratic and cubic equations, and the calculation of regular reciprocal pairs.[21] The tablets also include multiplication tables and methods for solving linear and quadratic equations. The Babylonian tablet YBC 7289 gives an approximation of $\sqrt{2}$ accurate to five decimal places.

Babylonian mathematics were written using a sexagesimal (base-60) numeral system. From this derives the modern day usage of 60 seconds in a minute, 60 minutes in an hour, and 360 (60 x 6) degrees in a circle, as well as the use of seconds and minutes of arc to denote fractions of a degree. Babylonian advances in mathematics were facilitated by the fact that 60 has many divisors. Also, unlike the Egyptians, Greeks, and Romans, the Babylonians had a true place-value system, where digits written in the left column represented larger values, much as in the decimal system. They lacked, however, an equivalent of the decimal point, and so the place value of a symbol often had to be inferred from the context. On the other hand, this "defect" is equivalent to the modern-day usage of floating point arithmetic; moreover, the use of base 60 means that any reciprocal of an integer which is a multiple of divisors of 60 necessarily has a finite expansion to the base 60. (In decimal arithmetic, only reciprocals of multiples of 2 and 5 have finite decimal expansions.) Accordingly, there is a strong argument that arithmetic Old Babylonian style is considerably more sophisticated than that of current usage.

The interpretation of Plimpton 322 was the source of controversy for many years after its significance in the context of Pythagorean triangles was realized. In historical context, inheritance problems involving equal-area subdivision of triangular and trapezoidal fields (with integer length sides) quickly convert into the need to calculate the square root of 2, or to solve the "Pythagorean equation" in integers.

Rather than considering a square as the sum of two squares, we can equivalently consider a square as a difference of two squares. Let a, b and c be integers that form a Pythagorean Triple: $a^2 + b^2 = c^2$. Then $c^2 - a^2 = b^2$, and using the expansion for the difference of two squares we get $(c-a)(c+a) = b^2$. Dividing by b^2, it becomes the product of two rational numbers giving 1: $(c/b - a/b)(c/b + a/b) = 1$. We require two rational numbers which are reciprocals and which differ by $2(a/b)$. This is easily solved by consulting a table of reciprocal pairs. E.g., $(1/2)$ $(2) = 1$ is a pair of reciprocals which differ by $3/2 = 2(a/b)$ Thus $a/b = 3/4$, giving a=3, b=4 and so c=5.

Solutions of the original equation are thus constructed by choosing a rational number x, from which Pythagorean-triples are $2x$, x^2-1, x^2+1. Other triples are made by scaling these by an integer (the scaling integer being half the difference between the largest and one other side). All Pythagorean triples arise in this way, and the examples provided in Plimpton 322 involve some quite large numbers, by modern standards, such as (4601, 4800, 6649) in decimal notation.

1.3   Egyptian mathematics

Main article: Egyptian mathematics
 Egyptian mathematics refers to mathematics written in the Egyptian language. From the Hellenistic period, Greek

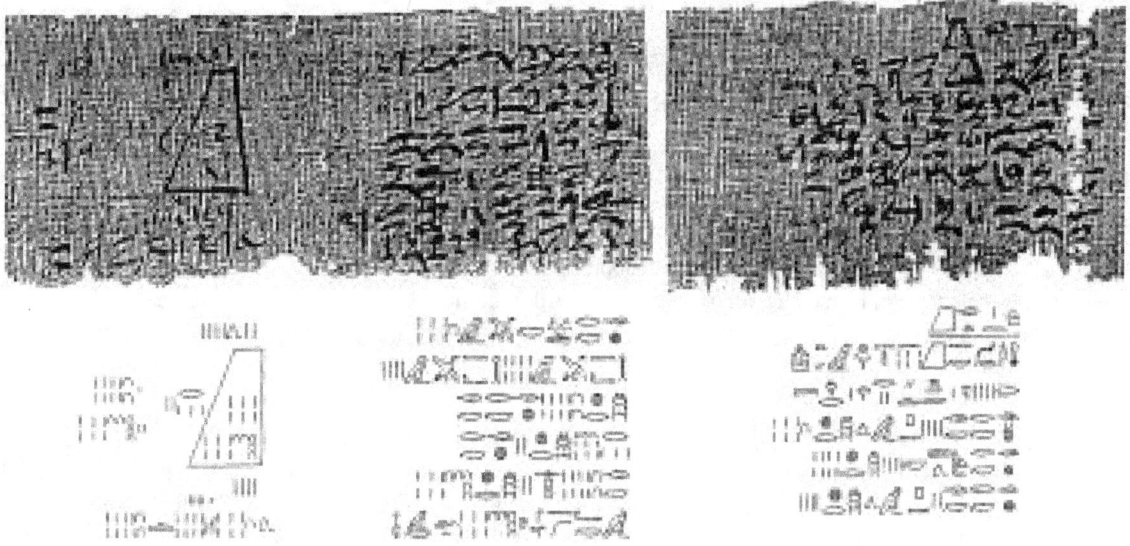

Image of Problem 14 from the Moscow Mathematical Papyrus. The problem includes a diagram indicating the dimensions of the truncated pyramid.

replaced Egyptian as the written language of Egyptian scholars. Mathematical study in Egypt later continued under the Arab Empire as part of Islamic mathematics, when Arabic became the written language of Egyptian scholars.

The most extensive Egyptian mathematical text is the Rhind papyrus (sometimes also called the Ahmes Papyrus after its author), dated to c. 1650 BC but likely a copy of an older document from the Middle Kingdom of about 2000-1800 BC.[22] It is an instruction manual for students in arithmetic and geometry. In addition to giving area formulas and methods for multiplication, division and working with unit fractions, it also contains evidence of other mathematical knowledge,[23] including composite and prime numbers; arithmetic, geometric and harmonic means; and simplistic understandings of both the Sieve of Eratosthenes and perfect number theory (namely, that of the number 6).[24] It also shows how to solve first order linear equations[25] as well as arithmetic and geometric series.[26]

Another significant Egyptian mathematical text is the Moscow papyrus, also from the Middle Kingdom period, dated to c. 1890 BC.[27] It consists of what are today called *word problems* or *story problems*, which were apparently intended as entertainment. One problem is considered to be of particular importance because it gives a method for finding the volume of a frustum: "If you are told: A truncated pyramid of 6 for the vertical height by 4 on the base by 2 on the top. You are to square this 4, result 16. You are to double 4, result 8. You are to square 2, result 4. You are to add the 16, the 8, and the 4, result 28. You are to take one third of 6, result 2. You are to take 28 twice, result 56. See, it is 56. You will find it right."

Finally, the Berlin Papyrus 6619 (c. 1800 BC) shows that ancient Egyptians could solve a second-order algebraic equation.[28]

1.4 Greek mathematics

Main article: Greek mathematics

Greek mathematics refers to the mathematics written in the Greek language from the time of Thales of Miletus (~600 BC) to the closure of the Academy of Athens in 529 AD.[29] Greek mathematicians lived in cities spread over the entire Eastern Mediterranean, from Italy to North Africa, but were united by culture and language. Greek mathematics of the period following Alexander the Great is sometimes called Hellenistic mathematics.[30]

Greek mathematics was much more sophisticated than the mathematics that had been developed by earlier cultures. All surviving records of pre-Greek mathematics show the use of inductive reasoning, that is, repeated observations used to establish rules of thumb. Greek mathematicians, by contrast, used deductive reasoning. The Greeks used logic to derive conclusions from definitions and axioms, and used mathematical rigor to prove them.[31]

Greek mathematics is thought to have begun with Thales of Miletus (c. 624–c.546 BC) and Pythagoras of Samos (c. 582–c. 507 BC). Although the extent of the influence is disputed, they were probably inspired by Egyptian and Babylonian mathematics. According to legend, Pythagoras traveled to Egypt to learn mathematics, geometry, and astronomy from Egyptian priests.

Thales used geometry to solve problems such as calculating the height of pyramids and the distance of ships from the shore. He is credited with the first use of deductive reasoning applied to geometry, by deriving four corollaries to Thales' Theorem. As a result, he has been hailed as the first true mathematician and the first known individual to whom a mathematical discovery has been attributed.[33] Pythagoras established the Pythagorean School, whose doctrine it was that mathematics ruled the universe and whose motto was "All is number".[34] It was the Pythagoreans who coined the term "mathematics", and with whom the study of mathematics for its own sake begins. The Pythagoreans are credited with the first proof of the Pythagorean theorem,[35] though the statement of the theorem has a long history, and with the proof of the existence of irrational numbers.[36][37]

Plato (428/427 BC – 348/347 BC) is important in the history of mathematics for inspiring and guiding others.[38] His Platonic Academy, in Athens, became the mathematical center of the world in the 4th century BC, and it was from this school that the leading mathematicians of the day, such as Eudoxus of Cnidus, came.[39] Plato also discussed the foundations of mathematics, clarified some of the definitions (e.g. that of a line as "breadthless length"), and reorganized the assumptions.[40] The analytic method is ascribed to Plato, while a formula for obtaining Pythagorean triples bears his name.[39]

Eudoxus (408–c.355 BC) developed the method of exhaustion, a precursor of modern integration[41] and a theory of ratios that avoided the problem of incommensurable magnitudes.[42] The former allowed the calculations of areas and volumes of curvilinear figures,[43] while the latter enabled subsequent geometers to make significant advances in geometry. Though he made no specific technical mathematical discoveries, Aristotle (384—c.322 BC) contributed significantly to the development of mathematics by laying the foundations of logic.[44]

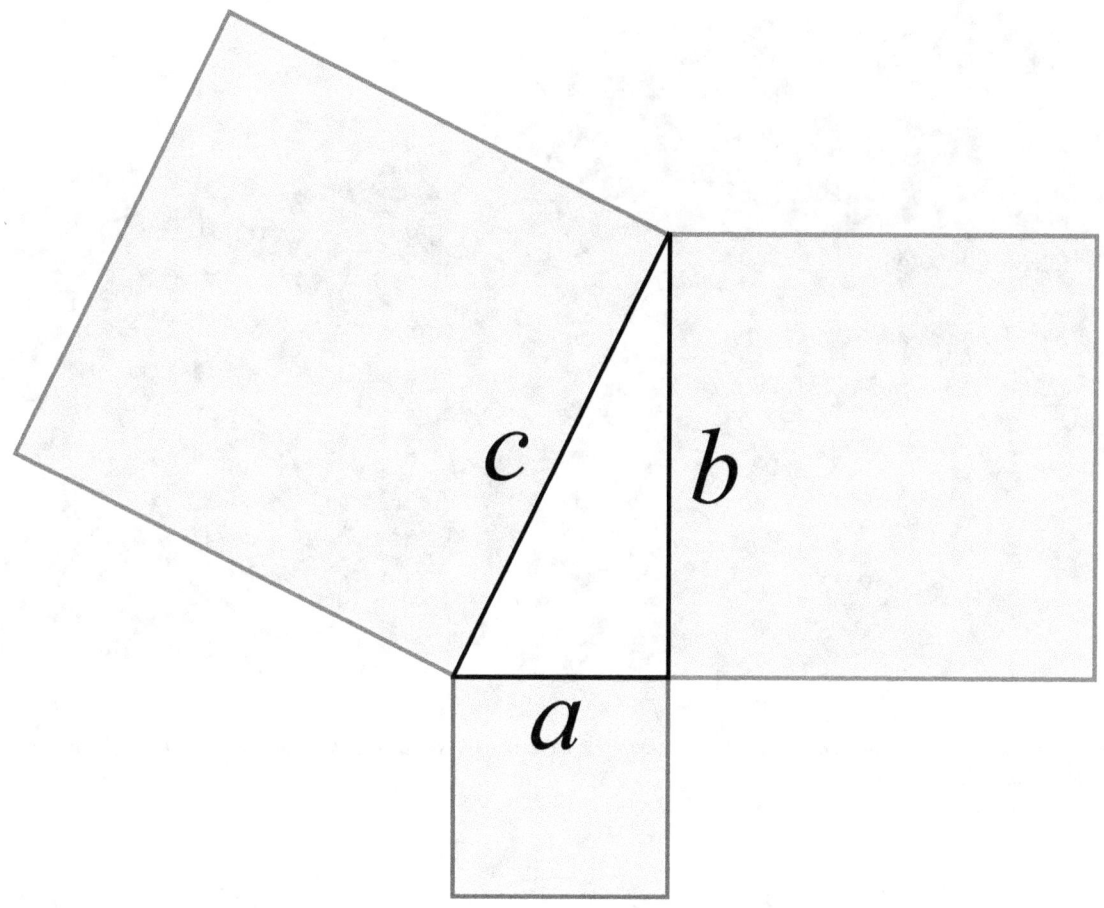

The Pythagorean theorem. The Pythagoreans are generally credited with the first proof of the theorem.

In the 3rd century BC, the premier center of mathematical education and research was the Musaeum of Alexandria.[45] It was there that Euclid (c. 300 BC) taught, and wrote the *Elements*, widely considered the most successful and influential textbook of all time.[1] The *Elements* introduced mathematical rigor through the axiomatic method and is the earliest example of the format still used in mathematics today, that of definition, axiom, theorem, and proof. Although most of the contents of the *Elements* were already known, Euclid arranged them into a single, coherent logical framework.[46] The *Elements* was known to all educated people in the West until the middle of the 20th century and its contents are still taught in geometry classes today.[47] In addition to the familiar theorems of Euclidean geometry, the *Elements* was meant as an introductory textbook to all mathematical subjects of the time, such as number theory, algebra and solid geometry,[46] including proofs that the square root of two is irrational and that there are infinitely many prime numbers. Euclid also wrote extensively on other subjects, such as conic sections, optics, spherical geometry, and mechanics, but only half of his writings survive.[48]

Archimedes (c.287–212 BC) of Syracuse, widely considered the greatest mathematician of antiquity,[49] used the method of exhaustion to calculate the area under the arc of a parabola with the summation of an infinite series, in a manner not too dissimilar from modern calculus.[50] He also showed one could use the method of exhaustion to calculate the value of π with as much precision as desired, and obtained the most accurate value of π then known, $3\frac{10}{71} < \pi < 3\frac{10}{70}$.[51] He also studied the spiral bearing his name, obtained formulas for the volumes of surfaces of revolution (paraboloid, ellipsoid, hyperboloid),[50] and an ingenious system for expressing very large numbers.[52] While he is also known for his contributions to physics and several advanced mechanical devices, Archimedes himself placed far greater value on the products of his thought and general mathematical principles.[53] He regarded as his greatest achievement his finding of the surface area and volume of a sphere, which he obtained by proving these are 2/3 the surface area and volume of a

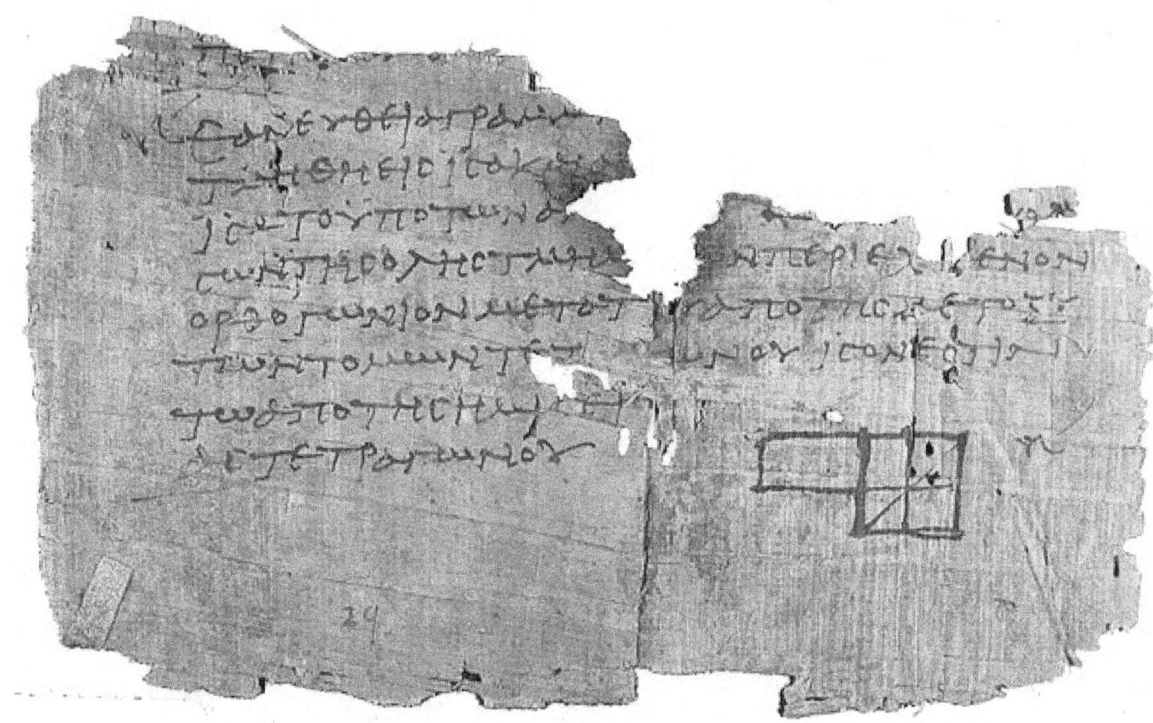

One of the oldest surviving fragments of Euclid's Elements, *found at Oxyrhynchus and dated to circa AD 100. The diagram accompanies Book II, Proposition 5.*[32]

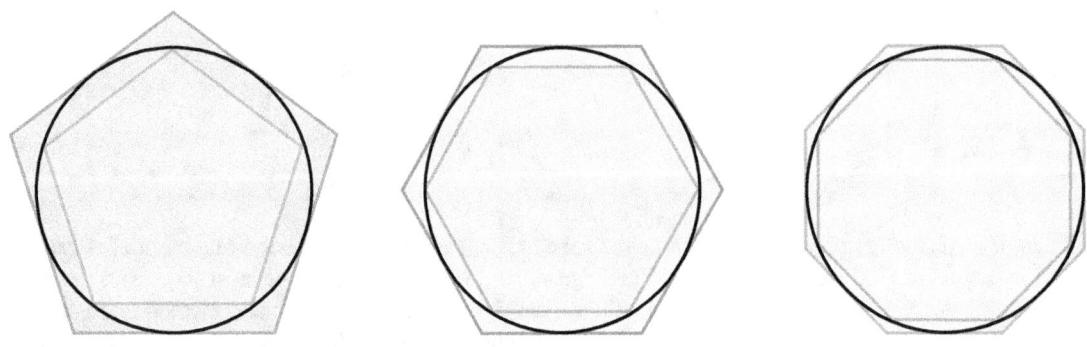

Archimedes used the method of exhaustion to approximate the value of pi.

cylinder circumscribing the sphere.[54]

Apollonius of Perga (c. 262-190 BC) made significant advances to the study of conic sections, showing that one can obtain all three varieties of conic section by varying the angle of the plane that cuts a double-napped cone.[55] He also coined the terminology in use today for conic sections, namely parabola ("place beside" or "comparison"), "ellipse" ("deficiency"), and "hyperbola" ("a throw beyond").[56] His work *Conics* is one of the best known and preserved mathematical works from antiquity, and in it he derives many theorems concerning conic sections that would prove invaluable to later mathematicians and astronomers studying planetary motion, such as Isaac Newton.[57] While neither Apollonius nor any other Greek mathematicians made the leap to coordinate geometry, Apollonius' treatment of curves is in some ways similar to the modern treatment, and some of his work seems to anticipate the development of analytical geometry by Descartes some 1800 years later.[58]

Around the same time, Eratosthenes of Cyrene (c. 276-194 BC) devised the Sieve of Eratosthenes for finding prime

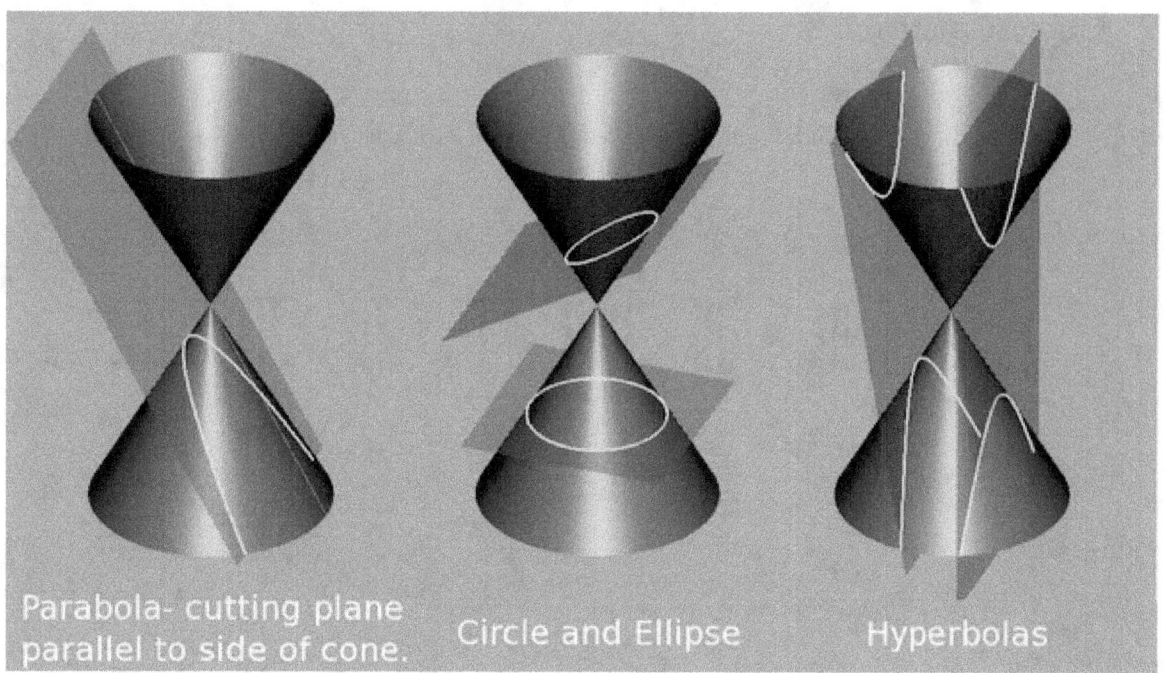

Apollonius of Perga made significant advances in the study of conic sections.

numbers.[59] The 3rd century BC is generally regarded as the "Golden Age" of Greek mathematics, with advances in pure mathematics henceforth in relative decline.[60] Nevertheless, in the centuries that followed significant advances were made in applied mathematics, most notably trigonometry, largely to address the needs of astronomers.[60] Hipparchus of Nicaea (c. 190-120 BC) is considered the founder of trigonometry for compiling the first known trigonometric table, and to him is also due the systematic use of the 360 degree circle.[61] Heron of Alexandria (c. 10–70 AD) is credited with Heron's formula for finding the area of a scalene triangle and with being the first to recognize the possibility of negative numbers possessing square roots.[62] Menelaus of Alexandria (c. 100 AD) pioneered spherical trigonometry through Menelaus' theorem.[63] The most complete and influential trigonometric work of antiquity is the *Almagest* of Ptolemy (c. AD 90-168), a landmark astronomical treatise whose trigonometric tables would be used by astronomers for the next thousand years.[64] Ptolemy is also credited with Ptolemy's theorem for deriving trigonometric quantities, and the most accurate value of π outside of China until the medieval period, 3.1416.[65]

Following a period of stagnation after Ptolemy, the period between 250 and 350 AD is sometimes referred to as the "Silver Age" of Greek mathematics.[66] During this period, Diophantus made significant advances in algebra, particularly indeterminate analysis, which is also known as "Diophantine analysis".[67] The study of Diophantine equations and Diophantine approximations is a significant area of research to this day. His main work was the *Arithmetica*, a collection of 150 algebraic problems dealing with exact solutions to determinate and indeterminate equations.[68] The *Arithmetica* had a significant influence on later mathematicians, such as Pierre de Fermat, who arrived at his famous Last Theorem after trying to generalize a problem he had read in the *Arithmetica* (that of dividing a square into two squares).[69] Diophantus also made significant advances in notation, the *Arithmetica* being the first instance of algebraic symbolism and syncopation.[68]

The first woman mathematician recorded by history was Hypatia of Alexandria (AD 350 - 415). She succeeded her father as Librarian at the Great Library and wrote many works on applied mathematics. Because of a political dispute, the Christian community in Alexandria punished her, presuming she was involved, by stripping her naked and scraping off her skin with clamshells (some say roofing tiles).[70]

1.5 Chinese mathematics

Main article: Chinese mathematics
Early Chinese mathematics is so different from that of other parts of the world that it is reasonable to assume independent

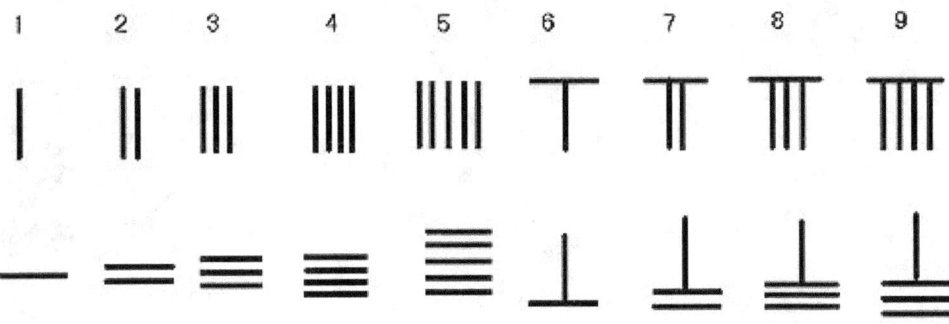

Counting rod numerals

development.[71] The oldest extant mathematical text from China is the *Chou Pei Suan Ching*, variously dated to between 1200 BC and 100 BC, though a date of about 300 BC appears reasonable.[72]

Of particular note is the use in Chinese mathematics of a decimal positional notation system, the so-called "rod numerals" in which distinct ciphers were used for numbers between 1 and 10, and additional ciphers for powers of ten.[73] Thus, the number 123 would be written using the symbol for "1", followed by the symbol for "100", then the symbol for "2" followed by the symbol for "10", followed by the symbol for "3". This was the most advanced number system in the world at the time, apparently in use several centuries before the common era and well before the development of the Indian numeral system.[74] Rod numerals allowed the representation of numbers as large as desired and allowed calculations to be carried out on the *suan pan*, or Chinese abacus. The date of the invention of the *suan pan* is not certain, but the earliest written mention dates from AD 190, in Xu Yue's *Supplementary Notes on the Art of Figures*.

The oldest existent work on geometry in China comes from the philosophical Mohist canon c. 330 BC, compiled by the followers of Mozi (470–390 BC). The *Mo Jing* described various aspects of many fields associated with physical science, and provided a small number of geometrical theorems as well.[75]

In 212 BC, the Emperor Qin Shi Huang (Shi Huang-ti) commanded all books in the Qin Empire other than officially sanctioned ones be burned. This decree was not universally obeyed, but as a consequence of this order little is known about ancient Chinese mathematics before this date. After the book burning of 212 BC, the Han dynasty (202 BC–220 AD) produced works of mathematics which presumably expanded on works that are now lost. The most important of these is *The Nine Chapters on the Mathematical Art*, the full title of which appeared by AD 179, but existed in part under other titles beforehand. It consists of 246 word problems involving agriculture, business, employment of geometry to figure height spans and dimension ratios for Chinese pagoda towers, engineering, surveying, and includes material on right triangles and values of π.[72] It created mathematical proof for the Pythagorean theorem, and a mathematical formula for Gaussian elimination. Liu Hui commented on the work in the 3rd century AD, and gave a value of π accurate to 5 decimal places.[76] Though more of a matter of computational stamina than theoretical insight, in the 5th century AD Zu Chongzhi computed the value of π to seven decimal places, which remained the most accurate value of π for almost the next 1000 years.[76] He also established a method which would later be called Cavalieri's principle to find the volume of a sphere.[77]

The high-water mark of Chinese mathematics occurs in the 13th century (latter part of the Song period), with the development of Chinese algebra. The most important text from that period is the *Precious Mirror of the Four Elements* by Chu Shih-chieh (fl. 1280-1303), dealing with the solution of simultaneous higher order algebraic equations using a method similar to Horner's method.[76] The *Precious Mirror* also contains a diagram of Pascal's triangle with coefficients of binomial expansions through the eighth power, though both appear in Chinese works as early as 1100.[78] The Chinese

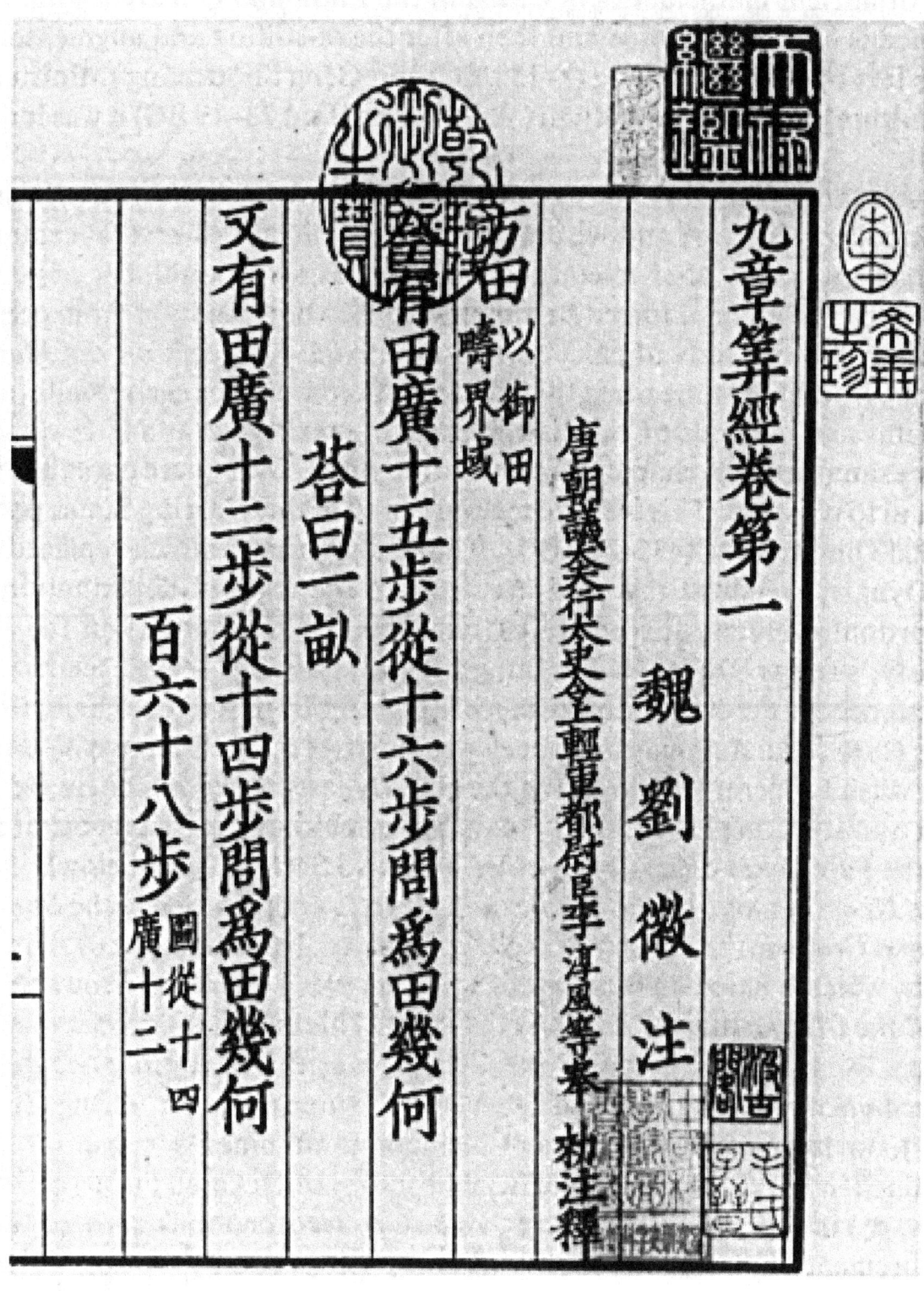

The Nine Chapters on the Mathematical Art, *one of the earliest surviving mathematical texts from China (2nd century AD).*

also made use of the complex combinatorial diagram known as the magic square and magic circles, described in ancient times and perfected by Yang Hui (AD 1238–1298).[78]

Even after European mathematics began to flourish during the Renaissance, European and Chinese mathematics were separate traditions, with significant Chinese mathematical output in decline from the 13th century onwards. Jesuit missionaries such as Matteo Ricci carried mathematical ideas back and forth between the two cultures from the 16th to 18th centuries, though at this point far more mathematical ideas were entering China than leaving.[78]

1.6   Indian mathematics

Main article: Indian mathematics
See also: History of the Hindu-Arabic numeral system
The earliest civilization on the Indian subcontinent is the Indus Valley Civilization that flourished between 2600 and 1900

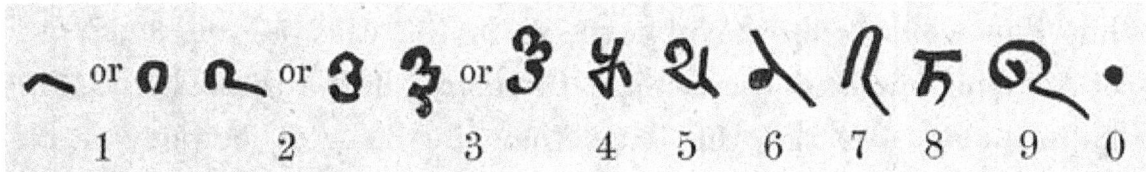

The numerals used in the Bakhshali manuscript, dated between the 2nd century BCE and the 2nd century CE.

1	2	3	4	5	6	7	8	9
—	=	≡	+	h	φ	?	⌐	?

Brahmi numerals (lower row) in India in the 1st century CE

BC in the Indus river basin. Their cities were laid out with geometric regularity, but no known mathematical documents survive from this civilization.[79]

The Hindu-Arabic numerals were invented by mathematicians in India. They were called "Hindu numerals". They were later called "Arabic" numerals by Europeans, because they were introduced in the West by Arab merchants.[80]

Various symbol sets are used to represent numbers in the Hindu–Arabic numeral system, all of which evolved from the Brahmi numerals. Each of the roughly dozen major scripts of India has its own numeral glyphs (as one will note when perusing Unicode character charts). This table shows two examples:

The oldest extant mathematical records from India are the Sulba Sutras (dated variously between the 8th century BC and the 2nd century AD),[81] appendices to religious texts which give simple rules for constructing altars of various shapes, such as squares, rectangles, parallelograms, and others.[82] As with Egypt, the preoccupation with temple functions points to an origin of mathematics in religious ritual.[81] The Sulba Sutras give methods for constructing a circle with approximately the same area as a given square, which imply several different approximations of the value of π.[83][84] In addition, they compute the square root of 2 to several decimal places, list Pythagorean triples, and give a statement of the Pythagorean theorem.[85] All of these results are present in Babylonian mathematics, indicating Mesopotamian influence.[81] It is not known to what extent the Sulba Sutras influenced later Indian mathematicians. As in China, there is a lack of continuity in Indian mathematics; significant advances are separated by long periods of inactivity.[81]

Pāṇini (c. 5th century BC) formulated the rules for Sanskrit grammar.[86] His notation was similar to modern mathematical notation, and used metarules, transformations, and recursion. Pingala (roughly 3rd-1st centuries BC) in his treatise of prosody uses a device corresponding to a binary numeral system.[87][88] His discussion of the combinatorics of meters

European (descended from the West Arabic)	0	1	2	3	4	5	6	7	8	9
Arabic-Indic	.	١	٢	٣	٤	٥	٦	٧	٨	٩
Eastern Arabic-Indic (Persian and Urdu)	.	١	٢	٣	۴	۵	۶	٧	٨	٩
Devanagari (Hindi)	०	१	२	३	४	५	६	७	८	९
Tamil		க	௨	௩	௪	௫	௬	௭	௮	௯

Table of numerals

corresponds to an elementary version of the binomial theorem. Pingala's work also contains the basic ideas of Fibonacci numbers (called *mātrāmeru*).[89]

The next significant mathematical documents from India after the *Sulba Sutras* are the *Siddhantas*, astronomical treatises from the 4th and 5th centuries AD (Gupta period) showing strong Hellenistic influence.[90] They are significant in that they contain the first instance of trigonometric relations based on the half-chord, as is the case in modern trigonometry, rather than the full chord, as was the case in Ptolemaic trigonometry.[91] Through a series of translation errors, the words "sine" and "cosine" derive from the Sanskrit "jiya" and "kojiya".[91]

In the 5th century AD, Aryabhata wrote the *Aryabhatiya*, a slim volume, written in verse, intended to supplement the rules of calculation used in astronomy and mathematical mensuration, though with no feeling for logic or deductive methodology.[92] Though about half of the entries are wrong, it is in the *Aryabhatiya* that the decimal place-value system first appears. Several centuries later, the Muslim mathematician Abu Rayhan Biruni described the *Aryabhatiya* as a "mix of common pebbles and costly crystals".[93]

In the 7th century, Brahmagupta identified the Brahmagupta theorem, Brahmagupta's identity and Brahmagupta's formula, and for the first time, in *Brahma-sphuta-siddhanta*, he lucidly explained the use of zero as both a placeholder and decimal digit, and explained the Hindu-Arabic numeral system.[94] It was from a translation of this Indian text on mathematics (c. 770) that Islamic mathematicians were introduced to this numeral system, which they adapted as Arabic numerals. Islamic scholars carried knowledge of this number system to Europe by the 12th century, and it has now displaced all older number systems throughout the world. In the 10th century, Halayudha's commentary on Pingala's work contains a study of the Fibonacci sequence and Pascal's triangle, and describes the formation of a matrix.

In the 12th century, Bhāskara II[95] lived in southern India and wrote extensively on all then known branches of mathematics. His work contains mathematical objects equivalent or approximately equivalent to infinitesimals, derivatives, the mean value theorem and the derivative of the sine function. To what extent he anticipated the invention of calculus is a controversial subject among historians of mathematics.[96]

In the 14th century, Madhava of Sangamagrama, the founder of the so-called Kerala School of Mathematics, found the Madhava–Leibniz series, and, using 21 terms, computed the value of π as 3.14159265359. Madhava also found the Madhava-Gregory series to determine the arctangent, the Madhava-Newton power series to determine sine and cosine and the Taylor approximation for sine and cosine functions.[97] In the 16th century, Jyesthadeva consolidated many of the Kerala School's developments and theorems in the *Yukti-bhāṣā*.[98] However, the Kerala School did not formulate a systematic theory of differentiation and integration, nor is there any direct evidence of their results being transmitted outside Kerala.[99][100][101][102]

1.7 Islamic mathematics

Main article: Mathematics in medieval Islam
See also: History of the Hindu-Arabic numeral system
 The Islamic Empire established across Persia, the Middle East, Central Asia, North Africa, Iberia, and in parts of India in the 8th century made significant contributions towards mathematics. Although most Islamic texts on mathematics were written in Arabic, most of them were not written by Arabs, since much like the status of Greek in the Hellenistic world, Arabic was used as the written language of non-Arab scholars throughout the Islamic world at the time. Persians contributed to the world of Mathematics alongside Arabs.

In the 9th century, the Persian mathematician Muḥammad ibn Mūsā al-Khwārizmī wrote several important books on the Hindu-Arabic numerals and on methods for solving equations. His book *On the Calculation with Hindu Numerals*, written about 825, along with the work of Al-Kindi, were instrumental in spreading Indian mathematics and Indian numerals to the West. The word *algorithm* is derived from the Latinization of his name, Algoritmi, and the word *algebra* from the title of one of his works, *Al-Kitāb al-mukhtaṣar fī hīsāb al-ğabr wa'l-muqābala* (*The Compendious Book on Calculation by Completion and Balancing*). He gave an exhaustive explanation for the algebraic solution of quadratic equations with positive roots,[103] and he was the first to teach algebra in an elementary form and for its own sake.[104] He also discussed the fundamental method of "reduction" and "balancing", referring to the transposition of subtracted terms to the other side of an equation, that is, the cancellation of like terms on opposite sides of the equation. This is the operation which al-Khwārizmī originally described as *al-jabr*.[105] His algebra was also no longer concerned "with a series of problems to be resolved, but an exposition which starts with primitive terms in which the combinations must give all possible prototypes for equations, which henceforward explicitly constitute the true object of study." He also studied an equation for its own sake and "in a generic manner, insofar as it does not simply emerge in the course of solving a problem, but is specifically called on to define an infinite class of problems."[106]

In Egypt, Abu Kamil extended algebra to the set of irrational numbers, accepting square roots and fourth roots as solutions and coefficients to quadratic equations. He also developed techniques used to solve three non-linear simultaneous equations with three unknown variables. One unique feature of his works was trying to find all the possible solutions to some of his problems, including one where he found 2676 solutions.[107] His works formed an important foundation for the development of algebra and influenced later mathematicians, such as al-Karaji and Fibonacci.

Further developments in algebra were made by Al-Karaji in his treatise *al-Fakhri*, where he extends the methodology to incorporate integer powers and integer roots of unknown quantities. Something close to a proof by mathematical induction appears in a book written by Al-Karaji around 1000 AD, who used it to prove the binomial theorem, Pascal's triangle, and the sum of integral cubes.[108] The historian of mathematics, F. Woepcke,[109] praised Al-Karaji for being "the first who introduced the theory of algebraic calculus." Also in the 10th century, Abul Wafa translated the works of Diophantus into Arabic. Ibn al-Haytham was the first mathematician to derive the formula for the sum of the fourth powers, using a method that is readily generalizable for determining the general formula for the sum of any integral powers. He performed an integration in order to find the volume of a paraboloid, and was able to generalize his result for the integrals of polynomials up to the fourth degree. He thus came close to finding a general formula for the integrals of polynomials, but he was not concerned with any polynomials higher than the fourth degree.[110]

In the late 11th century, Omar Khayyam wrote *Discussions of the Difficulties in Euclid*, a book about what he perceived as flaws in Euclid's *Elements*, especially the parallel postulate. He was also the first to find the general geometric solution to cubic equations. He was also very influential in calendar reform.

In the 13th century, Nasir al-Din Tusi (Nasireddin) made advances in spherical trigonometry. He also wrote influential work on Euclid's parallel postulate. In the 15th century, Ghiyath al-Kashi computed the value of π to the 16th decimal place. Kashi also had an algorithm for calculating nth roots, which was a special case of the methods given many centuries later by Ruffini and Horner.

Other achievements of Muslim mathematicians during this period include the addition of the decimal point notation to the Arabic numerals, the discovery of all the modern trigonometric functions besides the sine, al-Kindi's introduction of cryptanalysis and frequency analysis, the development of analytic geometry by Ibn al-Haytham, the beginning of algebraic geometry by Omar Khayyam and the development of an algebraic notation by al-Qalasādī.[111]

During the time of the Ottoman Empire and Safavid Empire from the 15th century, the development of Islamic mathematics became stagnant.

1.8   Medieval European mathematics

Medieval European interest in mathematics was driven by concerns quite different from those of modern mathematicians. One driving element was the belief that mathematics provided the key to understanding the created order of nature, frequently justified by Plato's *Timaeus* and the biblical passage (in the *Book of Wisdom*) that God had *ordered all things in measure, and number, and weight*.[112]

Boethius provided a place for mathematics in the curriculum in the 6th century when he coined the term *quadrivium* to describe the study of arithmetic, geometry, astronomy, and music. He wrote *De institutione arithmetica*, a free translation from the Greek of Nicomachus's *Introduction to Arithmetic*; *De institutione musica*, also derived from Greek sources; and a series of excerpts from Euclid's *Elements*. His works were theoretical, rather than practical, and were the basis of mathematical study until the recovery of Greek and Arabic mathematical works.[113][114]

In the 12th century, European scholars traveled to Spain and Sicily seeking scientific Arabic texts, including al-Khwārizmī's *The Compendious Book on Calculation by Completion and Balancing*, translated into Latin by Robert of Chester, and the complete text of Euclid's *Elements*, translated in various versions by Adelard of Bath, Herman of Carinthia, and Gerard of Cremona.[115][116]

See also: Latin translations of the 12th century

These new sources sparked a renewal of mathematics. Fibonacci, writing in the *Liber Abaci*, in 1202 and updated in 1254, produced the first significant mathematics in Europe since the time of Eratosthenes, a gap of more than a thousand years. The work introduced Hindu-Arabic numerals to Europe, and discussed many other mathematical problems.

The 14th century saw the development of new mathematical concepts to investigate a wide range of problems.[117] One important contribution was development of mathematics of local motion.

Thomas Bradwardine proposed that speed (V) increases in arithmetic proportion as the ratio of force (F) to resistance (R) increases in geometric proportion. Bradwardine expressed this by a series of specific examples, but although the logarithm had not yet been conceived, we can express his conclusion anachronistically by writing: $V = \log (F/R)$.[118] Bradwardine's analysis is an example of transferring a mathematical technique used by al-Kindi and Arnald of Villanova to quantify the nature of compound medicines to a different physical problem.[119]

One of the 14th-century Oxford Calculators, William Heytesbury, lacking differential calculus and the concept of limits, proposed to measure instantaneous speed "by the path that **would** be described by [a body] **if**... it were moved uniformly at the same degree of speed with which it is moved in that given instant".[120]

Heytesbury and others mathematically determined the distance covered by a body undergoing uniformly accelerated motion (today solved by integration), stating that "a moving body uniformly acquiring or losing that increment [of speed] will traverse in some given time a [distance] completely equal to that which it would traverse if it were moving continuously through the same time with the mean degree [of speed]".[121]

Nicole Oresme at the University of Paris and the Italian Giovanni di Casali independently provided graphical demonstrations of this relationship, asserting that the area under the line depicting the constant acceleration, represented the total distance traveled.[122] In a later mathematical commentary on Euclid's *Elements*, Oresme made a more detailed general analysis in which he demonstrated that a body will acquire in each successive increment of time an increment of any quality that increases as the odd numbers. Since Euclid had demonstrated the sum of the odd numbers are the square numbers, the total quality acquired by the body increases as the square of the time.[123]

1.9   Renaissance mathematics

During the Renaissance, the development of mathematics and of accounting were intertwined.[124] While there is no direct relationship between algebra and accounting, the teaching of the subjects and the books published often intended for the children of merchants who were sent to reckoning schools (in Flanders and Germany) or abacus schools (known as *abbaco* in Italy), where they learned the skills useful for trade and commerce. There is probably no need for algebra in performing bookkeeping operations, but for complex bartering operations or the calculation of compound interest, a basic knowledge of arithmetic was mandatory and knowledge of algebra was very useful.

Luca Pacioli's *Summa de Arithmetica, Geometria, Proportioni et Proportionalità* (Italian: "Review of Arithmetic, Geometry, Ratio and Proportion") was first printed and published in Venice in 1494. It included a 27-page treatise on bookkeeping, *"Particularis de Computis et Scripturis"* (Italian: "Details of Calculation and Recording"). It was written primarily for, and sold mainly to, merchants who used the book as a reference text, as a source of pleasure from the mathematical puzzles it contained, and to aid the education of their sons.[125] In *Summa Arithmetica*, Pacioli introduced symbols for plus and minus for the first time in a printed book, symbols that became standard notation in Italian Renaissance mathematics. *Summa Arithmetica* was also the first known book printed in Italy to contain algebra. It is important to note that Pacioli himself had borrowed much of the work of Piero Della Francesca whom he plagiarized.

In Italy, during the first half of the 16th century, Scipione del Ferro and Niccolò Fontana Tartaglia discovered solutions for cubic equations. Gerolamo Cardano published them in his 1545 book *Ars Magna*, together with a solution for the quartic equations, discovered by his student Lodovico Ferrari. In 1572 Rafael Bombelli published his *L'Algebra* in which he showed how to deal with the imaginary quantities that could appear in Cardano's formula for solving cubic equations.

Simon Stevin's book *De Thiende* ('the art of tenths'), first published in Dutch in 1585, contained the first systematic treatment of decimal notation, which influenced all later work on the real number system.

Driven by the demands of navigation and the growing need for accurate maps of large areas, trigonometry grew to be a major branch of mathematics. Bartholomaeus Pitiscus was the first to use the word, publishing his *Trigonometria* in 1595. Regiomontanus's table of sines and cosines was published in 1533.[126]

During the Renaissance the desire of artists to represent the natural world realistically, together with the rediscovered philosophy of the Greeks, led artists to study mathematics. They were also the engineers and architects of that time, and so had need of mathematics in any case. The art of painting in perspective, and the developments in geometry that involved, were studied intensely.[127]

1.10 Mathematics during the Scientific Revolution

1.10.1 17th century

The 17th century saw an unprecedented explosion of mathematical and scientific ideas across Europe. Galileo observed the moons of Jupiter in orbit about that planet, using a telescope based on a toy imported from Holland. Tycho Brahe had gathered an enormous quantity of mathematical data describing the positions of the planets in the sky. By his position as Brahe's assistant, Johannes Kepler was first exposed to and seriously interacted with the topic of planetary motion. Kepler's calculations were made simpler by the contemporaneous invention of logarithms by John Napier and Jost Bürgi. Kepler succeeded in formulating mathematical laws of planetary motion.[128] The analytic geometry developed by René Descartes (1596–1650) allowed those orbits to be plotted on a graph, in Cartesian coordinates. Simon Stevin (1585) created the basis for modern decimal notation capable of describing all numbers, whether rational or irrational.

Building on earlier work by many predecessors, Isaac Newton discovered the laws of physics explaining Kepler's Laws, and brought together the concepts now known as calculus. Independently, Gottfried Wilhelm Leibniz, who is arguably one of the most important mathematicians of the 17th century, developed calculus and much of the calculus notation still in use today. Science and mathematics had become an international endeavor, which would soon spread over the entire world.[129]

In addition to the application of mathematics to the studies of the heavens, applied mathematics began to expand into new areas, with the correspondence of Pierre de Fermat and Blaise Pascal. Pascal and Fermat set the groundwork for the investigations of probability theory and the corresponding rules of combinatorics in their discussions over a game of gambling. Pascal, with his wager, attempted to use the newly developing probability theory to argue for a life devoted to religion, on the grounds that even if the probability of success was small, the rewards were infinite. In some sense, this foreshadowed the development of utility theory in the 18th–19th century.

1.10.2   18th century

The most influential mathematician of the 18th century was arguably Leonhard Euler. His contributions range from founding the study of graph theory with the Seven Bridges of Königsberg problem to standardizing many modern mathematical terms and notations. For example, he named the square root of minus 1 with the symbol i, and he popularized the use of the Greek letter π to stand for the ratio of a circle's circumference to its diameter. He made numerous contributions to the study of topology, graph theory, calculus, combinatorics, and complex analysis, as evidenced by the multitude of theorems and notations named for him.

Other important European mathematicians of the 18th century included Joseph Louis Lagrange, who did pioneering work in number theory, algebra, differential calculus, and the calculus of variations, and Laplace who, in the age of Napoleon, did important work on the foundations of celestial mechanics and on statistics.

1.11   Modern mathematics

1.11.1   19th century

Throughout the 19th century mathematics became increasingly abstract. In the 19th century lived Carl Friedrich Gauss (1777–1855). Leaving aside his many contributions to science, in pure mathematics he did revolutionary work on functions of complex variables, in geometry, and on the convergence of series. He gave the first satisfactory proofs of the fundamental theorem of algebra and of the quadratic reciprocity law.

This century saw the development of the two forms of non-Euclidean geometry, where the parallel postulate of Euclidean geometry no longer holds. The Russian mathematician Nikolai Ivanovich Lobachevsky and his rival, the Hungarian mathematician János Bolyai, independently defined and studied hyperbolic geometry, where uniqueness of parallels no longer holds. In this geometry the sum of angles in a triangle add up to less than 180°. Elliptic geometry was developed later in the 19th century by the German mathematician Bernhard Riemann; here no parallel can be found and the angles in a triangle add up to more than 180°. Riemann also developed Riemannian geometry, which unifies and vastly generalizes the three types of geometry, and he defined the concept of a manifold, which generalizes the ideas of curves and surfaces.

The 19th century saw the beginning of a great deal of abstract algebra. Hermann Grassmann in Germany gave a first version of vector spaces, William Rowan Hamilton in Ireland developed noncommutative algebra. The British mathematician George Boole devised an algebra that soon evolved into what is now called Boolean algebra, in which the only numbers were 0 and 1. Boolean algebra is the starting point of mathematical logic and has important applications in computer science.

Augustin-Louis Cauchy, Bernhard Riemann, and Karl Weierstrass reformulated the calculus in a more rigorous fashion.

Also, for the first time, the limits of mathematics were explored. Niels Henrik Abel, a Norwegian, and Évariste Galois, a Frenchman, proved that there is no general algebraic method for solving polynomial equations of degree greater than four (Abel–Ruffini theorem). Other 19th-century mathematicians utilized this in their proofs that straightedge and compass alone are not sufficient to trisect an arbitrary angle, to construct the side of a cube twice the volume of a given cube, nor to construct a square equal in area to a given circle. Mathematicians had vainly attempted to solve all of these problems since the time of the ancient Greeks. On the other hand, the limitation of three dimensions in geometry was surpassed in the 19th century through considerations of parameter space and hypercomplex numbers.

Abel and Galois's investigations into the solutions of various polynomial equations laid the groundwork for further developments of group theory, and the associated fields of abstract algebra. In the 20th century physicists and other scientists have seen group theory as the ideal way to study symmetry.

In the later 19th century, Georg Cantor established the first foundations of set theory, which enabled the rigorous treatment of the notion of infinity and has become the common language of nearly all mathematics. Cantor's set theory, and the rise of mathematical logic in the hands of Peano, L. E. J. Brouwer, David Hilbert, Bertrand Russell, and A.N. Whitehead, initiated a long running debate on the foundations of mathematics.

The 19th century saw the founding of a number of national mathematical societies: the London Mathematical Society in 1865, the Société Mathématique de France in 1872, the Circolo Matematico di Palermo in 1884, the Edinburgh

Mathematical Society in 1883, and the American Mathematical Society in 1888. The first international, special-interest society, the Quaternion Society, was formed in 1899, in the context of a vector controversy.

In 1897, Hensel introduced p-adic numbers.

1.11.2   20th century

The 20th century saw mathematics become a major profession. Every year, thousands of new Ph.D.s in mathematics were awarded, and jobs were available in both teaching and industry. An effort to catalogue the areas and applications of mathematics was undertaken in Klein's encyclopedia.

In a 1900 speech to the International Congress of Mathematicians, David Hilbert set out a list of 23 unsolved problems in mathematics. These problems, spanning many areas of mathematics, formed a central focus for much of 20th-century mathematics. Today, 10 have been solved, 7 are partially solved, and 2 are still open. The remaining 4 are too loosely formulated to be stated as solved or not.

Notable historical conjectures were finally proven. In 1976, Wolfgang Haken and Kenneth Appel used a computer to prove the four color theorem. Andrew Wiles, building on the work of others, proved Fermat's Last Theorem in 1995. Paul Cohen and Kurt Gödel proved that the continuum hypothesis is independent of (could neither be proved nor disproved from) the standard axioms of set theory. In 1998 Thomas Callister Hales proved the Kepler conjecture.

Mathematical collaborations of unprecedented size and scope took place. An example is the classification of finite simple groups (also called the "enormous theorem"), whose proof between 1955 and 1983 required 500-odd journal articles by about 100 authors, and filling tens of thousands of pages. A group of French mathematicians, including Jean Dieudonné and André Weil, publishing under the pseudonym "Nicolas Bourbaki", attempted to exposit all of known mathematics as a coherent rigorous whole. The resulting several dozen volumes has had a controversial influence on mathematical education.[130]

Differential geometry came into its own when Einstein used it in general relativity. Entire new areas of mathematics such as mathematical logic, topology, and John von Neumann's game theory changed the kinds of questions that could be answered by mathematical methods. All kinds of structures were abstracted using axioms and given names like metric spaces, topological spaces etc. As mathematicians do, the concept of an abstract structure was itself abstracted and led to category theory. Grothendieck and Serre recast algebraic geometry using sheaf theory. Large advances were made in the qualitative study of dynamical systems that Poincaré had begun in the 1890s. Measure theory was developed in the late 19th and early 20th centuries. Applications of measures include the Lebesgue integral, Kolmogorov's axiomatisation of probability theory, and ergodic theory. Knot theory greatly expanded. Quantum mechanics led to the development of functional analysis. Other new areas include, Laurent Schwartz's distribution theory, fixed point theory, singularity theory and René Thom's catastrophe theory, model theory, and Mandelbrot's fractals. Lie theory with its Lie groups and Lie algebras became one of the major areas of study.

Non-standard analysis, introduced by Abraham Robinson, rehabillitated the infinitesimal approach to calculus, which had fallen into disrepute in favour of the theory of limits, by extending the field of real numbers to the Hyperreal numbers which include infinitesimal and infinite quantities. An even larger number system, the surreal numbers were discovered by John Horton Conway in connection with combinatorial games.

The development and continual improvement of computers, at first mechanical analog machines and then digital electronic machines, allowed industry to deal with larger and larger amounts of data to facilitate mass production and distribution and communication, and new areas of mathematics were developed to deal with this: Alan Turing's computability theory; complexity theory; Derrick Henry Lehmer's use of ENIAC to further number theory and the Lucas-Lehmer test; Claude Shannon's information theory; signal processing; data analysis; optimization and other areas of operations research. In the preceding centuries much mathematical focus was on calculus and continuous functions, but the rise of computing and communication networks led to an increasing importance of discrete concepts and the expansion of combinatorics including graph theory. The speed and data processing abilities of computers also enabled the handling of mathematical problems that were too time-consuming to deal with by pencil and paper calculations, leading to areas such as numerical analysis and symbolic computation. Some of the most important methods and algorithms of the 20th century are: the simplex algorithm, the Fast Fourier Transform, error-correcting codes, the Kalman filter from control theory and the RSA algorithm of public-key cryptography.

At the same time, deep insights were made about the limitations to mathematics. In 1929 and 1930, it was proved the truth or falsity of all statements formulated about the natural numbers plus one of addition and multiplication, was decidable, i.e. could be determined by some algorithm. In 1931, Kurt Gödel found that this was not the case for the natural numbers plus both addition and multiplication; this system, known as Peano arithmetic, was in fact incompletable. (Peano arithmetic is adequate for a good deal of number theory, including the notion of prime number.) A consequence of Gödel's two incompleteness theorems is that in any mathematical system that includes Peano arithmetic (including all of analysis and geometry), truth necessarily outruns proof, i.e. there are true statements that cannot be proved within the system. Hence mathematics cannot be reduced to mathematical logic, and David Hilbert's dream of making all of mathematics complete and consistent needed to be reformulated.

One of the more colorful figures in 20th-century mathematics was Srinivasa Aiyangar Ramanujan (1887–1920), an Indian autodidact who conjectured or proved over 3000 theorems, including properties of highly composite numbers, the partition function and its asymptotics, and mock theta functions. He also made major investigations in the areas of gamma functions, modular forms, divergent series, hypergeometric series and prime number theory.

Paul Erdős published more papers than any other mathematician in history, working with hundreds of collaborators. Mathematicians have a game equivalent to the Kevin Bacon Game, which leads to the Erdős number of a mathematician. This describes the "collaborative distance" between a person and Paul Erdős, as measured by joint authorship of mathematical papers.

Emmy Noether has been described by many as the most important woman in the history of mathematics,[131] she revolutionized the theories of rings, fields, and algebras.

As in most areas of study, the explosion of knowledge in the scientific age has led to specialization: by the end of the century there were hundreds of specialized areas in mathematics and the Mathematics Subject Classification was dozens of pages long.[132] More and more mathematical journals were published and, by the end of the century, the development of the world wide web led to online publishing.

1.11.3   21st century

In 2000, the Clay Mathematics Institute announced the seven Millennium Prize Problems, and in 2003 the Poincaré conjecture was solved by Grigori Perelman (who declined to accept an award, as he was critical of the mathematics establishment).

Most mathematical journals now have online versions as well as print versions, and many online-only journals are launched. There is an increasing drive towards open access publishing, first popularized by the arXiv.

1.12   Future of mathematics

Main article: Future of mathematics

There are many observable trends in mathematics, the most notable being that the subject is growing ever larger, computers are ever more important and powerful, the application of mathematics to bioinformatics is rapidly expanding, the volume of data to be analyzed being produced by science and industry, facilitated by computers, is explosively expanding.

1.13   See also

- History of algebra

- History of calculus

- History of combinatorics

- History of geometry

- History of logic

- History of mathematical notation

- History of number theory

- History of statistics

- History of trigonometry

- History of writing numbers

- Kenneth O. May Prize

- List of important publications in mathematics

- Lists of mathematicians

- List of mathematics history topics

- Timeline of mathematics

1.14   Notes

[1] (Boyer 1991, "Euclid of Alexandria" p. 119)

[2] J. Friberg, "Methods and traditions of Babylonian mathematics. Plimpton 322, Pythagorean triples, and the Babylonian triangle parameter equations", Historia Mathematica, 8, 1981, pp. 277—318.

[3] Neugebauer, Otto (1969) [1957]. *The Exact Sciences in Antiquity* (2 ed.). Dover Publications. ISBN 978-0-486-22332-2. Chap. IV "Egyptian Mathematics and Astronomy", pp. 71–96.

[4] Heath. *A Manual of Greek Mathematics.* p. 5.

[5] Sir Thomas L. Heath, *A Manual of Greek Mathematics*, Dover, 1963, p. 1: "In the case of mathematics, it is the Greek contribution which it is most essential to know, for it was the Greeks who first made mathematics a science."

[6] George Gheverghese Joseph, *The Crest of the Peacock: Non-European Roots of Mathematics*,Penguin Books, London, 1991, pp.140—148

[7] Georges Ifrah, *Universalgeschichte der Zahlen*, Campus, Frankfurt/New York, 1986, pp.428—437

[8] Robert Kaplan, "The Nothing That Is: A Natural History of Zero", Allen Lane/The Penguin Press, London, 1999

[9] "The ingenious method of expressing every possible number using a set of ten symbols (each symbol having a place value and an absolute value) emerged in India. The idea seems so simple nowadays that its significance and profound importance is no longer appreciated. Its simplicity lies in the way it facilitated calculation and placed arithmetic foremost amongst useful inventions. the importance of this invention is more readily appreciated when one considers that it was beyond the two greatest men of Antiquity, Archimedes and Apollonius." - Pierre Simon Laplace http://www-history.mcs.st-and.ac.uk/HistTopics/Indian_numerals.html

[10] A.P. Juschkewitsch, "Geschichte der Mathematik im Mittelalter", Teubner, Leipzig, 1964

[11] (Boyer 1991, "Origins" p. 3)

[12] Mathematics in (central) Africa before colonization

[13] Williams, Scott W. (2005). "The Oldest Mathematical Object is in Swaziland". *Mathematicians of the African Diaspora.* SUNY Buffalo mathematics department. Retrieved 2006-05-06.

[14] Marshack, Alexander (1991): *The Roots of Civilization*, Colonial Hill, Mount Kisco, NY.

[15] Rudman, Peter Strom (2007). *How Mathematics Happened: The First 50,000 Years.* Prometheus Books. p. 64. ISBN 978-1-59102-477-4.

[16] Marshack, A. 1972. The Roots of Civilization: the Cognitive Beginning of Man's First Art, Symbol and Notation. New York: McGraw-Hil

[17] Thom, Alexander, and Archie Thom, 1988, "The metrology and geometry of Megalithic Man", pp 132-151 in C.L.N. Ruggles, ed., *Records in Stone: Papers in memory of Alexander Thom*. Cambridge University Press. ISBN 0-521-33381-4.

[18] (Boyer 1991, "Mesopotamia" p. 24)

[19] (Boyer 1991, "Mesopotamia" p. 25)

[20] Duncan J. Melville (2003). Third Millennium Chronology, *Third Millennium Mathematics*. St. Lawrence University.

[21] Aaboe, Asger (1998). *Episodes from the Early History of Mathematics*. New York: Random House. pp. 30–31.

[22] (Boyer 1991, "Egypt" p. 11)

[23] Egyptian Unit Fractions at MathPages

[24] Egyptian Unit Fractions

[25] Egyptian Papyri

[26] Egyptian Algebra - Mathematicians of the African Diaspora

[27] (Boyer 1991, "Egypt" p. 19)

[28] Egyptian Mathematical Papyri - Mathematicians of the African Diaspora

[29] Howard Eves, *An Introduction to the History of Mathematics*, Saunders, 1990, ISBN 0-03-029558-0

[30] (Boyer 1991, "The Age of Plato and Aristotle" p. 99)

[31] Martin Bernal, "Animadversions on the Origins of Western Science", pp. 72–83 in Michael H. Shank, ed., *The Scientific Enterprise in Antiquity and the Middle Ages*, (Chicago: University of Chicago Press) 2000, p. 75.

[32] Bill Casselman. "One of the Oldest Extant Diagrams from Euclid". University of British Columbia. Retrieved 2008-09-26.

[33] (Boyer 1991, "Ionia and the Pythagoreans" p. 43)

[34] (Boyer 1991, "Ionia and the Pythagoreans" p. 49)

[35] Eves, Howard, An Introduction to the History of Mathematics, Saunders, 1990, ISBN 0-03-029558-0.

[36] Kurt Von Fritz (1945). "The Discovery of Incommensurability by Hippasus of Metapontum". *The Annals of Mathematics*.

[37] James R. Choike (1980). "The Pentagram and the Discovery of an Irrational Number". *The Two-Year College Mathematics Journal*.

[38] (Boyer 1991, "The Age of Plato and Aristotle" p. 86)

[39] (Boyer 1991, "The Age of Plato and Aristotle" p. 88)

[40] (Boyer 1991, "The Age of Plato and Aristotle" p. 87)

[41] (Boyer 1991, "The Age of Plato and Aristotle" p. 92)

[42] (Boyer 1991, "The Age of Plato and Aristotle" p. 93)

[43] (Boyer 1991, "The Age of Plato and Aristotle" p. 91)

[44] (Boyer 1991, "The Age of Plato and Aristotle" p. 98)

[45] (Boyer 1991, "Euclid of Alexandria" p. 100)

[46] (Boyer 1991, "Euclid of Alexandria" p. 104)

[47] Howard Eves, *An Introduction to the History of Mathematics*, Saunders, 1990, ISBN 0-03-029558-0 p. 141: "No work, except The Bible, has been more widely used...."

[48] (Boyer 1991, "Euclid of Alexandria" p. 102)

[49] (Boyer 1991, "Archimedes of Syracuse" p. 120)

[50] (Boyer 1991, "Archimedes of Syracuse" p. 130)

[51] (Boyer 1991, "Archimedes of Syracuse" p. 126)

[52] (Boyer 1991, "Archimedes of Syracuse" p. 125)

[53] (Boyer 1991, "Archimedes of Syracuse" p. 121)

[54] (Boyer 1991, "Archimedes of Syracuse" p. 137)

[55] (Boyer 1991, "Apollonius of Perga" p. 145)

[56] (Boyer 1991, "Apollonius of Perga" p. 146)

[57] (Boyer 1991, "Apollonius of Perga" p. 152)

[58] (Boyer 1991, "Apollonius of Perga" p. 156)

[59] (Boyer 1991, "Greek Trigonometry and Mensuration" p. 161)

[60] (Boyer 1991, "Greek Trigonometry and Mensuration" p. 175)

[61] (Boyer 1991, "Greek Trigonometry and Mensuration" p. 162)

[62] S.C. Roy. *Complex numbers: lattice simulation and zeta function applications*, p. 1 . Harwood Publishing, 2007, 131 pages. ISBN 1-904275-25-7

[63] (Boyer 1991, "Greek Trigonometry and Mensuration" p. 163)

[64] (Boyer 1991, "Greek Trigonometry and Mensuration" p. 164)

[65] (Boyer 1991, "Greek Trigonometry and Mensuration" p. 168)

[66] (Boyer 1991, "Revival and Decline of Greek Mathematics" p. 178)

[67] (Boyer 1991, "Revival and Decline of Greek Mathematics" p. 180)

[68] (Boyer 1991, "Revival and Decline of Greek Mathematics" p. 181)

[69] (Boyer 1991, "Revival and Decline of Greek Mathematics" p. 183)

[70] Ecclesiastical History,Bk VI: Chap. 15

[71] (Boyer 1991, "China and India" p. 201)

[72] (Boyer 1991, "China and India" p. 196)

[73] Katz 2007, pp. 194–199

[74] (Boyer 1991, "China and India" p. 198)

[75] Needham, Joseph (1986). "Science and Civilisation in China". 3, *Mathematics and the Sciences of the Heavens and the Earth*. Taipei: Caves Books Ltd.

[76] (Boyer 1991, "China and India" p. 202)

[77] Zill, Dennis G.; Wright, Scott; Wright, Warren S. (2009). *Calculus: Early Transcendentals* (3 ed.). Jones & Bartlett Learning. p. xxvii. ISBN 0-7637-5995-3., Extract of page 27

[78] (Boyer 1991, "China and India" p. 205)

[79] (Boyer 1991, "China and India" p. 206)

[80] Rowlett, Russ (2004-07-04), *Roman and "Arabic" Numerals*, University of North Carolina at Chapel Hill, retrieved 2009-06-22

[81] (Boyer 1991, "China and India" p. 207)

[82] T. K. Puttaswamy, "The Accomplishments of Ancient Indian Mathematicians", pp. 411–2, in Selin, Helaine; D'Ambrosio, Ubiratan, eds. (2000). *Mathematics Across Cultures: The History of Non-western Mathematics*. Springer. ISBN 1-4020-0260-2.

[83] R. P. Kulkarni, "The Value of π known to Śulbasūtras", *Indian Journal for the History of Science*, 13 **1** (1978): 32-41

[84] J.J. Connor, E.F. Robertson. *The Indian Sulba Sutras* Univ. of St. Andrew, Scotland The values for π are 4 x $(13/15)^2$ (3.0044...), 25/8 (3.125), 900/289 (3.11418685...), 1156/361 (3.202216...), and 339/108 (3.1389).

[85] J.J. Connor, E.F. Robertson. *The Indian Sulba Sutras* Univ. of St. Andrew, Scotland

[86] Bronkhorst, Johannes (2001). "Panini and Euclid: Reflections on Indian Geometry". *Journal of Indian Philosophy,* (Springer Netherlands) **29** (1–2): 43–80. doi:10.1023/A:1017506118885.

[87] Sanchez, Julio; Canton, Maria P. (2007). *Microcontroller programming : the microchip PIC*. Boca Raton, Florida: CRC Press. p. 37. ISBN 0-8493-7189-9.

[88] W. S. Anglin and J. Lambek, *The Heritage of Thales*, Springer, 1995, ISBN 0-387-94544-X

[89] Rachel W. Hall. Math for poets and drummers. *Math Horizons* **15** (2008) 10-11.

[90] (Boyer 1991, "China and India" p. 208)

[91] (Boyer 1991, "China and India" p. 209)

[92] (Boyer 1991, "China and India" p. 210)

[93] (Boyer 1991, "China and India" p. 211)

[94] Boyer (1991). "The Arabic Hegemony". *History of Mathematics*. p. 226. By 766 we learn that an astronomical-mathematical work, known to the Arabs as the *Sindhind*, was brought to Baghdad from India. It is generally thought that this was the *Brahmasphuta Siddhanta*, although it may have been the *Surya Siddhanata*. A few years later, perhaps about 775, this *Siddhanata* was translated into Arabic, and it was not long afterwards (ca. 780) that Ptolemy's astrological *Tetrabiblos* was translated into Arabic from the Greek.

[95] Plofker 2009 182-207

[96] Plofker 2009 pp 197 - 198; George Gheverghese Joseph, *The Crest of the Peacock: Non-European Roots of Mathematics*, Penguin Books, London, 1991 pp 298 - 300; Takao Hayashi, *Indian Mathematics*, pp 118 - 130 in *Companion History of the History and Philosophy of the Mathematical Sciences*, ed. I. Grattan.Guinness, Johns Hopkins University Press, Baltimore and London, 1994, p 126

[97] Plofker 2009 pp 217 - 253

[98] P. P. Divakaran, *The first textbook of calculus: Yukti-bhāṣā*, Journal of Indian Philosophy 35, 2007, pp 417 - 433.

[99] Pingree, David (December 1992), "Hellenophilia versus the History of Science", *Isis* **83** (4): 562, JSTOR 234257, One example I can give you relates to the Indian Mādhava's demonstration, in about 1400 A.D., of the infinite power series of trigonometrical functions using geometrical and algebraic arguments. When this was first described in English by Charles Whish, in the 1830s, it was heralded as the Indians' discovery of the calculus. This claim and Mādhava's achievements were ignored by Western historians, presumably at first because they could not admit that an Indian discovered the calculus, but later because no one read anymore the *Transactions of the Royal Asiatic Society*, in which Whish's article was published. The matter resurfaced in the 1950s, and now we have the Sanskrit texts properly edited, and we understand the clever way that Mādhava derived the series *without* the calculus; but many historians still find it impossible to conceive of the problem and its solution in terms of anything other than the calculus and proclaim that the calculus is what Mādhava found. In this case the elegance and brilliance of Mādhava's mathematics are being distorted as they are buried under the current mathematical solution to a problem to which he discovered an alternate and powerful solution.

[100] Bressoud, David (2002), "Was Calculus Invented in India?", *College Mathematics Journal* **33** (1): 2–13, doi:10.2307/1558972

[101] Plofker, Kim (November 2001), "The 'Error' in the Indian "Taylor Series Approximation" to the Sine", *Historia Mathematica* **28** (4): 293, doi:10.1006/hmat.2001.2331, It is not unusual to encounter in discussions of Indian mathematics such assertions as that 'the concept of differentiation was understood [in India] from the time of Manjula (... in the 10th century)' [Joseph 1991, 300], or that 'we may consider Madhava to have been the founder of mathematical analysis' (Joseph 1991, 293), or that Bhaskara II may claim to be 'the precursor of Newton and Leibniz in the discovery of the principle of the differential calculus' (Bag 1979, 294).... The points of resemblance, particularly between early European calculus and the Keralese work on power series, have even inspired suggestions of a possible transmission of mathematical ideas from the Malabar coast in or after the 15th century to the Latin scholarly world (e.g., in (Bag 1979, 285)).... It should be borne in mind, however, that such an emphasis on the similarity of Sanskrit (or Malayalam) and Latin mathematics risks diminishing our ability fully to see and comprehend the former. To speak of the Indian 'discovery of the principle of the differential calculus' somewhat obscures the fact that Indian techniques for expressing changes in the Sine by means of the Cosine or vice versa, as in the examples we have seen, remained within that specific trigonometric context. The differential 'principle' was not generalized to arbitrary functions—in fact, the explicit notion of an arbitrary function, not to mention that of its derivative or an algorithm for taking the derivative, is irrelevant here

[102] Katz, Victor J. (June 1995), "Ideas of Calculus in Islam and India" (PDF), *Mathematics Magazine* **68** (3): 163–174, JSTOR 2691411

[103] (Boyer 1991, "The Arabic Hegemony" p. 230) "The six cases of equations given above exhaust all possibilities for linear and quadratic equations having positive root. So systematic and exhaustive was al-Khwārizmī's exposition that his readers must have had little difficulty in mastering the solutions."

[104] Gandz and Saloman (1936), *The sources of Khwarizmi's algebra*, Osiris i, pp. 263–77: "In a sense, Khwarizmi is more entitled to be called "the father of algebra" than Diophantus because Khwarizmi is the first to teach algebra in an elementary form and for its own sake, Diophantus is primarily concerned with the theory of numbers".

[105] (Boyer 1991, "The Arabic Hegemony" p. 229) "It is not certain just what the terms *al-jabr* and *muqabalah* mean, but the usual interpretation is similar to that implied in the translation above. The word *al-jabr* presumably meant something like "restoration" or "completion" and seems to refer to the transposition of subtracted terms to the other side of an equation; the word *muqabalah* is said to refer to "reduction" or "balancing" - that is, the cancellation of like terms on opposite sides of the equation."

[106] Rashed, R.; Armstrong, Angela (1994). *The Development of Arabic Mathematics*. Springer. pp. 11–12. ISBN 0-7923-2565-6. OCLC 29181926.

[107] Sesiano, Jacques (1997-07-31). "Abū Kāmil". *Encyclopaedia of the history of science, technology, and medicine in non-western cultures*. Springer. pp. 4–5.

[108] Victor J. Katz (1998). *History of Mathematics: An Introduction*, pp. 255–59. Addison-Wesley. ISBN 0-321-01618-1.

[109] F. Woepcke (1853). *Extrait du Fakhri, traité d'Algèbre par Abou Bekr Mohammed Ben Alhacan Alkarkhi*. Paris.

[110] Victor J. Katz (1995), "Ideas of Calculus in Islam and India", *Mathematics Magazine* **68** (3): 163–74.

[111] O'Connor, John J.; Robertson, Edmund F., "Abu'l Hasan ibn Ali al Qalasadi", *MacTutor History of Mathematics archive*, University of St Andrews.

[112] *Wisdom*, 11:21

[113] Caldwell, John (1981) "The *De Institutione Arithmetica* and the *De Institutione Musica*", pp. 135–54 in Margaret Gibson, ed., *Boethius: His Life, Thought, and Influence,* (Oxford: Basil Blackwell).

[114] Folkerts, Menso, *"Boethius" Geometrie II*, (Wiesbaden: Franz Steiner Verlag, 1970).

[115] Marie-Thérèse d'Alverny, "Translations and Translators", pp. 421–62 in Robert L. Benson and Giles Constable, *Renaissance and Renewal in the Twelfth Century*, (Cambridge: Harvard University Press, 1982).

[116] Guy Beaujouan, "The Transformation of the Quadrivium", pp. 463–87 in Robert L. Benson and Giles Constable, *Renaissance and Renewal in the Twelfth Century*, (Cambridge: Harvard University Press, 1982).

[117] Grant, Edward and John E. Murdoch (1987), eds., *Mathematics and Its Applications to Science and Natural Philosophy in the Middle Ages,* (Cambridge: Cambridge University Press) ISBN 0-521-32260-X.

[118] Clagett, Marshall (1961) *The Science of Mechanics in the Middle Ages,* (Madison: University of Wisconsin Press), pp. 421–40.

[119] Murdoch, John E. (1969) "*Mathesis in Philosophiam Scholasticam Introducta:* The Rise and Development of the Application of Mathematics in Fourteenth Century Philosophy and Theology", in *Arts libéraux et philosophie au Moyen Âge* (Montréal: Institut d'Études Médiévales), at pp. 224–27.

[120] Clagett, Marshall (1961) *The Science of Mechanics in the Middle Ages,* (Madison: University of Wisconsin Press), pp. 210, 214–15, 236.

[121] Clagett, Marshall (1961) *The Science of Mechanics in the Middle Ages,* (Madison: University of Wisconsin Press), p. 284.

[122] Clagett, Marshall (1961) *The Science of Mechanics in the Middle Ages,* (Madison: University of Wisconsin Press), pp. 332–45, 382–91.

[123] Nicole Oresme, "Questions on the *Geometry* of Euclid" Q. 14, pp. 560–65, in Marshall Clagett, ed., *Nicole Oresme and the Medieval Geometry of Qualities and Motions,* (Madison: University of Wisconsin Press, 1968).

[124] Heeffer, Albrecht: *On the curious historical coincidence of algebra and double-entry bookkeeping*, Foundations of the Formal Sciences, Ghent University, November 2009, p.7

[125] Alan Sangster, Greg Stoner & Patricia McCarthy: "The market for Luca Pacioli's Summa Arithmetica" (Accounting, Business & Financial History Conference, Cardiff, September 2007) p. 1–2

[126] Grattan-Guinness, Ivor (1997). *The Rainbow of Mathematics: A History of the Mathematical Sciences.* W.W. Norton. ISBN 0-393-32030-8.

[127] Kline, Morris (1953). *Mathematics in Western Culture.* Great Britain: Pelican. pp. 150–151.

[128] Struik, Dirk (1987). *A Concise History of Mathematics* (3rd. ed.). Courier Dover Publications. p. 89. ISBN 9780486602554.

[129] Eves, Howard, An Introduction to the History of Mathematics, Saunders, 1990, ISBN 0-03-029558-0, p. 379, "...the concepts of calculus...(are) so far reaching and have exercised such an impact on the modern world that it is perhaps correct to say that without some knowledge of them a person today can scarcely claim to be well educated."

[130] Maurice Mashaal, 2006. *Bourbaki: A Secret Society of Mathematicians.* American Mathematical Society. ISBN 0-8218-3967-5, ISBN 978-0-8218-3967-6.

[131] Alexandrov, Pavel S. (1981), "In Memory of Emmy Noether", in Brewer, James W; Smith, Martha K, *Emmy Noether: A Tribute to Her Life and Work*, New York: Marcel Dekker, pp. 99–111, ISBN 0-8247-1550-0.

[132] Mathematics Subject Classification 2000

1.15   References

- Boyer, C.B. (1989), *A History of Mathematics* (2nd ed.), New York: Wiley, ISBN 0-471-09763-2 (1991 pbk ed. ISBN 0-471-54397-7)

- Eves, Howard (1990), *An Introduction to the History of Mathematics*, Saunders, ISBN 0-03-029558-0

- Katz, Victor J. (1998), *A History of Mathematics: An Introduction* (2nd ed.), Addison-Wesley, ISBN 978-0-321-01618-8

- Katz, Victor J., ed. (2007), *The Mathematics of Egypt, Mesopotamia, China, India, and Islam: A Sourcebook*, Princeton, NJ: Princeton University Press, ISBN 0-691-11485-4

- Plofker, Kim (2009), *Mathematics in India: 500 BCE–1800 CE*, Princeton, NJ: Princeton University Press, ISBN 0-691-12067-6

1.16 Further reading

1.16.1 General

- Aaboe, Asger (1964). *Episodes from the Early History of Mathematics*. New York: Random House.

- Bell, E. T. (1937). *Men of Mathematics*. Simon and Schuster.

- Burton, David M. *The History of Mathematics: An Introduction*. McGraw Hill: 1997.

- Grattan-Guinness, Ivor (2003). *Companion Encyclopedia of the History and Philosophy of the Mathematical Sciences*. The Johns Hopkins University Press. ISBN 0-8018-7397-5.

- Kline, Morris. *Mathematical Thought from Ancient to Modern Times*.

- Struik, D. J. (1987). *A Concise History of Mathematics*, fourth revised edition. Dover Publications, New York.

1.16.2 Books on a specific period

- Gillings, Richard J. (1972). *Mathematics in the Time of the Pharaohs*. Cambridge, MA: MIT Press.

- Heath, Sir Thomas (1981). *A History of Greek Mathematics*. Dover. ISBN 0-486-24073-8.

- Maier, Annaliese (1982), *At the Threshold of Exact Science: Selected Writings of Annaliese Maier on Late Medieval Natural Philosophy*, edited by Steven Sargent, Philadelphia: University of Pennsylvania Press.

- van der Waerden, B. L., *Geometry and Algebra in Ancient Civilizations*, Springer, 1983, ISBN 0-387-12159-5.

1.16.3 Books on a specific topic

- Hoffman, Paul, *The Man Who Loved Only Numbers: The Story of Paul Erdős and the Search for Mathematical Truth*. New York: Hyperion, 1998 ISBN 0-7868-6362-5.

- Menninger, Karl W. (1969). *Number Words and Number Symbols: A Cultural History of Numbers*. MIT Press. ISBN 0-262-13040-8.

- Stigler, Stephen M. (1990). *The History of Statistics: The Measurement of Uncertainty before 1900*. Belknap Press. ISBN 0-674-40341-X.

1.17 External links

1.17.1 Documentaries

- BBC (2008). *The Story of Maths*.

- MacTutor History of Mathematics archive (John J. O'Connor and Edmund F. Robertson; University of St Andrews, Scotland). An award-winning website containing detailed biographies on many historical and contemporary mathematicians, as well as information on notable curves and various topics in the history of mathematics.

- History of Mathematics Home Page (David E. Joyce; Clark University). Articles on various topics in the history of mathematics with an extensive bibliography.

- The History of Mathematics (David R. Wilkins; Trinity College, Dublin). Collections of material on the mathematics between the 17th and 19th century.

- History of Mathematics (Simon Fraser University).

- Earliest Known Uses of Some of the Words of Mathematics (Jeff Miller). Contains information on the earliest known uses of terms used in mathematics.

- Earliest Uses of Various Mathematical Symbols (Jeff Miller). Contains information on the history of mathematical notations.

- Mathematical Words: Origins and Sources (John Aldrich, University of Southampton) Discusses the origins of the modern mathematical word stock.

- Biographies of Women Mathematicians (Larry Riddle; Agnes Scott College).

- Mathematicians of the African Diaspora (Scott W. Williams; University at Buffalo).

- Fred Rickey's History of Mathematics Page

- A Bibliography of Collected Works and Correspondence of Mathematicians archive dated 2007/3/17 (Steven W. Rockey; Cornell University Library).

1.17.2   Organizations

- International Commission for the History of Mathematics

1.17.3   Journals

- *Historia Mathematica*

- Convergence, the Mathematical Association of America's online Math History Magazine

1.17.4   Directories

- Links to Web Sites on the History of Mathematics (The British Society for the History of Mathematics)

- History of Mathematics Math Archives (University of Tennessee, Knoxville)

- History/Biography The Math Forum (Drexel University)

- History of Mathematics (Courtright Memorial Library).

- History of Mathematics Web Sites (David Calvis; Baldwin-Wallace College)

- History of mathematics at DMOZ

- Historia de las Matemáticas (Universidad de La La guna)

- História da Matemática (Universidade de Coimbra)

- Using History in Math Class

- Mathematical Resources: History of Mathematics (Bruno Kevius)

- History of Mathematics (Roberta Tucci)

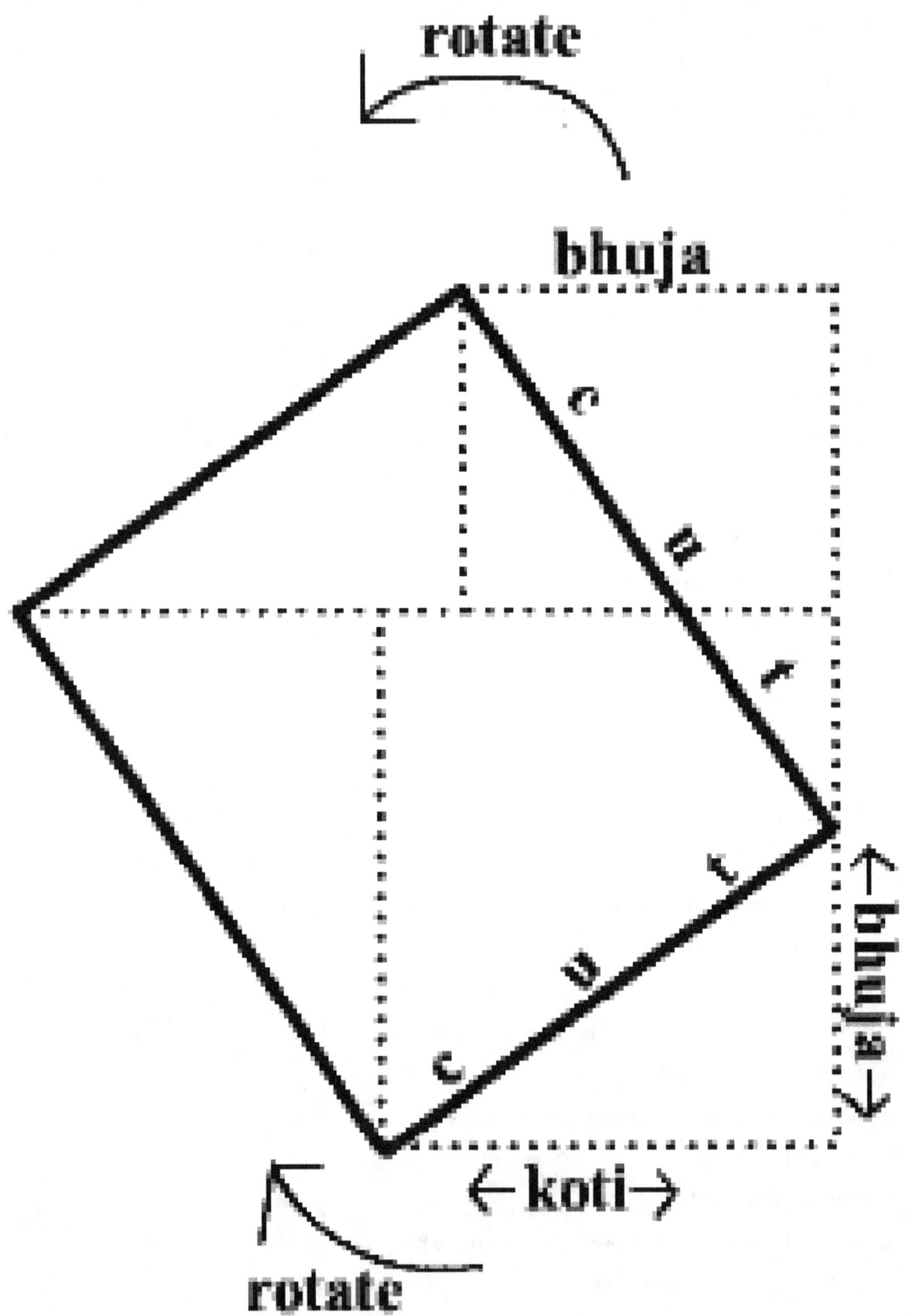

Explanation of the sine rule in Yuktibhāṣā

Portrait of Luca Pacioli, *a painting traditionally attributed to Jacopo de' Barbari, 1495, (Museo di Capodimonte).*

Gottfried Wilhelm Leibniz.

Leonhard Euler by Emanuel Handmann.

Carl Friedrich Gauss.

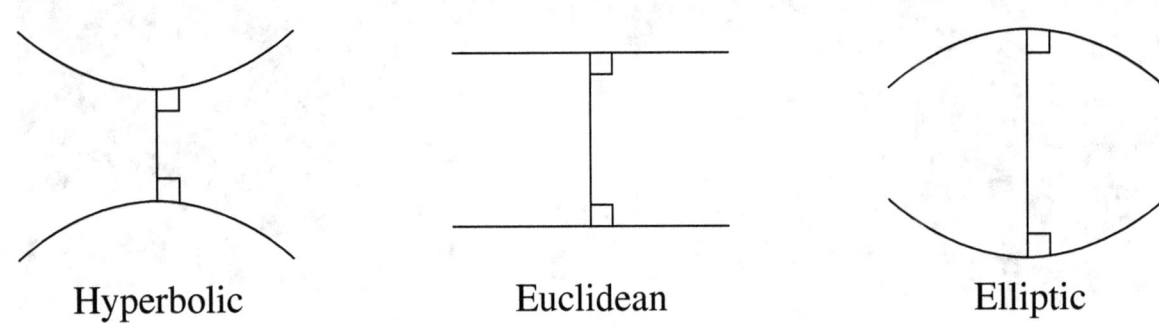

Hyperbolic   Euclidean   Elliptic

Behavior of lines with a common perpendicular in each of the three types of geometry

A map illustrating the Four Color Theorem

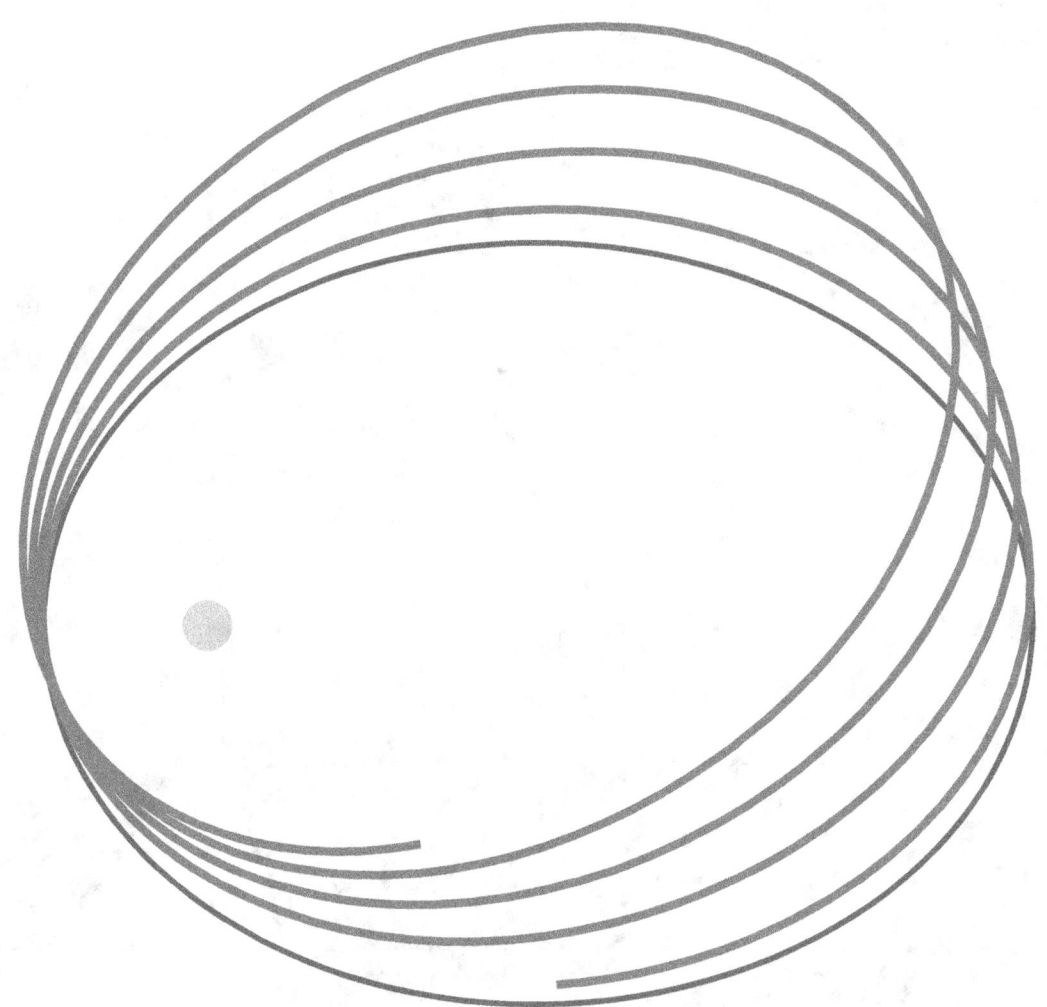

Newtonian (red) vs. Einsteinian orbit (blue) of a lone planet orbiting a star, with relativistic precession of apsides

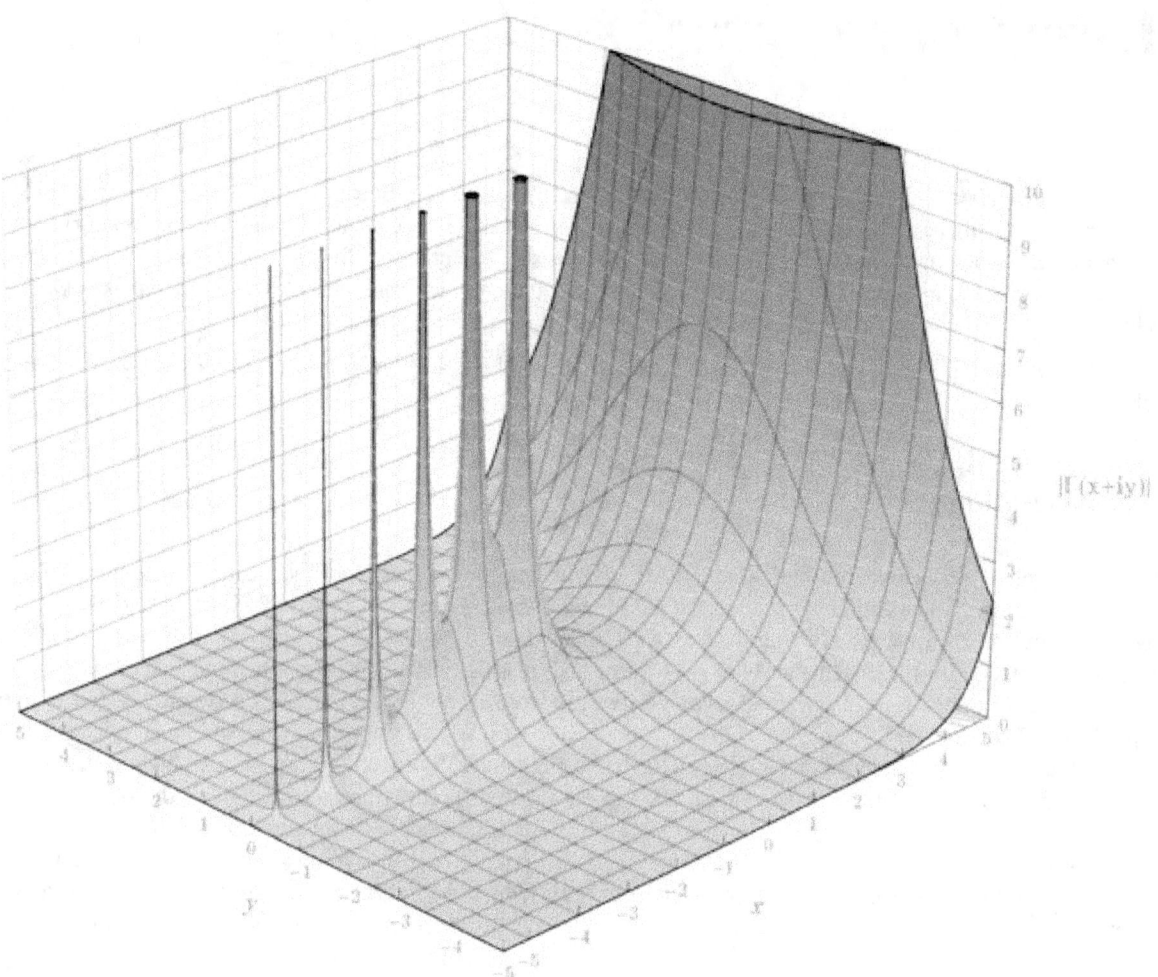

The absolute value of the Gamma function on the complex plane.

Chapter 2

History of algebra

As a branch of mathematics, **algebra** emerged at the end of 16th century in Europe, with the work of François Viète. Algebra can essentially be considered as doing computations similar to those of arithmetic but with non-numerical mathematical objects. However, until the 19th century, algebra consisted essentially of the theory of equations. For example, the fundamental theorem of algebra belongs to the theory of equations and is not, nowadays, considered as belonging to algebra.

This article describes the history of the theory of equations, called here "algebra", from the origins to the emergence of algebra as a separate area of mathematics.

2.1 Etymology

The word "algebra" is derived from the Arabic word *Al-Jabr*, and this comes from the treatise written in 820 by the medieval Persian mathematician, Muhammad ibn Mūsā al-Khwārizmī, entitled, in Arabic *Kitāb al-muḫtaṣar fī ḥisāb al-ǧabr wa-l-muqābala*, which can be translated as *The Compendious Book on Calculation by Completion and Balancing*. The treatise provided for the systematic solution of linear and quadratic equations. Although the exact meaning of the word *al-jabr* is still unknown, most historians agree that the word meant something like "restoration", "completion", (Boyer 1991, "The Arabic Hegemony" p. 229) "It is not certain just what the terms *al-jabr* and *muqabalah* mean, but the usual interpretation is similar to that implied in the translation above. The word *al-jabr* presumably meant something like "restoration" or "completion" and seems to refer to the transposition of subtracted terms to the other side of an equation, which is evident in the treatise; the word *muqabalah* is said to refer to "reduction" or "balancing"—that is, the cancellation of like terms on opposite sides of the equation." "reuniter of broken bones" or "bonesetter". The term is used by al-Khwarizmi to describe the operations that he introduced, "reduction" and "balancing", referring to the transposition of subtracted terms to the other side of an equation, that is, the cancellation of like terms on opposite sides of the equation.

2.2 Stages of algebra

See also: Timeline of algebra

2.2.1 Algebraic expression

Algebra did not always make use of the symbolism that is now ubiquitous in mathematics, rather, it went through three distinct stages. The stages in the development of symbolic algebra are roughly as follows:[1]

- **Rhetorical algebra**, where equations are written in full sentences. For example, the rhetorical form of $x + 1 = 2$

is "The thing plus one equals two" or possibly "The thing plus 1 equals 2". Rhetorical algebra was first developed by the ancient Babylonians and remained dominant up to the 16th century.

- **Syncopated algebra**, where some symbolism is used but which does not contain all of the characteristic of symbolic algebra. For instance, there may be a restriction that subtraction may be used only once within one side of an equation, which is not the case with symbolic algebra. Syncopated algebraic expression first appeared in Diophantus' *Arithmetica*, followed by Brahmagupta's *Brahma Sphuta Siddhanta*.

- **Symbolic algebra**, where full symbolism is used. Early steps toward this can be seen in the work of several Islamic mathematicians such as Ibn al-Banna and al-Qalasadi, though fully symbolic algebra has been developed by François Viète. Later, René Descartes has introduced the modern notation, and shown that the problems occurring in geometry may be expressed and (hopefully) solved in terms of algebra (Cartesian geometry).

As important as the symbolism, or lack thereof, that was used in algebra was the degree of the equations that were used. Quadratic equations played an important role in early algebra; and throughout most of history, until the early modern period, all quadratic equations were classified as belonging to one of three categories.

- $x^2 + px = q$

- $x^2 = px + q$

- $x^2 + q = px$

where p and q are positive. This trichotomy comes about because quadratic equations of the form $x^2 + px + q = 0$, with p and q positive, have no positive roots.[2]

In between the rhetorical and syncopated stages of symbolic algebra, a **geometric constructive algebra** was developed by classical Greek and Vedic Indian mathematicians in which algebraic equations were solved through geometry. For instance, an equation of the form $x^2 = A$ was solved by finding the side of a square of area A.

2.2.2 Conceptual stages

In addition to the three stages of expressing algebraic ideas, there were four conceptual stages in the development of algebra that occurred alongside the changes in expression. These four stages were as follows:[3]

- **Geometric stage**, where the concepts of algebra are largely geometric. This dates back to the Babylonians and continued with the Greeks, and was later revived by Omar Khayyám.

- **Static equation-solving stage**, where the objective is to find numbers satisfying certain relationships. The move away from geometric algebra dates back to Diophantus and Brahmagupta, but algebra didn't decisively move to the static equation-solving stage until Al-Khwarizmi's *Al-Jabr*.

- **Dynamic function stage**, where motion is an underlying idea. The idea of a function began emerging with Sharaf al-Dīn al-Tūsī, but algebra did not decisively move to the dynamic function stage until Gottfried Leibniz.

- **Abstract stage**, where mathematical structure plays a central role. Abstract algebra is largely a product of the 19th and 20th centuries.

2.3 Babylonian algebra

See also: Babylonian mathematics

The origins of algebra can be traced to the ancient Babylonians,[4] who developed a positional number system that greatly aided them in solving their rhetorical algebraic equations. The Babylonians were not interested in exact solutions but approximations, and so they would commonly use linear interpolation to approximate intermediate values.[5] One of the

The Plimpton 322 tablet.

most famous tablets is the Plimpton 322 tablet, created around 1900–1600 BCE, which gives a table of Pythagorean triples and represents some of the most advanced mathematics prior to Greek mathematics.[6]

Babylonian algebra was much more advanced than the Egyptian algebra of the time; whereas the Egyptians were mainly concerned with linear equations the Babylonians were more concerned with quadratic and cubic equations.[5] The Babylonians had developed flexible algebraic operations with which they were able to add equals to equals and multiply both sides of an equation by like quantities so as to eliminate fractions and factors.[5] They were familiar with many simple forms of factoring,[5] three-term quadratic equations with positive roots,[7] and many cubic equations[8] although it is not known if they were able to reduce the general cubic equation.[8]

2.4 Egyptian algebra

See also: Egyptian mathematics

Ancient Egyptian algebra dealt mainly with linear equations while the Babylonians found these equations too elementary and developed mathematics to a higher level than the Egyptians.[5]

The Rhind Papyrus, also known as the Ahmes Papyrus, is an ancient Egyptian papyrus written c. 1650 BCE by Ahmes, who transcribed it from an earlier work that he dated to between 2000 and 1800 BCE.[9] It is the most extensive ancient Egyptian mathematical document known to historians.[10] The Rhind Papyrus contains problems where linear equations of the form $x + ax = b$ and $x + ax + bx = c$ are solved, where a, b, and c are known and x, which is referred to as "aha" or heap, is the unknown.[11] The solutions were possibly, but not likely, arrived at by using the "method of false position", or *regula falsi*, where first a specific value is substituted into the left hand side of the equation, then the required arithmetic calculations are done, thirdly the result is compared to the right hand side of the equation, and finally the correct answer

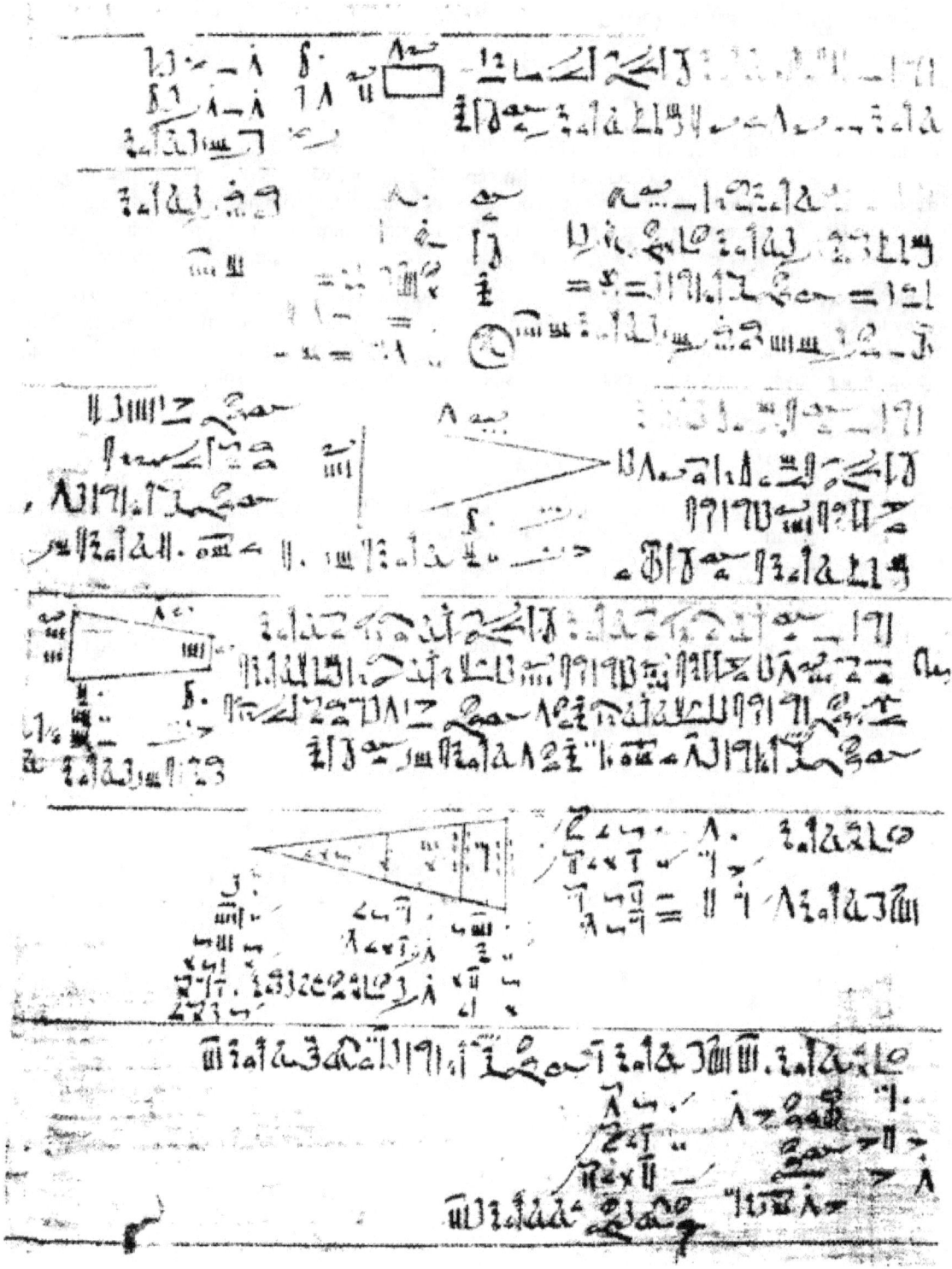

A portion of the Rhind Papyrus.

is found through the use of proportions. In some of the problems the author "checks" his solution, thereby writing one of the earliest known simple proofs.[11]

2.5 Greek geometric algebra

See also: Greek mathematics

It is sometimes alleged that the Greeks had no algebra, but this is inaccurate.[12] By the time of Plato, Greek mathematics had undergone a drastic change. The Greeks created a geometric algebra where terms were represented by sides of geometric objects,[13] usually lines, that had letters associated with them,[14] and with this new form of algebra they were able to find solutions to equations by using a process that they invented, known as "the application of areas".[13] "The application of areas" is only a part of geometric algebra and it is thoroughly covered in Euclid's *Elements*.

An example of geometric algebra would be solving the linear equation ax = bc. The ancient Greeks would solve this equation by looking at it as an equality of areas rather than as an equality between the ratios a:b and c:x. The Greeks would construct a rectangle with sides of length b and c, then extend a side of the rectangle to length a, and finally they would complete the extended rectangle so as to find the side of the rectangle that is the solution.[13]

2.5.1 Bloom of Thymaridas

Iamblichus in *Introductio arithmatica* tells us that Thymaridas (c. 400 BCE – c. 350 BCE) worked with simultaneous linear equations.[15] In particular, he created the then famous rule that was known as the "bloom of Thymaridas" or as the "flower of Thymaridas", which states that:

> If the sum of n quantities be given, and also the sum of every pair containing a particular quantity, then this particular quantity is equal to 1/ (n - 2) of the difference between the sums of these pairs and the first given sum.[16]

or using modern notion, the solution of the following system of n linear equations in n unknowns,[15]

$$x + x_1 + x_2 + ... + x_{n\text{-}1} = s$$
$$x + x_1 = m_1$$
$$x + x_2 = m_2$$
$$.$$
$$.$$
$$.$$
$$x + x_{n\text{-}1} = m_{n\text{-}1}$$

is,

$$x = \frac{(m_1 + m_2 + ... + m_{n-1}) - s}{n - 2} = \frac{\left(\sum_{i=1}^{n-1} m_i\right) - s}{n - 2}$$

Iamblichus goes on to describe how some systems of linear equations that are not in this form can be placed into this form.[15]

2.5.2 Euclid of Alexandria

Euclid (Greek: Εὐκλείδης) was a Greek mathematician who flourished in Alexandria, Egypt, almost certainly during the reign of Ptolemy I (323–283 BCE).[17][18] Neither the year nor place of his birth[17] have been established, nor the circumstances of his death.

Euclid is regarded as the "father of geometry". His *Elements* is the most successful textbook in the history of mathematics.[17] Although he is one of the most famous mathematicians in history there are no new discoveries attributed to him, rather he is remembered for his great explanatory skills.[19] The *Elements* is not, as is sometimes thought, a collection of all Greek mathematical knowledge to its date, rather, it is an elementary introduction to it.[20]

Ἐπὶ τῆς δοθείσης εὐθείας πεπερασμένης τρίγωνον ἰσόπλευρον συστήσασθαι.

Ἔστω ἡ δοθεῖσα εὐθεῖα πεπερασμένη ἡ AB.

Δεῖ δὴ ἐπὶ τῆς AB εὐθείας τρίγωνον ἰσόπλευρον συστήσασθαι.

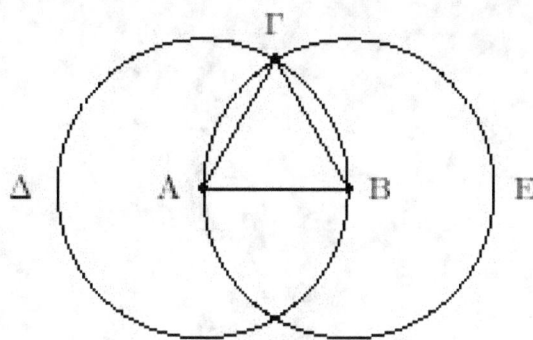

Κέντρῳ μὲν τῷ A διαστήματι δὲ τῷ AB κύκλος γεγράφθω ὁ ΒΓΔ, καὶ πάλιν κέντρῳ μὲν τῷ B διαστήματι δὲ τῷ BA κύκλος γεγράφθω ὁ ΑΓΕ, καὶ ἀπὸ τοῦ Γ σημείου, καθ' ὃ τέμνουσιν ἀλλήλους οἱ κύκλοι, ἐπὶ τὰ A, B σημεῖα ἐπεζεύχθωσαν εὐθεῖαι αἱ ΓΑ, ΓΒ.

Καὶ ἐπεὶ τὸ A σημεῖον κέντρον ἐστὶ τοῦ ΓΔΒ κύκλου, ἴση ἐστὶν ἡ ΑΓ τῇ AB· πάλιν, ἐπεὶ τὸ B σημεῖον κέντρον ἐστὶ τοῦ ΓΑΕ κύκλου, ἴση ἐστὶν ἡ ΒΓ τῇ ΒΑ. ἐδείχθη δὲ καὶ ἡ ΓΑ τῇ AB ἴση· ἑκατέρα ἄρα τῶν ΓΑ, ΓΒ τῇ AB ἐστὶν ἴση. τὰ δὲ τῷ αὐτῷ ἴσα καὶ ἀλλήλοις ἐστὶν ἴσα· καὶ ἡ ΓΑ ἄρα τῇ ΓΒ ἐστὶν ἴση· αἱ τρεῖς ἄρα αἱ ΓΑ, AB, ΒΓ ἴσαι ἀλλήλαις εἰσίν.

Ἰσόπλευρον ἄρα ἐστὶ τὸ ΑΒΓ τρίγωνον, καὶ συνέσταται ἐπὶ τῆς δοθείσης εὐθείας πεπερασμένης τῆς AB.

[Ἐπὶ τῆς δοθείσης ἄρα εὐθείας πεπερασμένης τρίγωνον ἰσόπλευρον συνέσταται]· ὅπερ ἔδει ποιῆσαι.

A proof from Euclid's Elements *that, given a line segment, an equilateral triangle exists that includes the segment as one of its sides.*

Elements

The geometric work of the Greeks, typified in Euclid's *Elements*, provided the framework for generalizing formulae beyond the solution of particular problems into more general systems of stating and solving equations.

Book II of the *Elements* contains fourteen propositions, which in Euclid's time were extremely significant for doing geometric algebra. These propositions and their results are the geometric equivalents of our modern symbolic algebra and trigonometry.[12] Today, using modern symbolic algebra, we let symbols represent known and unknown magnitudes (i.e. numbers) and then apply algebraic operations on them. While in Euclid's time magnitudes were viewed as line segments and then results were deduced using the axioms or theorems of geometry.[12]

Many basic laws of addition and multiplication are included or proved geometrically in the *Elements*. For instance, proposition 1 of Book II states:

Hellenistic mathematician Euclid details geometrical algebra.

If there be two straight lines, and one of them be cut into any number of segments whatever, the rectangle contained by the two straight lines is equal to the rectangles contained by the uncut straight line and each of the segments.

But this is nothing more than the geometric version of the (left) distributive law, $a(b + c + d) = ab + ac + ad$; and in Books V and VII of the *Elements* the commutative and associative laws for multiplication are demonstrated.[12]

Many basic equations were also proved geometrically. For instance, proposition 5 in Book II proves that $a^2 - b^2 = (a + b)(a - b)$,[21] and proposition 4 in Book II proves that $(a + b)^2 = a^2 + 2ab + b^2$.[12]

Furthermore, there are also geometric solutions given to many equations. For instance, proposition 6 of Book II gives the solution to the quadratic equation $ax + x^2 = b^2$, and proposition 11 of Book II gives a solution to $ax + x^2 = a^2$.[22]

Data

Data is a work written by Euclid for use at the schools of Alexandria and it was meant to be used as a companion volume to the first six books of the *Elements*. The book contains some fifteen definitions and ninety-five statements, of which there are about two dozen statements that serve as algebraic rules or formulas.[23] Some of these statements are geometric equivalents to solutions of quadratic equations.[23] For instance, *Data* contains the solutions to the equations $dx^2 - adx + b^2c = 0$ and the familiar Babylonian equation $xy = a^2$, x ± y = b.[23]

2.5.3 Conic sections

A conic section is a curve that results from the intersection of a cone with a plane. There are three primary types of conic sections: ellipses (including circles), parabolas, and hyperbolas. The conic sections are reputed to have been discovered by Menaechmus[24] (c. 380 BCE – c. 320 BCE) and since dealing with conic sections is equivalent to dealing with their respective equations, they played geometric roles equivalent to cubic equations and other higher order equations.

Menaechmus knew that in a parabola, the equation y² = *l*x holds, where *l* is a constant called the latus rectum, although he was not aware of the fact that any equation in two unknowns determines a curve.[25] He apparently derived these properties of conic sections and others as well. Using this information it was now possible to find a solution to the problem of the duplication of the cube by solving for the points at which two parabolas intersect, a solution equivalent to solving a cubic equation.[25]

We are informed by Eutocius that the method he used to solve the cubic equation was due to Dionysodorus (250 BCE – 190 BCE). Dionysodorus solved the cubic by means of the intersection of a rectangular hyperbola and a parabola. This was related to a problem in Archimedes' *On the Sphere and Cylinder*. Conic sections would be studied and used for thousands of years by Greek, and later Islamic and European, mathematicians. In particular Apollonius of Perga's famous *Conics* deals with conic sections, among other topics.

2.6 Chinese algebra

See also: Chinese mathematics

Chinese Mathematics dates to at least 300 BCE with the *Chou Pei Suan Ching*, generally considered to be one of the oldest Chinese mathematical documents.[26]

2.6.1 *Nine Chapters on the Mathematical Art*

Chiu-chang suan-shu or *The Nine Chapters on the Mathematical Art*, written around 250 BCE, is one of the most influential of all Chinese math books and it is composed of some 246 problems. Chapter eight deals with solving determinate and indeterminate simultaneous linear equations using positive and negative numbers, with one problem dealing with solving four equations in five unknowns.[26]

2.6.2 *Sea-Mirror of the Circle Measurements*

Ts'e-yuan hai-ching, or *Sea-Mirror of the Circle Measurements*, is a collection of some 170 problems written by Li Zhi (or Li Ye) (1192 – 1272 CE). He used *fan fa*, or Horner's method, to solve equations of degree as high as six, although

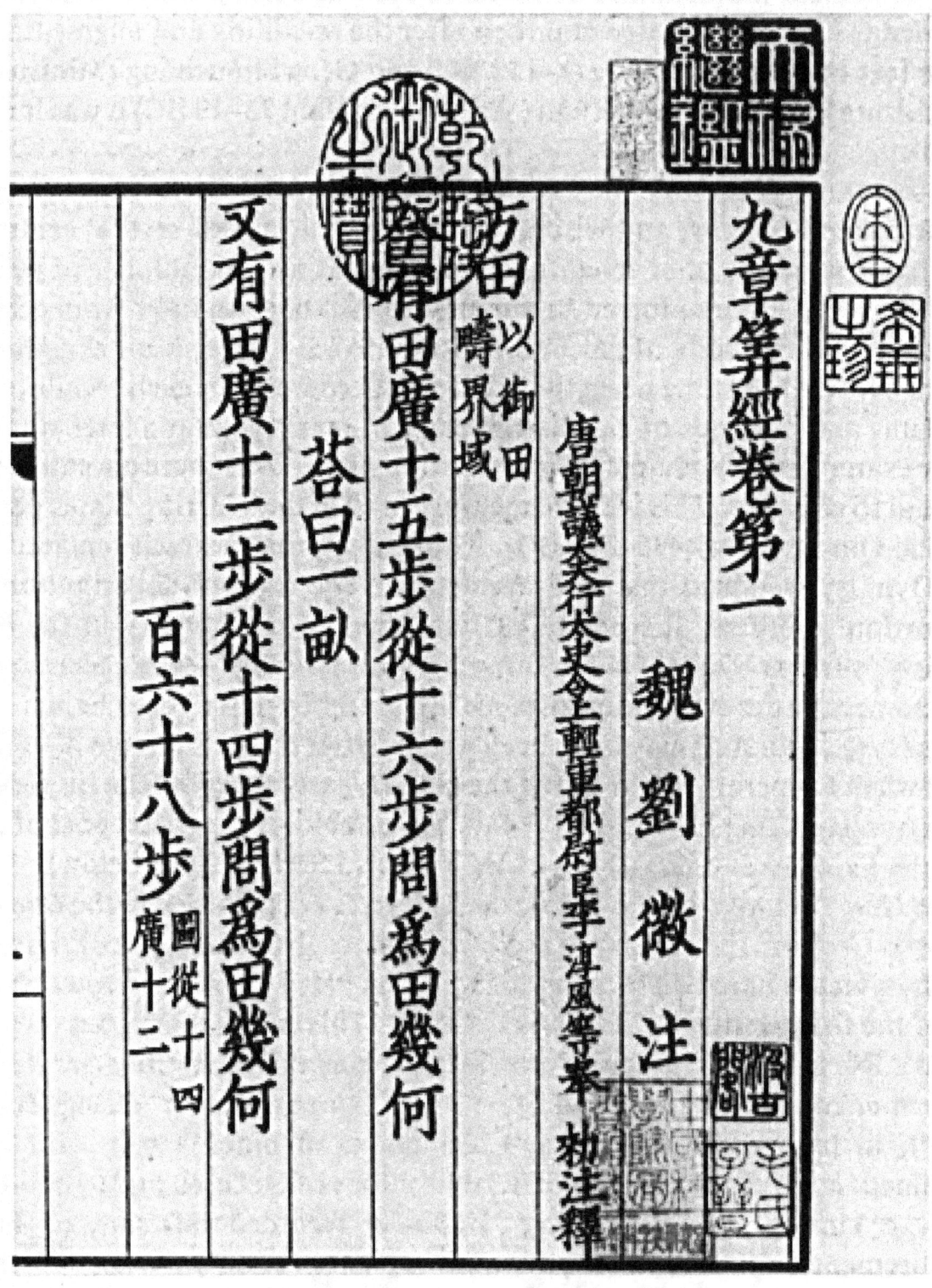

Nine Chapters on the Mathematical Art

he did not describe his method of solving equations.[27]

2.6.3 *Mathematical Treatise in Nine Sections*

Shu-shu chiu-chang, or *Mathematical Treatise in Nine Sections*, was written by the wealthy governor and minister Ch'in Chiu-shao (c. 1202 – c. 1261 CE) and with the invention of a method of solving simultaneous congruences, now called Chinese remainder theorem, it marks the high point in Chinese indeterminate analysis.[27]

2.6.4 Magic squares

The earliest known magic squares appeared in China.[28] In *Nine Chapters* the author solves a system of simultaneous linear equations by placing the coefficients and constant terms of the linear equations into a magic square (i.e. a matrix) and performing column reducing operations on the magic square.[28] The earliest known magic squares of order greater than three are attributed to Yang Hui (fl. c. 1261 – 1275), who worked with magic squares of order as high as ten.[29]

2.6.5 *Precious Mirror of the Four Elements*

Ssy-yüan yü-chien▢▢▢▢▢▢, or *Precious Mirror of the Four Elements*, was written by Chu Shih-chieh in 1303 and it marks the peak in the development of Chinese algebra. The four elements, called heaven, earth, man and matter, represented the four unknown quantities in his algebraic equations. The *Ssy-yüan yü-chien* deals with simultaneous equations and with equations of degrees as high as fourteen. The author uses the method of *fan fa*, today called Horner's method, to solve these equations.[30]

The *Precious Mirror* opens with a diagram of the arithmetic triangle (Pascal's triangle) using a round zero symbol, but Chu Shih-chieh denies credit for it. A similar triangle appears in Yang Hui's work, but without the zero symbol.[31]

There are many summation series equations given without proof in the *Precious mirror*. A few of the summation series are:[31]

$$1^2 + 2^2 + 3^2 + \cdots + n^2 = \frac{n(n+1)(2n+1)}{3!}$$

$$1 + 8 + 30 + 80 + \cdots + \frac{n^2(n+1)(n+2)}{3!} = \frac{n(n+1)(n+2)(n+3)(4n+1)}{5!}$$

2.7 Diophantine algebra

Diophantus was a Hellenistic mathematician who lived c. 250 CE, but the uncertainty of this date is so great that it may be off by more than a century. He is known for having written *Arithmetica*, a treatise that was originally thirteen books but of which only the first six have survived.[32] *Arithmetica* has very little in common with traditional Greek mathematics since it is divorced from geometric methods, and it is different from Babylonian mathematics in that Diophantus is concerned primarily with exact solutions, both determinate and indeterminate, instead of simple approximations.[33]

In *Arithmetica*, Diophantus is the first to use symbols for unknown numbers as well as abbreviations for powers of numbers, relationships, and operations;[33] thus he used what is now known as *syncopated* algebra. The main difference between Diophantine syncopated algebra and modern algebraic notation is that the former lacked special symbols for operations, relations, and exponentials.[34] So, for example, what we would write as

$$x^3 - 2x^2 + 10x - 1 = 5$$

Diophantus would have written this as

$$\mathrm{K}^{\Upsilon} \ \bar{\alpha}\varsigma\bar{\iota} \ ▢ \ \Delta^{\Upsilon} \ \bar{\beta} \ \mathbf{M} \ \bar{\alpha} \ \mathring{\iota}\sigma \ \mathbf{M} \ \bar{\varepsilon}$$

where the symbols represent the following:[35][36]

Note that the coefficients come after the variables and that addition is represented by the juxtaposition of terms. A literal symbol-for-symbol translation of Diophantus's syncopated equation into a modern symbolic equation would be the following:[35]

$$x^3 1 x 10 - x^2 2 x^0 1 = x^0 5$$

and, to clarify, if the modern parentheses and plus are used then the above equation can be rewritten as:[35]

$$(x^3 1 + x 10) - (x^2 2 + x^0 1) = x^0 5$$

Arithmetica is a collection of some 150 solved problems with specific numbers and there is no postulational development nor is a general method explicitly explained, although generality of method may have been intended and there is no attempt to find all of the solutions to the equations.[33] *Arithmetica* does contain solved problems involving several unknown quantities, which are solved, if possible, by expressing the unknown quantities in terms of only one of them.[33] *Arithmetica* also makes use of the identities:[37]

2.8 Indian algebra

See also: Indian mathematics

The Indian mathematicians were active in studying about number systems. Not much verifiable account remains of their contribution to algebra. However, The earliest known Indian mathematical documents are dated to around the middle of the first millennium BCE (around the 6th century BCE).[38]

The recurring themes in Indian mathematics are, among others, determinate and indeterminate linear and quadratic equations, simple mensuration, and Pythagorean triples.[39]

2.8.1 *Aryabhata*

Aryabhata (476–550 CE) was an Indian mathematician who authored *Aryabhatiya*. In it he gave the rules,[40]

$$1^2 + 2^2 + \cdots + n^2 = \frac{n(n+1)(2n+1)}{6}$$

and

$$1^3 + 2^3 + \cdots + n^3 = (1 + 2 + \cdots + n)^2$$

2.8.2 *Brahma Sphuta Siddhanta*

Brahmagupta (fl. 628) was an Indian mathematician who authored *Brahma Sphuta Siddhanta*. In his work Brahmagupta solves the general quadratic equation for both positive and negative roots.[41] In indeterminate analysis Brahmagupta gives the Pythagorean triads m, $\frac{1}{2}(\frac{m^2}{n} - n)$, $\frac{1}{2}(\frac{m^2}{n} + n)$, but this is a modified form of an old Babylonian rule that Brahmagupta may have been familiar with.[42] He was the first to give a general solution to the linear Diophantine equation ax + by = c, where a, b, and c are integers. Unlike Diophantus who only gave one solution to an indeterminate equation, Brahmagupta gave *all* integer solutions; but that Brahmagupta used some of the same examples as Diophantus has led some historians to consider the possibility of a Greek influence on Brahmagupta's work, or at least a common Babylonian source.[43]

Like the algebra of Diophantus, the algebra of Brahmagupta was syncopated. Addition was indicated by placing the numbers side by side, subtraction by placing a dot over the subtrahend, and division by placing the divisor below the dividend, similar to our notation but without the bar. Multiplication, evolution, and unknown quantities were represented by abbreviations of appropriate terms.[43] The extent of Greek influence on this syncopation, if any, is not known and it is possible that both Greek and Indian syncopation may be derived from a common Babylonian source.[43]

2.8.3 Bhāskara II

Bhāskara II (1114 – c. 1185) was the leading mathematician of the 12th century. In Algebra, he gave the general solution of the Pell equation.[43] He is the author of *Lilavati* and *Vija-Ganita*, which contain problems dealing with determinate and indeterminate linear and quadratic equations, and Pythagorean triples[39] and he fails to distinguish between exact and approximate statements.[44] Many of the problems in *Lilavati* and *Vija-Ganita* are derived from other Hindu sources, and so Bhaskara is at his best in dealing with indeterminate analysis.[44]

Bhaskara uses the initial symbols of the names for colors as the symbols of unknown variables. So, for example, what we would write today as

$$(-x - 1) + (2x - 8) = x - 9$$

Bhaskara would have written as

       . _ .

    ya 1 *ru* 1

       .

    ya 2 *ru* 8

      .

   Sum *ya* 1 ru *9*

where *ya* indicates the first syllable of the word for *black*, and *ru* is taken from the word *species*. The dots over the numbers indicate subtraction.

2.9 Islamic algebra

See also: Islamic mathematics

The first century of the Islamic Arab Empire saw almost no scientific or mathematical achievements since the Arabs, with their newly conquered empire, had not yet gained any intellectual drive and research in other parts of the world had faded. In the second half of the 8th century, Islam had a cultural awakening, and research in mathematics and the sciences increased.[45] The Muslim Abbasid caliph al-Mamun (809–833) is said to have had a dream where Aristotle appeared to him, and as a consequence al-Mamun ordered that Arabic translation be made of as many Greek works as possible, including Ptolemy's *Almagest* and Euclid's *Elements*. Greek works would be given to the Muslims by the Byzantine Empire in exchange for treaties, as the two empires held an uneasy peace.[45] Many of these Greek works were translated by Thabit ibn Qurra (826–901), who translated books written by Euclid, Archimedes, Apollonius, Ptolemy, and Eutocius.[46]

There are three theories about the origins of Arabic Algebra. The first emphasizes Hindu influence, the second emphasizes Mesopotamian or Persian-Syriac influence and the third emphasizes Greek influence. Many scholars believe that it is the result of a combination of all three sources.[47]

Throughout their time in power, before the fall of Islamic civilization, the Arabs used a fully rhetorical algebra, where often even the numbers were spelled out in words. The Arabs would eventually replace spelled out numbers (e.g. twenty-two) with Arabic numerals (e.g. 22), but the Arabs did not adopt or develop a syncopated or symbolic algebra[46] until the work of Ibn al-Banna in the 13th century and Abū al-Hasan ibn Alī al-Qalasādī in the 15th century.

2.9.1 *Al-jabr wa'l muqabalah*

See also: The Compendious Book on Calculation by Completion and Balancing

The Muslim[48] Persian mathematician Muhammad ibn Mūsā al-Khwārizmī was a faculty member of the "House of Wisdom" (*Bait al-Hikma*) in Baghdad, which was established by Al-Mamun. Al-Khwarizmi, who died around 850 CE, wrote more than half a dozen mathematical and astronomical works; some of which were based on the Indian *Sindhind*.[45] One of al-Khwarizmi's most famous books is entitled *Al-jabr wa'l muqabalah* or *The Compendious Book on Calculation by Completion and Balancing*, and it gives an exhaustive account of solving polynomials up to the second degree.[49] The book also introduced the fundamental concept of "reduction" and "balancing", referring to the transposition of subtracted terms to the other side of an equation, that is, the cancellation of like terms on opposite sides of the equation. This is the operation which Al-Khwarizmi originally described as *al-jabr*.[50]

R. Rashed and Angela Armstrong write:

> "Al-Khwarizmi's text can be seen to be distinct not only from the Babylonian tablets, but also from Diophantus' *Arithmetica*. It no longer concerns a series of problems to be resolved, but an exposition which starts with primitive terms in which the combinations must give all possible prototypes for equations, which henceforward explicitly constitute the true object of study. On the other hand, the idea of an equation for its own sake appears from the beginning and, one could say, in a generic manner, insofar as it does not simply emerge in the course of solving a problem, but is specifically called on to define an infinite class of problems."[51]

Al-Jabr is divided into six chapters, each of which deals with a different type of formula. The first chapter of *Al-Jabr* deals with equations whose squares equal its roots ($ax^2 = bx$), the second chapter deals with squares equal to number ($ax^2 = c$), the third chapter deals with roots equal to a number ($bx = c$), the fourth chapter deals with squares and roots equal a number ($ax^2 + bx = c$), the fifth chapter deals with squares and number equal roots ($ax^2 + c = bx$), and the sixth and final chapter deals with roots and number equal to squares ($bx + c = ax^2$).[52]

In *Al-Jabr*, al-Khwarizmi uses geometric proofs,[14] he does not recognize the root $x = 0$,[52] and he only deals with positive roots.[53] He also recognizes that the discriminant must be positive and described the method of completing the square, though he does not justify the procedure.[54] The Greek influence is shown by *Al-Jabr*'s geometric foundations[47][55] and by one problem taken from Heron.[56] He makes use of lettered diagrams but all of the coefficients in all of his equations are specific numbers since he had no way of expressing with parameters what he could express geometrically; although generality of method is intended.[14]

Al-Khwarizmi most likely did not know of Diophantus's *Arithmetica*,[57] which became known to the Arabs sometime before the 10th century.[58] And even though al-Khwarizmi most likely knew of Brahmagupta's work, *Al-Jabr* is fully rhetorical with the numbers even being spelled out in words.[57] So, for example, what we would write as

$$x^2 + 10x = 39$$

Diophantus would have written as[59]

$$\Delta^\Upsilon \bar{\alpha} \; \varsigma\bar{\iota} \; '\iota\sigma \; \mathbf{M} \; \lambda\vartheta$$

And al-Khwarizmi would have written as[59]

> One square and ten roots of the same amount to thirty-nine *dirhems*; that is to say, what must be the square which, when increased by ten of its own roots, amounts to thirty-nine?

2.9.2 *Logical Necessities in Mixed Equations*

'Abd al-Hamīd ibn Turk authored a manuscript entitled *Logical Necessities in Mixed Equations*, which is very similar to al-Khwarzimi's *Al-Jabr* and was published at around the same time as, or even possibly earlier than, *Al-Jabr*.[58] The manuscript gives exactly the same geometric demonstration as is found in *Al-Jabr*, and in one case the same example as found in *Al-Jabr*, and even goes beyond *Al-Jabr* by giving a geometric proof that if the discriminant is negative then the quadratic equation has no solution.[58] The similarity between these two works has led some historians to conclude that Arabic algebra may have been well developed by the time of al-Khwarizmi and 'Abd al-Hamid.[58]

2.9.3 Abu Kamil and al-Karkhi

Arabic mathematicians treated irrational numbers as algebraic objects.[60] The Egyptian mathematician Abū Kāmil Shujā ibn Aslam (c. 850–930) was the first to accept irrational numbers (often in the form of a square root, cube root or fourth root) as solutions to quadratic equations or as coefficients in an equation.[61] He was also the first to solve three non-linear simultaneous equations with three unknown variables.[62]

Al-Karkhi (953–1029), also known as Al-Karaji, was the successor of Abū al-Wafā' al-Būzjānī (940–998) and he discovered the first numerical solution to equations of the form $ax^{2n} + bx^n = c$.[63] Al-Karkhi only considered positive roots.[63] Al-Karkhi is also regarded as the first person to free algebra from geometrical operations and replace them with the type of arithmetic operations which are at the core of algebra today. His work on algebra and polynomials, gave the rules for arithmetic operations to manipulate polynomials. The historian of mathematics F. Woepcke, in *Extrait du Fakhri, traité d'Algèbre par Abou Bekr Mohammed Ben Alhacan Alkarkhi* (Paris, 1853), praised Al-Karaji for being "the first who introduced the theory of algebraic calculus". Stemming from this, Al-Karaji investigated binomial coefficients and Pascal's triangle.[64]

2.9.4 Omar Khayyám, Sharaf al-Dīn, and al-Kashi

Omar Khayyám (c. 1050 – 1123) wrote a book on Algebra that went beyond *Al-Jabr* to include equations of the third degree.[65] Omar Khayyám provided both arithmetic and geometric solutions for quadratic equations, but he only gave geometric solutions for general cubic equations since he mistakenly believed that arithmetic solutions were impossible.[65] His method of solving cubic equations by using intersecting conics had been used by Menaechmus, Archimedes, and Ibn al-Haytham (Alhazen), but Omar Khayyám generalized the method to cover all cubic equations with positive roots.[65] He only considered positive roots and he did not go past the third degree.[65] He also saw a strong relationship between Geometry and Algebra.[65]

In the 12th century, Sharaf al-Dīn al-Tūsī (1135–1213) wrote the *Al-Mu'adalat* (*Treatise on Equations*), which dealt with eight types of cubic equations with positive solutions and five types of cubic equations which may not have positive solutions. He used what would later be known as the "Ruffini-Horner method" to numerically approximate the root of a cubic equation. He also developed the concepts of the maxima and minima of curves in order to solve cubic equations which may not have positive solutions.[66] He understood the importance of the discriminant of the cubic equation and used an early version of Cardano's formula[67] to find algebraic solutions to certain types of cubic equations. Some scholars, such as Roshdi Rashed, argue that Sharaf al-Din discovered the derivative of cubic polynomials and realized its significance, while other scholars connect his solution to the ideas of Euclid and Archimedes.[68]

Sharaf al-Din also developed the concept of a function. In his analysis of the equation $x^3 + d = bx^2$ for example, he begins by changing the equation's form to $x^2(b - x) = d$. He then states that the question of whether the equation has a solution depends on whether or not the "function" on the left side reaches the value d. To determine this, he finds a maximum value for the function. He proves that the maximum value occurs when $x = \frac{2b}{3}$, which gives the functional value $\frac{4b^3}{27}$. Sharaf al-Din then states that if this value is less than d , there are no positive solutions; if it is equal to d , then there is one solution at $x = \frac{2b}{3}$; and if it is greater than d , then there are two solutions, one between 0 and $\frac{2b}{3}$ and one between $\frac{2b}{3}$ and b .[69]

In the early 15th century, Jamshīd al-Kāshī developed an early form of Newton's method to numerically solve the equation $x^P - N = 0$ to find roots of N .[70] Al-Kāshī also developed decimal fractions and claimed to have discovered it himself. However, J. Lennart Berggrenn notes that he was mistaken, as decimal fractions were first used five centuries before him by the Baghdadi mathematician Abu'l-Hasan al-Uqlidisi as early as the 10th century.[62]

2.9.5 Al-Hassār, Ibn al-Banna, and al-Qalasadi

Al-Hassār, a mathematician from Morocco specializing in Islamic inheritance jurisprudence during the 12th century, developed the modern symbolic mathematical notation for fractions, where the numerator and denominator are separated by a horizontal bar. This same fractional notation appeared soon after in the work of Fibonacci in the 13th century.

Abū al-Hasan ibn Alī al-Qalasādī (1412–1486) was the last major medieval Arab algebraist, who made the first attempt at creating an algebraic notation since Ibn al-Banna two centuries earlier, who was himself the first to make such an attempt since Diophantus and Brahmagupta in ancient times.[71] The syncopated notations of his predecessors, however, lacked symbols for mathematical operations.[34] Al-Qalasadi "took the first steps toward the introduction of algebraic symbolism by using letters in place of numbers"[71] and by "using short Arabic words, or just their initial letters, as mathematical symbols."[71]

2.10 European algebra

2.10.1 Dark Ages

Just as the death of Hypatia signals the close of the Library of Alexandria as a mathematical center, so does the death of Boethius signal the end of mathematics in the Western Roman Empire. Although there was some work being done at Athens, it came to a close when in 529 the Byzantine emperor Justinian closed the pagan philosophical schools. The year 529 is now taken to be the beginning of the medieval period. Scholars fled the West towards the more hospitable East, particularly towards Persia, where they found haven under King Chosroes and established what might be termed an "Athenian Academy in Exile".[72] Under a treaty with Justinian, Chosroes would eventually return the scholars to the Eastern Empire. During the Dark Ages, European mathematics was at its nadir with mathematical research consisting mainly of commentaries on ancient treatises; and most of this research was centered in the Byzantine Empire. The end of the medieval period is set as the fall of Constantinople to the Turks in 1453.

2.10.2 Late Middle Ages

The 12th century saw a flood of translations from Arabic into Latin and by the 13th century, European mathematics was beginning to rival the mathematics of other lands. In the 13th century, the solution of a cubic equation by Fibonacci is representative of the beginning of a revival in European algebra.

As the Islamic world was declining after the 15th century, the European world was ascending. And it is here that Algebra was further developed.

2.11 Modern algebra

Another key event in the further development of algebra was the general algebraic solution of the cubic and quartic equations, developed in the mid-16th century. The idea of a determinant was developed by Japanese mathematician Kowa Seki in the 17th century, followed by Gottfried Leibniz ten years later, for the purpose of solving systems of simultaneous linear equations using matrices. Gabriel Cramer also did some work on matrices and determinants in the 18th century.

The symbol x commonly denotes an unknown variable. Even though any letter can be used, x is the most common choice. The tradition of using x to represent unknowns was started by René Descartes in his *La geometrie* (1637).[73] In mathematics, an "italicized x" (x) is often used to avoid potential confusion with the multiplication symbol.

2.11.1   Gottfried Leibniz

Although the mathematical notion of function was implicit in trigonometric and logarithmic tables, which existed in his day, Gottfried Leibniz was the first, in 1692 and 1694, to employ it explicitly, to denote any of several geometric concepts derived from a curve, such as abscissa, ordinate, tangent, chord, and the perpendicular.[74] In the 18th century, "function" lost these geometrical associations.

Leibniz realized that the coefficients of a system of linear equations could be arranged into an array, now called a matrix, which can be manipulated to find the solution of the system, if any. This method was later called Gaussian elimination. Leibniz also discovered Boolean algebra and symbolic logic, also relevant to algebra.

2.11.2   Abstract algebra

The ability to do algebra is a skill cultivated in mathematics education. As explained by Andrew Warwick, Cambridge University students in the early 19th century practiced "mixed mathematics",[75] doing exercises based on physical variables such as space, time, and weight. Over time the association of variables with physical quantities faded away as mathematical technique grew. Eventually mathematics was concerned completely with abstract polynomials, complex numbers, hypercomplex numbers and other concepts. Application to physical situations was then called applied mathematics or mathematical physics, and the field of mathematics expanded to include abstract algebra. For instance, the issue of constructible numbers showed some mathematical limitations, and the field of Galois theory was developed.

2.12   The father of algebra

The Hellenistic mathematician Diophantus has traditionally been known as "the father of algebra"[76][77] but debate now exists as to whether or not Al-Khwarizmi deserves this title instead.[76] Those who support Diophantus point to the fact that the algebra found in *Al-Jabr* is more elementary than the algebra found in *Arithmetica* and that *Arithmetica* is syncopated while *Al-Jabr* is fully rhetorical.[76]

Those who support Al-Khwarizmi point to the fact that he gave an exhaustive explanation for the algebraic solution of quadratic equations with positive roots,[78] and was the first to teach algebra in an elementary form and for its own sake, whereas Diophantus was primarily concerned with the theory of numbers.[79] Al-Khwarizmi also introduced the fundamental concept of "reduction" and "balancing" (which he originally used the term *al-jabr* to refer to), referring to the transposition of subtracted terms to the other side of an equation, that is, the cancellation of like terms on opposite sides of the equation.[50] Other supporters of Al-Khwarizmi point to his algebra no longer being concerned "with a series of problems to be resolved, but an exposition which starts with primitive terms in which the combinations must give all possible prototypes for equations, which henceforward explicitly constitute the true object of study." They also point to his treatment of an equation for its own sake and "in a generic manner, insofar as it does not simply emerge in the course of solving a problem, but is specifically called on to define an infinite class of problems."[51]

2.13   See also

- Algebra
- Timeline of algebra
- History of Mathematics

2.14 Footnotes and citations

[1] (Boyer 1991, "Revival and Decline of Greek Mathematics" p.180) "It has been said that three stages of in the historical development of algebra can be recognized: (1) the rhetorical or early stage, in which everything is written out fully in words; (2) a syncopated or intermediate state, in which some abbreviations are adopted; and (3) a symbolic or final stage. Such an arbitrary division of the development of algebra into three stages is, of course, a facile oversimplification; but it can serve effectively as a first approximation to what has happened""

[2] (Boyer 1991, "Mesopotamia" p. 32) "Until modern times there was no thought of solving a quadratic equation of the form $x^2 + px + q = 0$, where p and q are positive, for the equation has no positive root. Consequently, quadratic equations in ancient and Medieval times—and even in the early modern period—were classified under three types: (1) $x^2 + px = q$ (2) $x^2 = px + q$ (3) $x^2 + q = px$ "

[3] Victor J. Katz, Bill Barton (October 2007), "Stages in the History of Algebra with Implications for Teaching", *Educational Studies in Mathematics* (Springer Netherlands) **66** (2): 185–201, doi:10.1007/s10649-006-9023-7

[4] Struik, Dirk J. (1987). *A Concise History of Mathematics*. New York: Dover Publications.

[5] (Boyer 1991, "Mesopotamia" p. 30) "Babylonian mathematicians did not hesitate to interpolate by proportional parts to approximate intermediate values. Linear interpolation seems to have been a commonplace procedure in ancient Mesopotamia, and the positional notation lent itself conveniently to the rile of three. [...] a table essential in Babylonian algebra; this subject reached a considerably higher level in Mesopotamia than in Egypt. Many problem texts from the Old Babylonian period show that the solution of the complete three-term quadratic equation afforded the Babylonians no serious difficulty, for flexible algebraic operations had been developed. They could transpose terms in an equations by adding equals to equals, and they could multiply both sides by like quantities to remove fractions or to eliminate factors. By adding 4ab to $(a - b)^2$ they could obtain $(a + b)^2$ for they were familiar with many simple forms of factoring. [...]Egyptian algebra had been much concerned with linear equations, but the Babylonians evidently found these too elementary for much attention. [...] In another problem in an Old Babylonian text we find two simultaneous linear equations in two unknown quantities, called respectively the "first silver ring" and the "second silver ring.""

[6] Joyce, David E. (1995). "Plimpton 322". The clay tablet with the catalog number 322 in the G. A. Plimpton Collection at Columbia University may be the most well known mathematical tablet, certainly the most photographed one, but it deserves even greater renown. It was scribed in the Old Babylonian period between -1900 and -1600 and shows the most advanced mathematics before the development of Greek mathematics.

[7] (Boyer 1991, "Mesopotamia" p. 31) "The solution of a three-term quadratic equation seems to have exceeded by far the algebraic capabilities of the Egyptians, but Neugebauer in 1930 disclosed that such equations had been handled effectively by the Babylonians in some of the oldest problem texts."

[8] (Boyer 1991, "Mesopotamia" p. 33) "There is no record in Egypt of the solution of a cubic equations, but among the Babylonians there are many instances of this. [...] Whether or not the Babylonians were able to reduce the general four-term cubic, $ax^3 + bx^2 + cx = d$, to their normal form is not known."

[9] (Boyer 1991, "Egypt" p. 11) "It had been bought in 1959 in a Nile resort town by a Scottish antiquary, Henry Rhind; hence, it often is known as the Rhind Papyrus or, less frequently, as the Ahmes Papyrus in honor of the scribe by whose hand it had been copied in about 1650 BCE. The scribe tells us that the material is derived from a prototype from the Middle Kingdom of about 2000 to 1800 BCE."

[10] (Boyer 1991, "Egypt" p. 19) "Much of our information about Egyptian mathematics has been derived from the Rhind or Ahmes Papyrus, the most extensive mathematical document from ancient Egypt; but there are other sources as well."

[11] (Boyer 1991, "Egypt" pp. 15–16) "The Egyptian problems so far described are best classified as arithmetic, but there are others that fall into a class to which the term algebraic is appropriately applied. These do not concern specific concrete objects such as bread and beer, nor do they call for operations on known numbers. Instead they require the equivalent of solutions of linear equations of the form $x + ax = b$ or $x + ax + bx = c$, where a and b and c are known and x is unknown. The unknown is referred to as "aha," or heap. [...] The solution given by Ahmes is not that of modern textbooks, but one proposed characteristic of a procedure now known as the "method of false position," or the "rule of false." A specific false value has been proposed by 1920s scholars and the operations indicated on the left hand side of the equality sign are performed on this assumed number. Recent scholarship shows that scribes had not guessed in these situations. Exact rational number answers written in Egyptian fraction series had confused the 1920s scholars. The attested result shows that Ahmes "checked" result by showing that 16 + 1/2 + 1/8 exactly added to a seventh of this (which is 2 + 1/4 + 1/8), does obtain 19. Here we see another significant step in the development of mathematics, for the check is a simple instance of a proof."

[12] (Boyer 1991, "Euclid of Alexandria" p.109) "Book II of the *Elements* is a short one, containing only fourteen propositions, not one of which plays any role in modern textbooks; yet in Euclid's day this book was of great significance. This sharp discrepancy between ancient and modern views is easily explained—today we have symbolic algebra and trigonometry that have replaced the geometric equivalents from Greece. For instance, Proposition 1 of Book II states that "If there be two straight lines, and one of them be cut into any number of segments whatever, the rectangle contained by the two straight lines is equal to the rectangles contained by the uncut straight line and each of the segments." This theorem, which asserts (Fig. 7.5) that AD (AP + PR + RB) = AD·AP + AD·PR + AD·RB, is nothing more than a geometric statement of one of the fundamental laws of arithmetic known today as the distributive law: a (b + c + d) = ab + ac + ad. In later books of the *Elements* (V and VII) we find demonstrations of the commutative and associative laws for multiplication. Whereas in our time magnitudes are represented by letters that are understood to be numbers (either known or unknown) on which we operate with algorithmic rules of algebra, in Euclid's day magnitudes were pictured as line segments satisfying the axions and theorems of geometry. It is sometimes asserted that the Greeks had no algebra, but this is patently false. They had Book II of the *Elements*, which is geometric algebra and served much the same purpose as does our symbolic algebra. There can be little doubt that modern algebra greatly facilitates the manipulation of relationships among magnitudes. But it is undoubtedly also true that a Greek geometer versed in the fourteen theorems of Euclid's "algebra" was far more adept in applying these theorems to practical mensuration than is an experienced geometer of today. Ancient geometric "algebra" was not an ideal tool, but it was far from ineffective. Euclid's statement (Proposition 4), "If a straight line be cut at random, the square on the while is equal to the squares on the segments and twice the rectangle contained by the segments, is a verbose way of saying that $(a+b)^2 = a^2 + 2ab + b^2$,"

[13] (Boyer 1991, "The Heroic Age" pp. 77–78) "Whether deduction came into mathematics in the sixth century BCE or the fourth and whether incommensurability was discovered before or after 400 BCE, there can be no doubt that Greek mathematics had undergone drastic changes by the time of Plato. [...] A "geometric algebra" had to take the place of the older "arithmetic algebra," and in this new algebra there could be no adding of lines to areas or of areas to volumes. From now on there had to be strict homogeneity of terms in equations, and the Mesopotamian normal form, xy = A, x ± y = b, were to be interpreted geometrically. [...] In this way the Greeks built up the solution of quadratic equations by their process known as "the application of areas," a portion of geometric algebra that is fully covered by Euclid's *Elements*. [...] The linear equation ax = bc, for example, was looked upon as an equality of the areas ax and bc, rather than as a proportion—an equality between the two ratios a:b and c:x. Consequently, in constructing the fourth proportion *x* in this case, it was usual to construct a rectangle OCDB with the sides b = OB and c = OC (Fig 5.9) and then along OC to lay off OA = a. One completes the rectangle OCDB and draws the diagonal OE cutting CD in P. It is now clear that CP is the desired line x, for rectangle OARS is equal in area to rectangle OCDB"

[14] (Boyer 1991, "Europe in the Middle Ages" p. 258) "In the arithmetical theorems in Euclid's *Elements* VII–IX, numbers had been represented by line segments to which letters had been attached, and the geometric proofs in al-Khwarizmi's *Algebra* made use of lettered diagrams; but all coefficients in the equations used in the *Algebra* are specific numbers, whether represented by numerals or written out in words. The idea of generality is implied in al-Khwarizmi's exposition, but he had no scheme for expressing algebraically the general propositions that are so readily available in geometry."

[15] (Heath 1981a, "The ('Bloom') of Thymaridas" pp. 94–96) Thymaridas of Paros, an ancient Pythagorean already mentioned (p. 69), was the author of a rule for solving a certain set of *n* simultaneous simple equations connecting *n* unknown quantities. The rule was evidently well known, for it was called by the special name [...] the 'flower' or 'bloom' of Thymaridas. [...] The rule is very obscurely worded, but it states in effect that, if we have the following *n* equations connecting *n* unknown quantities $x, x_1, x_2 ... x_{n-1}$, namely [...] Iamblichus, our informant on this subject, goes on to show that other types of equations can be reduced to this, so that they rule does not 'leave us in the lurch' in those cases either."

[16] (Flegg 1983, "Unknown Numbers" p. 205) "Thymaridas (fourth century) is said to have had this rule for solving a particular set of *n* linear equations in *n* unknowns:
If the sum of *n* quantities be given, and also the sum of every pair containing a particular quantity, then this particular quantity is equal to 1/ (n - 2) of the difference between the sums of these pairs and the first given sum."

[17] (Boyer 1991, "Euclid of Alexandria" p. 100) "but by 306 BCE control of the Egyptian portion of the empire was firmly in the hands of Ptolemy I, and this enlightened ruler was able to turn his attention to constructive efforts. Among his early acts was the establishment at Alexandria of a school or institute, known as the Museum, second to none in its day. As teachers at the school he called a band of leading scholars, among whom was the author of the most fabulously successful mathematics textbook ever written—the *Elements* (*Stoichia*) of Euclid. Considering the fame of the author and of his best seller, remarkably little is known of Euclid's life. So obscure was his life that no birthplace is associated with his name."

[18] (Boyer 1991, "Euclid of Alexandria" p. 101) "The tale related above in connection with a request of Alexander the Great for an easy introduction to geometry is repeated in the case of Ptolemy, who Euclid is reported to have assured that "there is no royal road to geometry.""

[19] (Boyer 1991, "Euclid of Alexandria" p. 104) "Some of the faculty probably excelled in research, others were better fitted to be administrators, and still some others were noted for teaching ability. It would appear, from the reports we have, that Euclid very definitely fitted into the last category. There is no new discovery attributed to him, but he was noted for expository skills."

[20] (Boyer 1991, "Euclid of Alexandria" p. 104) "The *Elements* was not, as is sometimes thought, a compendium of all geometric knowledge; it was instead an introductory textbook covering all *elementary* mathematics."

[21] (Boyer 1991, "Euclid of Alexandria" p. 110) "The same holds true for *Elements* II.5, which contains what we should regard as an impractical circumlocution for $a^2 - b^2 = (a + b)(a - b)$ "

[22] (Boyer 1991, "Euclid of Alexandria" p. 111) "In an exactly analogous manner the quadratic equation $ax + x^2 = b^2$ is solved through the use of II.6: If a straight line be bisected and a straight line be added to it in a straight line, the rectangle contained by the whole (with the added straight line) and the added straight line together with the square on the half is equal to the square on the straight line made up of the half and the added straight line. [...] with II.11 being an important special case of II.6. Here Euclid solves the equation $ax + x^2 = a^2$ "

[23] (Boyer 1991, "Euclid of Alexandria" p. 103) "Euclid's *Data*, a work that has come down to us through both Greek and the Arabic. It seems to have been composed for use at the schools of Alexandria, serving as a companion volume to the first six books of the *Elements* in much the same way that a manual of tables supplements a textbook. [...] It opens with fifteen definitions concerning magnitudes and loci. The body of the text comprises ninety-five statements concerning the implications of conditions and magnitudes that may be given in a problem. [...] There are about two dozen similar statements serving as algebraic rules or formulas. [...] Some of the statements are geometric equivalents of the solution of quadratic equations. For example[...] Eliminating y we have $(a - x)dx = b^2 c$ or $dx^2 - adx + b^2 c = 0$, from which $x = a/2 + / - sqrt((a/2)^2 - b^2 c/d)$. The geometric solution given by Euclid is equivalent to this, except that the negative sign before the radical us used. Statements 84 and 85 in the Data are geometric replacements of the familiar Babylonian algebraic solutions of the systems $xy = a^2$, x ± y = b., which again are the equivalents of solutions of simultaneous equations."

[24] (Boyer 1991, "The Euclidean Synthesis" p. 103) "Eutocius and Proclus both attribute the discovery of the conic sections to Menaechmus, who lived in Athens in the late fourth century BCE. Proclus, quoting Eratosthenes, refers to "the conic section triads of Menaechmus." Since this quotation comes just after a discussion of "the section of a right-angled cone" and "the section of an acute-angled cone," it is inferred that the conic sections were produced by cutting a cone with a plane perpendicular to one of its elements. Then if the vertex angle of the cone is acute, the resulting section (called *oxytome*) is an ellipse. If the angle is right, the section (*orthotome*) is a parabola, and if the angle is obtuse, the section (*amblytome*) is a hyperbola (see Fig. 5.7)."

[25] (Boyer 1991, "The age of Plato and Aristotle" p. 94–95) "If OP=y and OD = x are coordinates of point P, we have y<sup2 = R).OV, or, on substituting equals,
y^2=R'D.OV=AR'.BC/AB.DO.BC/AB=AR'.BC^2/AB^2.x
Inasmuch as segments AR', BC, and AB are the same for all points P on the curve EQDPG, we can write the equation of the curve, a "section of a right-angled cone," as y^2=lx, where l is a constant, later to be known as the latus rectum of the curve. [...] Menaechmus apparently derived these properties of the conic sections and others as well. Since this material has a string resemblance to the use of coordinates, as illustrated above, it has sometimes been maintains that Menaechmus had analytic geometry. Such a judgment is warranted only in part, for certainly Menaechmus was unaware that any equation in two unknown quantities determines a curve. In fact, the general concept of an equation in unknown quantities was alien to Greek thought. [...] He had hit upon the conics in a successful search for curves with the properties appropriate to the duplication of the cube. In terms of modern notation the solution is easily achieved. By shifting the curring plane (Gig. 6.2), we can find a parabola with any latus rectum. If, then, we wish to duplicate a cube of edge a, we locate on a right-angled cone two parabolas, one with latus rectum *a* and another with latus rectum 2*a*. [...] It is probable that Menaechmus knew that the duplication could be achieved also by the use of a rectangular hyperbola and a parabola."

[26] (Boyer 1991, "China and India" pp. 195–197) "estimates concerning the *Chou Pei Suan Ching*, generally considered to be the oldest of the mathematical classics, differ by almost a thousand years. [...] A date of about 300 B.C. would appear reasonable, thus placing it in close competition with another treatise, the *Chiu-chang suan-shu*, composed about 250 B.C., that is, shortly before the Han dynasty (202 B.C.). [...] Almost as old at the *Chou Pei*, and perhaps the most influential of all Chinese mathematical books, was the *Chui-chang suan-shu*, or *Nine Chapters on the Mathematical Art*. This book includes 246 problems on surveying, agriculture, partnerships, engineering, taxation, calculation, the solution of equations, and the properties of right triangles. [...] Chapter eight of the *Nine chapters* is significant for its solution of problems of simultaneous linear equations, using both positive and negative numbers. The last problem int the chapter involves four equations in five unknowns, and the topic of indeterminate equations was to remain a favorite among Oriental peoples."

[27] (Boyer 1991, "China and India" p. 204) "Li Chih (or Li Yeh, 1192–1279), a mathematician of Peking who was offered a government post by Khublai Khan in 1206, but politely found an excuse to decline it. His *Ts'e-yuan hai-ching* (*Sea-Mirror of*

the Circle Measurements) includes 170 problems dealing with[...]some of the problems leading to equations of fourth degree. Although he did not describe his method of solution of equations, including some of sixth degree, it appears that it was not very different form that used by Chu Shih-chieh and Horner. Others who used the Horner method were Ch'in Chiu-shao (c. 1202 – c. 1261) and Yang Hui (fl. c. 1261 – 1275). The former was an unprincipled governor and minister who acquired immense wealth within a hundred days of assuming office. His *Shu-shu chiu-chang* (*Mathematical Treatise in Nine Sections*) marks the high point of Chinese indeterminate analysis, with the invention of routines for solving simultaneous congruences."

[28] (Boyer 1991, "China and India" p. 197) "The Chinese were especially fond of patters; hence, it is not surprising that the first record (of ancient but unknown origin) of a magic square appeared there. [...] The concern for such patterns left the author of the *Nine Chapters* to solve the system of simultaneous linear equations [...] by performing column operations on the matrix [...] to reduce it to [...] The second form represented the equations $36z = 99$, $5y + z = 24$, and $3x + 2y + z = 39$ from which the values of z, y, and x are successively found with ease."

[29] (Boyer 1991, "China and India" pp. 204–205) "The same "Horner" device was used by Yang Hui, about whose life almost nothing is known and who work has survived only in part. Among his contributions that are extant are the earliest Chinese magic squares of order greater than three, including two each of orders four through eight and one each of orders nine and ten."

[30] (Boyer 1991, "China and India" p. 203) "The last and greatest of the Sung mathematicians was Chu Chih-chieh (fl. 1280–1303), yet we known little about him-, [...] Of greater historical and mathematical interest is the *Ssy-yüan yü-chien* (*Precious Mirror of the Four Elements*) of 1303. In the eighteenth century this, too, disappeared in China, only to be rediscovered in the next century. The four elements, called heaven, earth, man, and matter, are the representations of four unknown quantities in the same equation. The book marks the peak in the development of Chinese algebra, for it deals with simultaneous equations and with equations of degrees as high as fourteen. In it the author describes a transformation method that he calls *fan fa*, the elements of which to have arisen long before in China, but which generally bears the name of Horner, who lived half a millennium later."

[31] (Boyer 1991, "China and India" p. 205) "A few of the many summations of series found in the *Precious Mirror* are the following:[...] However, no proofs are given, nor does the topic seem to have been continued again in China until about the nineteenth century. [...] The *Precious Mirror* opens with a diagram of the arithmetic triangle, inappropriately known in the West as "pascal's triangle." (See illustration.) [...] Chu disclaims credit for the triangle, referring to it as a "diagram of the old method for finding eighth and lower powers." A similar arrangement of coefficients through the sixth power had appeared in the work of Yang Hui, but without the round zero symbol."

[32] (Boyer 1991, "Revival and Decline of Greek Mathematics" p. 178) Uncertainty about the life of Diophantus is so great that we do not know definitely in which century he lived. Generally he is assumed to have flourished about 250 CE, but dates a century or more earlier or later are sometimes suggested[...] If this conundrum is historically accurate, Diophantus lived to be eighty-four-years old. [...] The chief Diophantine work known to us is the *Arithmetica*, a treatise originally in thirteen books, only the first six of which have survived."

[33] (Boyer 1991, "Revival and Decline of Greek Mathematics" pp. 180–182) "In this respect it can be compared with the great classics of the earlier Alexandrian Age; yet it has practically nothing in common with these or, in fact, with any traditional Greek mathematics. It represents essentially a new branch and makes use of a different approach. Being divorced from geometric methods, it resembles Babylonian algebra to a large extent. But whereas Babylonian mathematicians had been concerned primarily with *approximate* solutions of *determinate* equations as far as the third degree, the *Arithmetica* of Diophantus (such as we have it) is almost entirely devoted to the *exact* solution of equations, both *determinate* and *indeterminate*. [...] Throughout the six surviving books of *Arithmetica* there is a systematic use of abbreviations for powers of numbers and for relationships and operations. An unknown number is represented by a symbol resembling the Greek letter ζ (perhaps for the last letter of arithmos). [...] It is instead a collection of some 150 problems, all worked out in terms of specific numerical examples, although perhaps generality of method was intended. There is no postulation development, nor is an effort made to find all possible solutions. In the case of quadratic equations with two positive roots, only the larger is give, and negative roots are not recognized. No clear-cut distinction is made between determinate and indeterminate problems, and even for the latter for which the number of solutions generally is unlimited, only a single answer is given. Diophantus solved problems involving several unknown numbers by skillfully expressing all unknown quantities, where possible, in terms of only one of them."

[34] (Boyer 1991, "Revival and Decline of Greek Mathematics" p. 178) "The chief difference between Diophantine syncopation and the modern algebraic notation is the lack of special symbols for operations and relations, as well as of the exponential notation."

[35] (Derbyshire 2006, "The Father of Algebra" pp. 35–36)

[36] (Cooke 1997, "Mathematics in the Roman Empire" pp. 167–168)

[37] (Boyer 1991, "Europe in the Middle Ages" p. 257) "The book makes frequent use of the identities [...] which had appeared in Diophantus and had been widely used by the Arabs."

[38] (Boyer 1991, "The Mathematics of the Hindus" p. 197) "The oldest surviving documents on Hindu mathematics are copies of works written in the middle of the first millennium B.C.E., approximately the time during which Thales and Pythagoras lived. [...] from the sixth century B.C.E."

[39] (Boyer 1991, "China and India" p. 222) "The *Livavanti*, like the *Vija-Ganita*, contains numerous problems dealing with favorite Hindu topics; linear and quadratic equations, both determinate and indeterminate, simple mensuration, arithmetic and geometric progretions, surds, Pythagorean triads, and others."

[40] (Boyer 1991, "The Mathematics of the Hindus" p. 207) "He gave more elegant rules for the sum of the squares and cubes of an initial segment of the positive integers. The sixth part of the product of three quantities consisting of the number of terms, the number of terms plus one, and twice the number of terms plus one is the sum of the squares. The square of the sum of the series is the sum of the cubes."

[41] (Boyer 1991, "China and India" p. 219) "Brahmagupta (fl. 628), who lived in Central India somewhat more than a century after Aryabhata [...] in the trigonometry of his best-known work, the *Brahmasphuta Siddhanta*, [...] here we find general solutions of quadratic equations, including two roots even in cases in which one of them is negative."

[42] (Boyer 1991, "China and India" p. 220) "Hindu algebra is especially noteworthy in its development of indeterminate analysis, to which Brahmagupta made several contributions. For one thing, in his work we find a rule for the formation of Pythagorean triads expressed in the form m, $1/2 \, (m^2/n - n)$, $1/2 \, (m^2/n + n)$; but this is only a modified form of the old Babylonian rule, with which he may have become familiar."

[43] (Boyer 1991, "China and India" p. 221) "he was the first one to give a *general* solution of the linear Diophantine equation $ax + by = c$, where a, b, and c are integers. [...] It is greatly to the credit of Brahmagupta that he gave *all* integral solutions of the linear Diophantine equation, whereas Diophantus himself had been satisfied to give one particular solution of an indeterminate equation. Inasmuch as Brahmagupta used some of the same examples as Diophantus, we see again the likelihood of Greek influence in India—or the possibility that they both made use of a common source, possibly from Babylonia. It is interesting to note also that the algebra of Brahmagupta, like that of Diophantus, was syncopated. Addition was indicated by juxtaposition, subtraction by placing a dot over the subtrahend, and division by placing the divisor below the dividend, as in our fractional notation but without the bar. The operations of multiplication and evolution (the taking of roots), as well as unknown quantities, were represented by abbreviations of appropriate words. [...] Bhaskara (1114 – c. 1185), the leading mathematician of the twelfth century. It was he who filled some of the gaps in Brahmagupta's work, as by giving a general solution of the Pell equation and by considering the problem of division by zero."

[44] (Boyer 1991, "China and India" pp. 222–223) "In treating of the circle and the sphere the *Lilavati* fails also to distinguish between exact and approximate statements. [...] Many of Bhaskara's problems in the *Livavati* and the *Vija-Ganita* evidently were derived from earlier Hindu sources; hence, it is no surprise to note that the author is at his best in dealing with indeterminate analysis."

[45] (Boyer 1991, "The Arabic Hegemony" p. 227) "The first century of the Muslim empire had been devoid of scientific achievement. This period (from about 650 to 750) had been, in fact, perhaps the nadir in the development of mathematics, for the Arabs had not yet achieved intellectual drive, and concern for learning in other parts of the world had faded. Had it not been for the sudden cultural awakening in Islam during the second half of the eighth century, considerably more of ancient science and mathematics would have been lost. [...] It was during the caliphate of al-Mamun (809–833), however, that the Arabs fully indulged their passion for translation. The caliph is said to have had a dream in which Aristotle appeared, and as a consequence al-Mamun determined to have Arabic versions made of all the Greek works that he could lay his hands on, including Ptolemy's *Almagest* and a complete version of Euclid's *Elements*. From the Byzantine Empire, with which the Arabs maintained an uneasy peace, Greek manuscripts were obtained through peace treaties. Al-Mamun established at Baghdad a "House of Wisdom" (Bait al-hikma) comparable to the ancient Museum at Alexandria. Among the faculty members was a mathematician and astronomer, Mohammed ibn-Musa al-Khwarizmi, whose name, like that of Euclid, later was to become a household word in Western Europe. The scholar, who died sometime before 850, wrote more than half a dozen astronomical and mathematical works, of which the earliest were probably based on the *Sindhad* derived from India."

[46] (Boyer 1991, "The Arabic Hegemony" p. 234) "but al-Khwarizmi's work had a serious deficiency that had to be removed before it could serve its purpose effectively in the modern world: a symbolic notation had to be developed to replace the rhetorical form. This step the Arabs never took, except for the replacement of number words by number signs. [...] Thabit was the founder of a school of translators, especially from Greek and Syriac, and to him we owe an immense debt for translations into Arabic of works by Euclid, Archimedes, Apollonius, Ptolemy, and Eutocius."

[47] (Boyer 1991, "The Arabic Hegemony" p. 230) "Al-Khwarizmi continued: "We have said enough so far as numbers are concerned, about the six types of equations. Now, however, it is necessary that we should demonstrate geometrically the truth of the same problems which we have explained in numbers." The ring of this passage is obviously Greek rather than Babylonian

or Indian. There are, therefore, three main schools of thought on the origin of Arabic algebra: one emphasizes Hindu influence, another stresses the Mesopotamian, or Syriac-Persian, tradition, and the third points to Greek inspiration. The truth is probably approached if we combine the three theories."

[48] (Boyer 1991, "The Arabic Hegemony" pp. 228–229) "the author's preface in Arabic gave fulsome praise to Mohammed, the prophet, and to al-Mamun, "the Commander of the Faithful.""

[49] (Boyer 1991, "The Arabic Hegemony" p. 228) "The Arabs in general loved a good clear argument from premise to conclusion, as well as systematic organization—respects in which neither Diophantus nor the Hindus excelled."

[50] (Boyer 1991, "The Arabic Hegemony" p. 229) "It is not certain just what the terms *al-jabr* and *muqabalah* mean, but the usual interpretation is similar to that implied in the translation above. The word *al-jabr* presumably meant something like "restoration" or "completion" and seems to refer to the transposition of subtracted terms to the other side of an equation, which is evident in the treatise; the word *muqabalah* is said to refer to "reduction" or "balancing"—that is, the cancellation of like terms on opposite sides of the equation."

[51] Rashed, R.; Armstrong, Angela (1994), *The Development of Arabic Mathematics*, Springer, pp. 11–2, ISBN 0-7923-2565-6, OCLC 29181926

[52] (Boyer 1991, "The Arabic Hegemony" p. 229) "in six short chapters, of the six types of equations made up from the three kinds of quantities: roots, squares, and numbers (that is x, x^2, and numbers). Chapter I, in three short paragraphs, covers the case of squares equal to roots, expressed in modern notation as $x^2 = 5x$, $x^2/3 = 4x$, and $5x^2 = 10x$, giving the answers x = 5, x = 12, and x = 2 respectively. (The root x = 0 was not recognized.) Chapter II covers the case of squares equal to numbers, and Chapter III solves the cases of roots equal to numbers, again with three illustrations per chapter to cover the cases in which the coefficient of the variable term is equal to, more than, or less than one. Chapters IV, V, and VI are more interesting, for they cover in turn the three classical cases of three-term quadratic equations: (1) squares and roots equal to numbers, (2) squares and numbers equal to roots, and (3) roots and numbers equal to squares."

[53] (Boyer 1991, "The Arabic Hegemony" pp. 229–230) "The solutions are "cookbook" rules for "completing the square" applied to specific instances. [...] In each case only the positive answer is give. [...] Again only one root is given for the other is negative. [...]The six cases of equations given above exhaust all possibilities for linear and quadratic equations having positive roots."

[54] (Boyer 1991, "The Arabic Hegemony" p. 230) "Al-Khwarizmi here calls attention to the fact that what we designate as the discriminant must be positive: "You ought to understand also that when you take the half of the roots in this form of equation and then multiply the half by itself; if that which proceeds or results from the multiplication is less than the units above mentioned as accompanying the square, you have an equation." [...] Once more the steps in completing the square are meticulously indicated, without justification,"

[55] (Boyer 1991, "The Arabic Hegemony" p. 231) "The *Algebra* of al-Khwarizmi betrays unmistakable Hellenic elements,"

[56] (Boyer 1991, "The Arabic Hegemony" p. 233) "A few of al-Khwarizmi's problems give rather clear evidence of Arabic dependence on the Babylonian-Heronian stream of mathematics. One of them presumably was taken directly from Heron, for the figure and dimensions are the same."

[57] (Boyer 1991, "The Arabic Hegemony" p. 228) "the algebra of al-Khwarizmi is thoroughly rhetorical, with none of the syncopation found in the Greek *Arithmetica* or in Brahmagupta's work. Even numbers were written out in words rather than symbols! It is quite unlikely that al-Khwarizmi knew of the work of Diophantus, but he must have been familiar with at least the astronomical and computational portions of Brahmagupta; yet neither al-Khwarizmi nor other Arabic scholars made use of syncopation or of negative numbers."

[58] (Boyer 1991, "The Arabic Hegemony" p. 234) "The *Algebra* of al-Khwarizmi usually is regarded as the first work on the subject, but a recent publication in Turkey raises some questions about this. A manuscript of a work by 'Abd-al-Hamid ibn-Turk, entitled "Logical Necessities in Mixed Equations," was part of a book on *Al-jabr wa'l muqabalah* which was evidently very much the same as that by al-Khwarizmi and was published at about the same time—possibly even earlier. The surviving chapters on "Logical Necessities" give precisely the same type of geometric demonstration as al-Khwarizmi's *Algebra* and in one case the same illustrative example $x^2 + 21 = 10x$. In one respect 'Abd-al-Hamad's exposition is more thorough than that of al-Khwarizmi for he gives geometric figures to prove that if the discriminant is negative, a quadratic equation has no solution. Similarities in the works of the two men and the systematic organization found in them seem to indicate that algebra in their day was not so recent a development as has usually been assumed. When textbooks with a conventional and well-ordered exposition appear simultaneously, a subject is likely to be considerably beyond the formative stage. [...] Note the omission of Diophantus and Pappus, authors who evidently were not at first known in Arabia, although the Diophantine *Arithmetica* became familiar before the end of the tenth century."

[59] (Derbyshire 2006, "The Father of Algebra" p. 49)

[60] O'Connor, John J.; Robertson, Edmund F., "Arabic mathematics: forgotten brilliance?", *MacTutor History of Mathematics archive*, University of St Andrews. "Algebra was a unifying theory which allowed rational numbers, irrational numbers, geometrical magnitudes, etc., to all be treated as "algebraic objects"."

[61] Jacques Sesiano, "Islamic mathematics", p. 148, in Selin, Helaine; D'Ambrosio, Ubiratan, eds. (2000), *Mathematics Across Cultures: The History of Non-Western Mathematics*, Springer, ISBN 1-4020-0260-2

[62] Berggren, J. Lennart (2007). "Mathematics in Medieval Islam". *The Mathematics of Egypt, Mesopotamia, China, India, and Islam: A Sourcebook*. Princeton University Press. p. 518. ISBN 978-0-691-11485-9.

[63] (Boyer 1991, "The Arabic Hegemony" p. 239) "Abu'l Wefa was a capable algebraist as well as a trionometer. [...] His successor al-Karkhi evidently used this translation to become an Arabic disciple of Diophantus—but without Diophantine analysis! [...] In particular, to al-Karkhi is attributed the first numerical solution of equations of the form $ax^{2n} + bx^n = c$ (only equations with positive roots were considered),"

[64] O'Connor, John J.; Robertson, Edmund F., "Abu Bekr ibn Muhammad ibn al-Husayn Al-Karaji", *MacTutor History of Mathematics archive*, University of St Andrews.

[65] (Boyer 1991, "The Arabic Hegemony" pp. 241–242) "Omar Khayyam (c. 1050 – 1123), the "tent-maker," wrote an *Algebra* that went beyond that of al-Khwarizmi to include equations of third degree. Like his Arab predecessors, Omar Khayyam provided for quadratic equations both arithmetic and geometric solutions; for general cubic equations, he believed (mistakenly, as the sixteenth century later showed), arithmetic solutions were impossible; hence he gave only geometric solutions. The scheme of using intersecting conics to solve cubics had been used earlier by Menaechmus, Archimedes, and Alhazan, but Omar Khayyam took the praiseworthy step of generalizing the method to cover all third-degree equations (having positive roots). .. For equations of higher degree than three, Omar Khayyam evidently did not envision similar geometric methods, for space does not contain more than three dimensions, [...] One of the most fruitful contributions of Arabic eclecticism was the tendency to close the gap between numerical and geometric algebra. The decisive step in this direction came much later with Descartes, but Omar Khayyam was moving in this direction when he wrote, "Whoever thinks algebra is a trick in obtaining unknowns has thought it in vain. No attention should be paid to the fact that algebra and geometry are different in appearance. Algebras are geometric facts which are proved.""

[66] O'Connor, John J.; Robertson, Edmund F., "Sharaf al-Din al-Muzaffar al-Tusi", *MacTutor History of Mathematics archive*, University of St Andrews.

[67] Rashed, Roshdi; Armstrong, Angela (1994), *The Development of Arabic Mathematics*, Springer, pp. 342–3, ISBN 0-7923-2565-6

[68] Berggren, J. L. (1990), "Innovation and Tradition in Sharaf al-Din al-Tusi's Muadalat", *Journal of the American Oriental Society* **110** (2): 304–9, doi:10.2307/604533, Rashed has argued that Sharaf al-Din discovered the derivative of cubic polynomials and realized its significance for investigating conditions under which cubic equations were solvable; however, other scholars have suggested quite difference explanations of Sharaf al-Din's thinking, which connect it with mathematics found in Euclid or Archimedes.

[69] Victor J. Katz, Bill Barton (October 2007), "Stages in the History of Algebra with Implications for Teaching", *Educational Studies in Mathematics* (Springer Netherlands) **66** (2): 185–201 [192], doi:10.1007/s10649-006-9023-7

[70] Tjalling J. Ypma (1995), "Historical development of the Newton-Raphson method", *SIAM Review* **37** (4): 531–51, doi:10.1137/1037125

[71] O'Connor, John J.; Robertson, Edmund F., "Abu'l Hasan ibn Ali al Qalasadi", *MacTutor History of Mathematics archive*, University of St Andrews.

[72] (Boyer 1991, "Euclid of Alexandria pp. 192–193) "The death of Boethius may be taken to mark the end of ancient mathematics in the Western Roman Empire, as the death of Hypatia had marked the close of Alexandria as a mathematical center; but work continued for a few years longer at Athens. [...] When in 527 Justinian became emperor in the East, he evidently felt that the pagan learning of the Academy and other philosophical schools at Athens was a threat to orthodox Christianity; hence, in 529 the philosophical schools were closed and the scholars dispersed. Rome at the time was scarcely a very hospitable home for scholars, and Simplicius and some of the other philosophers looked to the East for haven. This they found in Persia, where under King Chosroes they established what might be called the "Athenian Academy in Exile."(Sarton 1952; p. 400)."

[73] Cajori, F. (1928). *A History of Mathematical Notations* (v. 1-2). Dover Publications. p. 382. ISBN 9780486677668. Retrieved 2015-06-20.

[74] Struik (1969), 367

[75] Andrew Warwick (2003) *Masters of Theory: Cambridge and the Rise of Mathematical Physics*, Chicago: University of Chicago Press ISBN 0-226-87374-9

[76] (Boyer 1991, "The Arabic Hegemony" p. 228) "Diophantus sometimes is called "the father of algebra," but this title more appropriately belongs to Abu Abdullah bin mirsmi al-Khwarizmi. It is true that in two respects the work of al-Khwarizmi represented a retrogression from that of Diophantus. First, it is on a far more elementary level than that found in the Diophantine problems and, second, the algebra of al-Khwarizmi is thoroughly rhetorical, with none of the syncopation found in the Greek *Arithmetica* or in Brahmagupta's work. Even numbers were written out in words rather than symbols! It is quite unlikely that al-Khwarizmi knew of the work of Diophantus, but he must have been familiar with at least the astronomical and computational portions of Brahmagupta; yet neither al-Khwarizmi nor other Arabic scholars made use of syncopation or of negative numbers."

[77] (Derbyshire 2006, "The Father of Algebra" p. 31) "Diophantus, the father of algebra, in whose honor I have named this chapter, lived in Alexandria, in Roman Egypt, in either the 1st, the 2nd, or the 3rd century CE."

[78] (Boyer 1991, "The Arabic Hegemony" p. 230) "The six cases of equations given above exhaust all possibilities for linear and quadratic equations having positive root. So systematic and exhaustive was al-Khwarizmi's exposition that his readers must have had little difficulty in mastering the solutions."

[79] Gandz and Saloman (1936), *The sources of al-Khwarizmi's algebra*, Osiris i, p. 263–277: "In a sense, Khwarizmi is more entitled to be called "the father of algebra" than Diophantus because Khwarizmi is the first to teach algebra in an elementary form and for its own sake, Diophantus is primarily concerned with the theory of numbers".

2.15   References

- Bashmakova, I, and Smirnova, G. (2000) *The Beginnings and Evolution of Algebra*, Dolciani Mathematical Expositions 23. Translated by Abe Shenitzer. The Mathematical Association of America.

- Boyer, Carl B. (1991), *A History of Mathematics* (Second Edition ed.), John Wiley & Sons, Inc., ISBN 0-471-54397-7

- Cooke, Roger (1997), *The History of Mathematics: A Brief Course*, Wiley-Interscience, ISBN 0-471-18082-3

- Derbyshire, John (2006), *Unknown Quantity: A Real And Imaginary History of Algebra*, Joseph Henry Press, ISBN 0-309-09657-X

- Stillwell, John (2004), *Mathematics and its History* (Second Edition ed.), Springer Science + Business Media Inc., ISBN 0-387-95336-1

- Burton, David M. (1997), *The History of Mathematics: An Introduction* (Third Edition ed.), The McGraw-Hill Companies, Inc., ISBN 0-07-009465-9

- Heath, Thomas Little (1981a), *A History of Greek Mathematics, Volume I*, Dover publications, ISBN 0-486-24073-8

- Heath, Thomas Little (1981b), *A History of Greek Mathematics, Volume II*, Dover publications, ISBN 0-486-24074-6

- Flegg, Graham (1983), *Numbers: Their History and Meaning*, Dover publications, ISBN 0-486-42165-1

2.16   External links

- "Commentary by Islam's Sheikh Zakariyya al-Ansari on Ibn al-Hā'im's Poem on the Science of Algebra and Balancing Called the Creator's Epiphany in Explaining the Cogent" featuring the basic concepts of algebra dating back to the 15th century, from the World Digital Library.

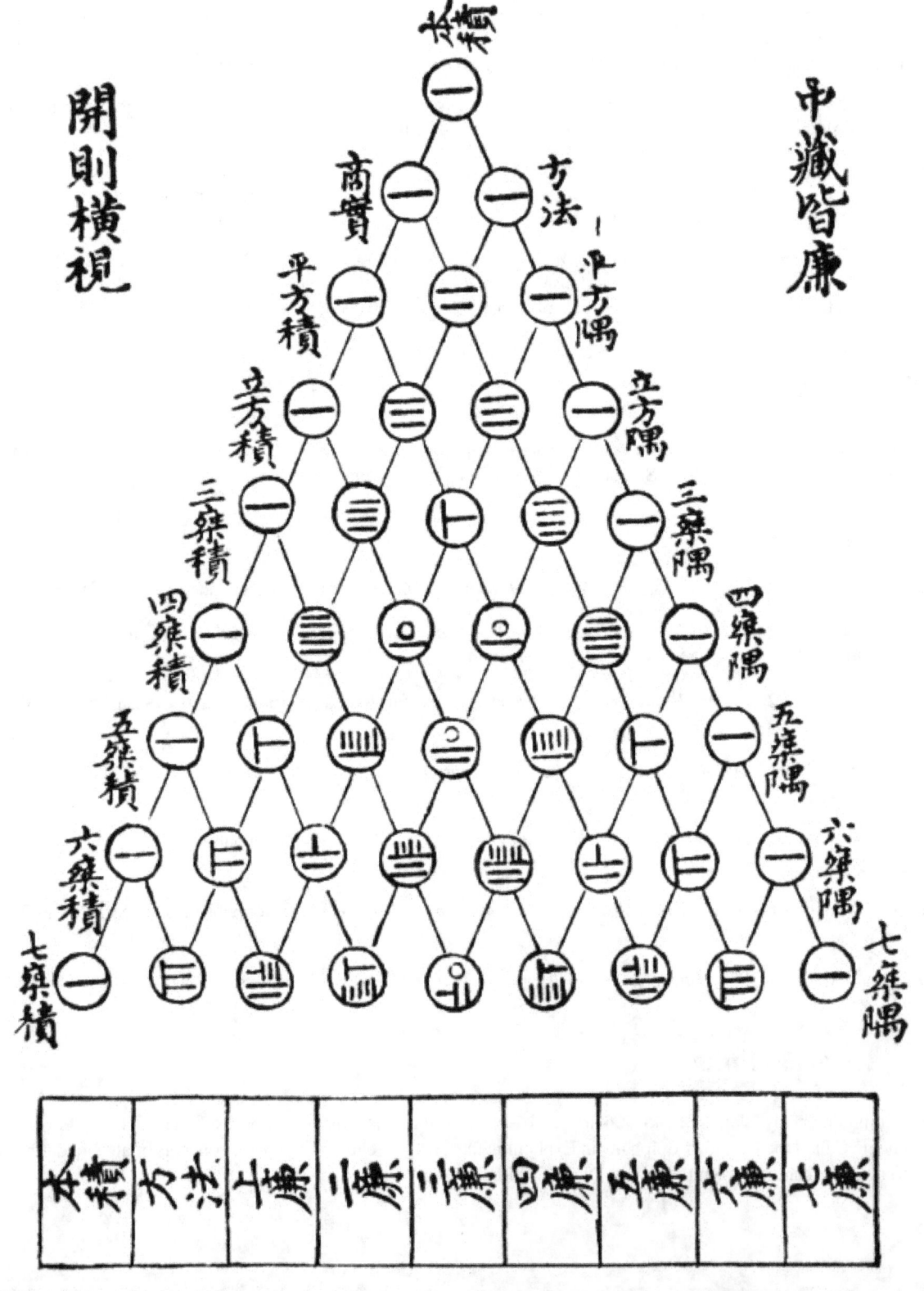

Yang Hui (Pascal's) triangle, as depicted by the ancient Chinese using rod numerals.

DIOPHANTI
ALEXANDRINI
ARITHMETICORVM
LIBRI SEX,

ET DE NVMERIS MVLTANGVLIS,

LIBER VNVS.

Nunc primùm Græcè & Latinè editi, atque absolutissimis
Commentariis illustrati.

AVCTORE CLAVDIO GASPARE BACHETO
MEZIRIACO SEBVSIANO. V. C.

LVTETIAE PARISIORVM,

Sumptibus Sebastiani Cramoisy, via
Iacobæa, sub Ciconiis.

M. DC. XXI.

CVM PRIVILEGIO REGIS.

Omar Khayyam is credited with identifying the foundations of algebraic geometry and found the general geometric solution of the cubic equation.

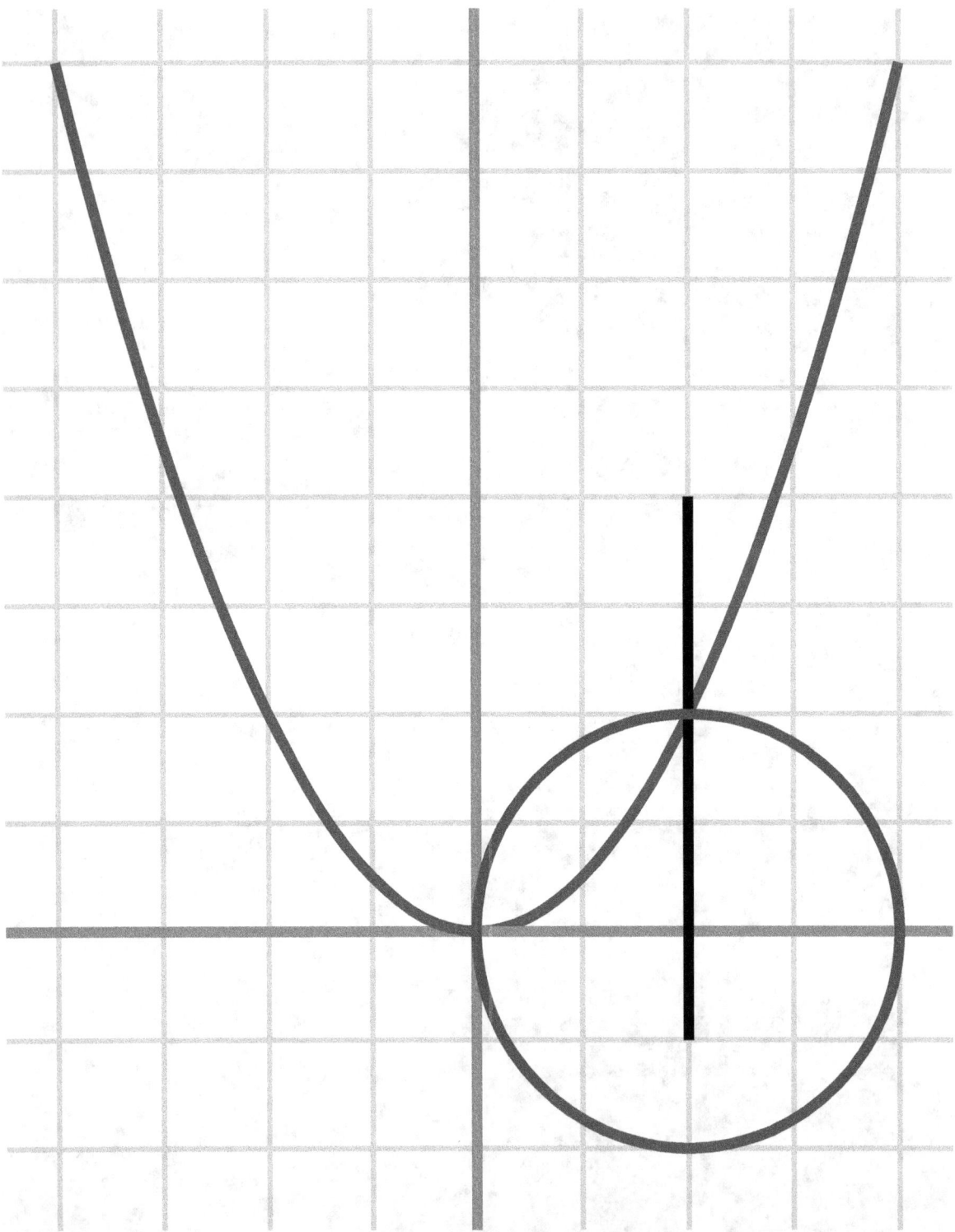

To solve the third-degree equation $x^3 + a^2x = b$ Khayyám constructed the parabola $x^2 = ay$, a circle with diameter b/a^2, and a vertical line through the intersection point. The solution is given by the length of the horizontal line segment from the origin to the intersection of the vertical line and the x-axis.

Chapter 3

History of calculus

Calculus, known in its early history as *infinitesimal calculus*, is a mathematical discipline focused on limits, functions, derivatives, integrals, and infinite series. Isaac Newton and Gottfried Leibniz independently invented calculus in the mid-17th century. However, each inventor claimed that the other one stole his work in a bitter dispute that continued until the end of their lives.

3.1 Precursors of calculus

3.1.1 Ancient

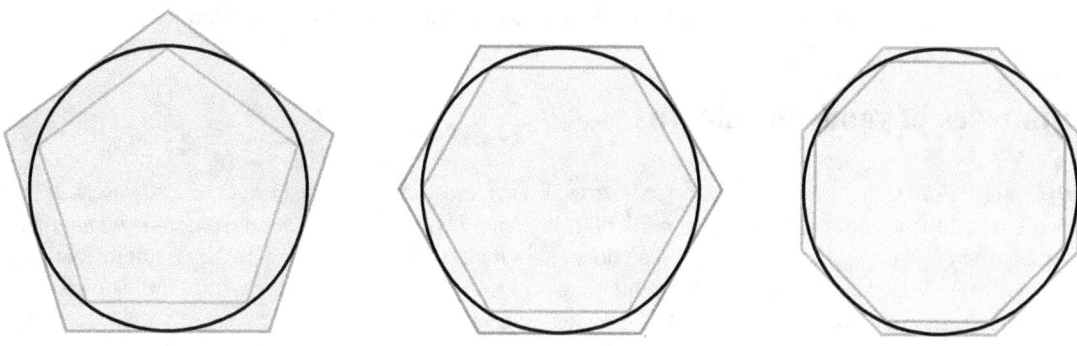

Archimedes used the method of exhaustion to compute the area inside a circle

The ancient period introduced some of the ideas that led to integral calculus, but does not seem to have developed these ideas in a rigorous and systematic way. Calculations of volumes and areas, one goal of integral calculus, can be found in the Egyptian Moscow papyrus (c. 1820 BC), but the formulas are only given for concrete numbers, some are only approximately true, and they are not derived by deductive reasoning.[1]

From the age of Greek mathematics, Eudoxus (c. 408–355 BC) used the method of exhaustion, which foreshadows the concept of the limit, to calculate areas and volumes, while Archimedes (c. 287–212 BC) developed this idea further, inventing heuristics which resemble the methods of integral calculus.[2] The method of exhaustion was later reinvented in China by Liu Hui in the 3rd century AD in order to find the area of a circle.[3] In the 5th century AD, Zu Chongzhi established a method that would later be called Cavalieri's principle to find the volume of a sphere.[4] Greek mathematicians are also credited with a significant use of infinitesimals. Democritus is the first person recorded to consider seriously the division of objects into an infinite number of cross-sections, but his inability to rationalize discrete cross-sections with a

66

cone's smooth slope prevented him from accepting the idea. At approximately the same time, Zeno of Elea discredited infinitesimals further by his articulation of the paradoxes which they create.

Archimedes of Syracuse developed this method further, while also inventing heuristic methods which resemble modern day concepts somewhat in his *The Quadrature of the Parabola*, *The Method*, and *Archimedes on Spheres & Cylinders*.[5] It should not be thought that infinitesimals were put on a rigorous footing during this time, however. Only when it was supplemented by a proper geometric proof would Greek mathematicians accept a proposition as true. It was not until the 17th century that the method was formalized by Cavalieri as the method of Indivisibles and eventually incorporated by Newton into a general framework of integral calculus. Archimedes was the first to find the tangent to a curve, other than a circle, in a method akin to differential calculus. While studying the spiral, he separated a point's motion into two components, one radial motion component and one circular motion component, and then continued to add the two component motions together, thereby finding the tangent to the curve.[6] The pioneers of the calculus such as Isaac Barrow and Johann Bernoulli were diligent students of Archimedes; see for instance C. S. Roero (1983).

3.1.2 Medieval

India had a long history of trigonometry as witnessed by the 8th century BCE treatise Sulba Sutras, or rules of the chord, where the sine, cosine, and tangent were conceived. Indian mathematicians gave a semi-rigorous method of differentiation of some trigonometric functions. In the Middle East, Alhazen derived a formula for the sum of fourth powers. He used the results to carry out what would now be called an integration, where the formulas for the sums of integral squares and fourth powers allowed him to calculate the volume of a paraboloid.[7] In the 14th century, Indian mathematician Madhava of Sangamagrama and the Kerala school of astronomy and mathematics stated components of calculus such as the Taylor series and infinite series approximations.[8] However, they were not able to combine many differing ideas under the two unifying themes of the derivative and the integral, show the connection between the two, and turn calculus into the powerful problem-solving tool we have today.[7]

The mathematical study of continuity was revived in the 14th century by the Oxford Calculators and French collaborators such as Nicole Oresme. They proved the "Merton mean speed theorem": that a uniformly accelerated body travels the same distance as a body with uniform speed whose speed is half the final velocity of the accelerated body.[9]

3.2 Pioneers of modern calculus

In the 17th century, European mathematicians Isaac Barrow, René Descartes, Pierre de Fermat, Blaise Pascal, John Wallis and others discussed the idea of a derivative. In particular, in *Methodus ad disquirendam maximam et minima* and in *De tangentibus linearum curvarum*, Fermat developed an adequality method for determining maxima, minima, and tangents to various curves that was closely related to differentiation.[10] Isaac Newton would later write that his own early ideas about calculus came directly from "Fermat's way of drawing tangents."[11]

On the integral side, Cavalieri developed his method of indivisibles in the 1630s and 1640s, providing a more modern form of the ancient Greek method of exhaustion, and computing Cavalieri's quadrature formula, the area under the curves x^n of higher degree, which had previously only been computed for the parabola, by Archimedes. Torricelli extended this work to other curves such as the cycloid, and then the formula was generalized to fractional and negative powers by Wallis in 1656. In a 1659 treatise, Fermat is credited with an ingenious trick for evaluating the integral of any power function directly.[12] Fermat also obtained a technique for finding the centers of gravity of various plane and solid figures, which influenced further work in quadrature. James Gregory, influenced by Fermat's contributions both to tangency and to quadrature, was then able to prove a restricted version of the second fundamental theorem of calculus in the mid-17th century. The first full proof of the fundamental theorem of calculus was given by Isaac Barrow.[13]

Newton and Leibniz, building on this work, independently developed the surrounding theory of infinitesimal calculus in the late 17th century. Also, Leibniz did a great deal of work with developing consistent and useful notation and concepts. Newton provided some of the most important applications to physics, especially of integral calculus.

The first proof of Rolle's theorem was given by Michel Rolle in 1691 using methods developed by the Dutch mathematician Johann van Waveren Hudde.[14] The mean value theorem in its modern form was stated by Bernard Bolzano and Augustin-Louis Cauchy (1789–1857) also after the founding of modern calculus. Important contributions were also made by

Barrow, Huygens, and many others.

3.3   Newton and Leibniz

See also: Leibniz–Newton calculus controversy

Before Newton and Leibniz, the word "calculus" was a general term used to refer to any body of mathematics, but in the following years, "calculus" became a popular term for a field of mathematics based upon their insights.[15] The purpose of this section is to examine Newton and Leibniz's investigations into the developing field of infinitesimal calculus. Specific importance will be put on the justification and descriptive terms which they used in an attempt to understand calculus as they themselves conceived it.

By the middle of the 17th century, European mathematics had changed its primary repository of knowledge. In comparison to the last century which maintained Hellenistic mathematics as the starting point for research, Newton, Leibniz and their contemporaries increasingly looked towards the works of more modern thinkers.[16] Europe had become home to a burgeoning mathematical community and with the advent of enhanced institutional and organizational bases a new level of organization and academic integration was being achieved. Importantly, however, the community lacked formalism; instead it consisted of a disordered mass of various methods, techniques, notations, theories, and paradoxes.

Newton came to calculus as part of his investigations in physics and geometry. He viewed calculus as the scientific description of the generation of motion and magnitudes. In comparison, Leibniz focused on the tangent problem and came to believe that calculus was a metaphysical explanation of change. Importantly, the core of their insight was the formalization of the inverse properties between the integral and the differential of a function. This insight had been anticipated by their predecessors, but they were the first to conceive calculus as a system in which new rhetoric and descriptive terms were created.[17] Their unique discoveries lay not only in their imagination, but also in their ability to synthesize the insights around them into a universal algorithmic process, thereby forming a new mathematical system.

3.3.1   Newton

Newton completed no definitive publication formalizing his Fluxional Calculus; rather, many of his mathematical discoveries were transmitted through correspondence, smaller papers or as embedded aspects in his other definitive compilations, such as the *Principia* and *Opticks*. Newton would begin his mathematical training as the chosen heir of Isaac Barrow in Cambridge. His incredible aptitude was recognized early and he quickly learned the current theories. By 1664 Newton had made his first important contribution by advancing the binomial theorem, which he had extended to include fractional and negative exponents. Newton succeeded in expanding the applicability of the binomial theorem by applying the algebra of finite quantities in an analysis of infinite series. He showed a willingness to view infinite series not only as approximate devices, but also as alternative forms of expressing a term.[18]

Many of Newton's critical insights occurred during the plague years of 1665-1666[19] which he later described as, "the prime of my age for invention and minded mathematics and [natural] philosophy more than at any time since." It was during his plague-induced isolation that the first written conception of Fluxionary Calculus was recorded in the unpublished *De Analysi per Aequationes Numero Terminorum Infinitas*. In this paper, Newton determined the area under a curve by first calculating a momentary rate of change and then extrapolating the total area. He began by reasoning about an indefinitely small triangle whose area is a function of x and y. He then reasoned that the infinitesimal increase in the abscissa will create a new formula where $x = x + o$ (importantly, o is the letter, not the digit 0). He then recalculated the area with the aid of the binomial theorem, removed all quantities containing the letter o and re-formed an algebraic expression for the area. Significantly, Newton would then "blot out" the quantities containing o because terms "multiplied by it will be nothing in respect to the rest".

At this point Newton had begun to realize the central property of inversion. He had created an expression for the area under a curve by considering a momentary increase at a point. In effect, the fundamental theorem of calculus was built into his calculations. While his new formulation offered incredible potential, Newton was well aware of its logical limitations at the time. He admits that "errors are not to be disregarded in mathematics, no matter how small" and that what he had achieved was "shortly explained rather than accurately demonstrated."

In an effort to give calculus a more rigorous explication and framework, Newton compiled in 1671 the *Methodus Fluxionum et Serierum Infinitarum*. In this book, Newton's strict empiricism shaped and defined his Fluxional Calculus. He exploited instantaneous motion and infinitesimals informally. He used math as a methodological tool to explain the physical world. The base of Newton's revised Calculus became continuity; as such he redefined his calculations in terms of continual flowing motion. For Newton, variable magnitudes are not aggregates of infinitesimal elements, but are generated by the indisputable fact of motion. As with many of his works, Newton delayed publication. *Methodus Fluxionum* was not published until 1736.[20]

Newton attempted to avoid the use of the infinitesimal by forming calculations based on ratios of changes. In the *Methodus Fluxionum* he defined the rate of generated change as a fluxion, which he represented by a dotted letter, and the quantity generated he defined as a fluent. For example, if x and y are fluents, then $\dot{x}$ and $\dot{y}$ are their respective fluxions. This revised calculus of ratios continued to be developed and was maturely stated in the 1676 text *De Quadratura Curvarum* where Newton came to define the present day derivative as the ultimate ratio of change, which he defined as the ratio between evanescent increments (the ratio of fluxions) purely at the moment in question. Essentially, the ultimate ratio is the ratio as the increments vanish into nothingness. Importantly, Newton explained the existence of the ultimate ratio by appealing to motion;

"For by the ultimate velocity is meant that, with which the body is moved, neither before it arrives at its last place, when the motion ceases nor after but at the very instant when it arrives... the ultimate ratio of evanescent quantities is to be understood, the ratio of quantities not before they vanish, not after, but with which they vanish"[21]

Newton developed his Fluxional Calculus in an attempt to evade the informal use of infinitesimals in his calculations.

3.3.2 Leibniz

While Newton began development of his fluxional calculus in 1665-1666 his findings did not become widely circulated until later. In the intervening years Leibniz also strove to create his calculus. In comparison to Newton who came to math at an early age, Leibniz began his rigorous math studies with a mature intellect. He was a polymath, and his intellectual interests and achievements involved metaphysics, law, economics, politics, logic, and mathematics. In order to understand Leibniz's reasoning in calculus his background should be kept in mind. Particularly, his metaphysics which considered the world as an infinite aggregate of indivisible monads. and his plans of creating a precise formal logic whereby, "a general method in which all truths of the reason would be reduced to a kind of calculation." In 1672 Leibniz met the mathematician Huygens who convinced Leibniz to dedicate significant time to the study of mathematics. By 1673 he had progressed to reading Pascal's *Traité des Sinus du Quarte Cercle* and it was during his largely autodidactic research that Leibniz said "a light turned on" . Like Newton, Leibniz, saw the tangent as a ratio but declared it as simply the ratio between ordinates and abscissas. He continued this reasoning to argue that the integral was in fact the sum of the ordinates for infinitesimal intervals in the abscissa; in effect, the sum of an infinite number of rectangles. From these definitions the inverse relationship or differential became clear and Leibniz quickly realized the potential to form a whole new system of mathematics. Where Newton over the course of his career used several approaches in addition to an approach using infinitesimals, Leibniz made it the cornerstone of his notation and calculus.

In the manuscripts of 25 October to 11 November 1675, Leibniz recorded his discoveries and experiments with various forms of notation. He was acutely aware of the notational terms used and his earlier plans to form a precise logical symbolism became evident. Eventually, Leibniz denoted the infinitesimal increments of abscissas and ordinates dx and dy, and the summation of infinitely many infinitesimally thin rectangles as a long s ($\int$), which became the present integral symbol $\int$.

While Leibniz's notation is used by modern mathematics, his logical base was different from our current one. Leibniz embraced infinitesimals and wrote extensively so as, "not to make of the infinitely small a mystery, as had Pascal." According to Deleuze, Leibniz's zeroes "are nothings, but they are not absolute nothings, they are nothings respectively" (quoting Leibniz' text "Justification of the calculus of infinitesimals by the calculus of ordinary algebra.") [22] Alternatively, he defines them as, "less than any given quantity." For Leibniz, the world was an aggregate of infinitesimal points and the lack of scientific proof for their existence did not trouble him. Infinitesimals to Leibniz were ideal quantities of a different type from appreciable numbers. The truth of continuity was proven by existence itself. For Leibniz the principle of continuity and thus the validity of his Calculus was assured. Three hundred years after Leibniz's work, Abraham Robinson showed that using infinitesimal quantities in calculus could be given a solid foundation.

3.3.3   Legacy

The rise of Calculus stands out as a unique moment in mathematics. Calculus is the math of motion and change, and as such, its invention required the creation of a new mathematical system. Importantly, Newton and Leibniz did not create the same Calculus and they did not conceive of modern Calculus. While they were both involved in the process of creating a mathematical system to deal with variable quantities their elementary base was different. For Newton, change was a variable quantity over time and for Leibniz it was the difference ranging over a sequence of infinitely close values. Notably, the descriptive terms each system created to describe change was different.

Historically, there was much debate over whether it was Newton or Leibniz who first "invented" calculus. This argument, the Leibniz and Newton calculus controversy, involving Leibniz, who was German, and the Englishman Newton, led to a rift in the European mathematical community lasting over a century. Leibniz was the first to publish his investigations; however, it is well established that Newton had started his work several years prior to Leibniz and had already developed a theory of tangents by the time Leibniz became interested in the question. Much of the controversy centers on the question whether Leibniz had seen certain early manuscripts of Newton before publishing his own memoirs on the subject. Newton began his work on calculus no later than 1666, and Leibniz did not begin his work until 1673. Leibniz visited England in 1673 and again in 1676, and was shown some of Newton's unpublished writings. He also corresponded with several English scientists (as well as with Newton himself), and may have gained access to Newton's manuscripts through them. It is not known how much this may have influenced Leibniz. The initial accusations were made by students and supporters of the two great scientists at the turn of the century, but after 1711 both of them became personally involved, accusing each other of plagiarism.

The priority dispute had an effect of separating English-speaking mathematicians from those in the continental Europe for many years. Only in the 1820s, due to the efforts of the Analytical Society, did Leibnizian analytical calculus become accepted in England. Today, both Newton and Leibniz are given credit for independently developing the basics of calculus. It is Leibniz, however, who is credited with giving the new discipline the name it is known by today: "calculus". Newton's name for it was "the science of fluents and fluxions".

The work of both Newton and Leibniz is reflected in the notation used today. Newton introduced the notation $\dot{f}$ for the derivative of a function f.[23] Leibniz introduced the symbol $\int$ for the integral and wrote the derivative of a function y of the variable x as $\frac{dy}{dx}$, both of which are still in use.

3.4   Integrals

Niels Henrik Abel seems to have been the first to consider in a general way the question as to what differential expressions can be integrated in a finite form by the aid of ordinary functions, an investigation extended by Liouville. Cauchy early undertook the general theory of determining definite integrals, and the subject has been prominent during the 19th century. Frullani's theorem (1821), Bierens de Haan's work on the theory (1862) and his elaborate tables (1867), Lejeune Dirichlet's lectures (1858) embodied in Meyer's treatise (1871), and numerous memoirs of Legendre, Poisson, Plana, Raabe, Sohncke, Schlömilch, Elliott, Leudesdorf, and Kronecker are among the noteworthy contributions.

Eulerian integrals were first studied by Euler and afterwards investigated by Legendre, by whom they were classed as Eulerian integrals of the first and second species, as follows:

$$\int_0^1 x^{n-1}(1-x)^{n-1}\, dx$$

$$\int_0^\infty e^{-x} x^{n-1}\, dx$$

although these were not the exact forms of Euler's study.

If n is an integer, it follows that:

$$\int_0^\infty e^{-x} x^{n-1} dx = (n-1)!,$$

but the integral converges for all positive real n and defines an analytic continuation of the factorial function to all of the complex plane except for poles at zero and the negative integers. To it Legendre assigned the symbol Γ, and it is now called the gamma function. Besides being analytic over positive reals $\mathbb{R}^+$, Γ also enjoys the uniquely defining property that $\log \Gamma$ is convex, which aesthetically justifies this analytic continuation of the factorial function over any other analytic continuation. To the subject Lejeune Dirichlet has contributed an important theorem (Liouville, 1839), which has been elaborated by Liouville, Catalan, Leslie Ellis, and others. On the evaluation of $\Gamma(x)$ and $\log \Gamma(x)$ Raabe (1843–44), Bauer (1859), and Gudermann (1845) have written. Legendre's great table appeared in 1816.

3.5   Symbolic methods

Symbolic methods may be traced back to Taylor, and the much debated analogy between successive differentiation and ordinary exponentials had been observed by numerous writers before the 19th century. Arbogast (1800) was the first, however, to separate the symbol of operation from that of quantity in a differential equation. François (1812) and Servois (1814) seem to have been the first to give correct rules on the subject. Hargreave (1848) applied these methods in his memoir on differential equations, and Boole freely employed them. Grassmann and Hermann Hankel made great use of the theory, the former in studying equations, the latter in his theory of complex numbers.

3.6   Calculus of variations

The calculus of variations may be said to begin with a problem of Johann Bernoulli's (1696). It immediately occupied the attention of Jakob Bernoulli and the Marquis de l'Hôpital, but Euler first elaborated the subject. His contributions began in 1733, and his *Elementa Calculi Variationum* gave to the science its name. Lagrange contributed extensively to the theory, and Legendre (1786) laid down a method, not entirely satisfactory, for the discrimination of maxima and minima. To this discrimination Brunacci (1810), Gauss (1829), Poisson (1831), Ostrogradsky (1834), and Jacobi (1837) have been among the contributors. An important general work is that of Sarrus (1842) which was condensed and improved by Cauchy (1844). Other valuable treatises and memoirs have been written by Strauch (1849), Jellett (1850), Hesse (1857), Clebsch (1858), and Carll (1885), but perhaps the most important work of the century is that of Weierstrass. His course on the theory may be asserted that he was the first to place calculus on a firm and rigorous foundation.

3.7   Applications

The application of the infinitesimal calculus to problems in physics and astronomy was contemporary with the origin of the science. All through the 18th century these applications were multiplied, until at its close Laplace and Lagrange had brought the whole range of the study of forces into the realm of analysis. To Lagrange (1773) we owe the introduction of the theory of the potential into dynamics, although the name "potential function" and the fundamental memoir of the subject are due to Green (1827, printed in 1828). The name "potential" is due to Gauss (1840), and the distinction between potential and potential function to Clausius. With its development are connected the names of Lejeune Dirichlet, Riemann, von Neumann, Heine, Kronecker, Lipschitz, Christoffel, Kirchhoff, Beltrami, and many of the leading physicists of the century.

It is impossible in this place to enter into the great variety of other applications of analysis to physical problems. Among them are the investigations of Euler on vibrating chords; Sophie Germain on elastic membranes; Poisson, Lamé, Saint-Venant, and Clebsch on the elasticity of three-dimensional bodies; Fourier on heat diffusion; Fresnel on light; Maxwell, Helmholtz, and Hertz on electricity; Hansen, Hill, and Gyldén on astronomy; Maxwell on spherical harmonics; Lord Rayleigh on acoustics; and the contributions of Lejeune Dirichlet, Weber, Kirchhoff, F. Neumann, Lord Kelvin, Clausius, Bjerknes, MacCullagh, and Fuhrmann to physics in general. The labors of Helmholtz should be especially mentioned, since he contributed to the theories of dynamics, electricity, etc., and brought his great analytical powers to bear on the fundamental axioms of mechanics as well as on those of pure mathematics.

Furthermore, infinitesimal calculus was introduced into the social sciences, starting with Neoclassical economics. Today, it is a valuable tool in mainstream economics.

3.8 Non-European antecedents of the calculus

3.8.1 Indian mathematics

Main article: Kerala school of astronomy and mathematics
See also: Yuktibhāṣā

3.8.2 Islamic mathematics

In the 11th century, when Ibn al-Haytham (known as *Alhacen* in Europe), an Iraqi mathematician working in Egypt, devised what is now known as "Alhazen's problem", which leads to an equation of the fourth degree, in his *Book of Optics*. While solving this problem, he was the first mathematician to derive the formula for the sum of the fourth powers, using a method that is readily generalizable for determining the general formula for the sum of any integral powers. He performed an integration in order to find the volume of a paraboloid, and was able to generalize his result for the integrals of polynomials up to the fourth degree. He thus came close to finding a general formula for the integrals of polynomials, but he was not concerned with any polynomials higher than the fourth degree.[24]

3.9 See also

- Analytic geometry

- Non-standard calculus

- History of mathematics

3.10 Notes

[1] Kline, Morris. *Mathematical thought from ancient to modern times* **1**. Oxford University Press. pp. 18–21. ISBN 978-0-19-506135-2.

[2] Archimedes, *Method*, in *The Works of Archimedes* ISBN 978-0-521-66160-7

[3] Dun, Liu; Fan, Dainian; Cohen, Robert Sonné (1966). "A comparison of Archimdes' and Liu Hui's studies of circles". Chinese studies in the history and philosophy of science and technology **130**. Springer. p. 279. ISBN 0-7923-3463-9., Chapter , p. 279

[4] Zill, Dennis G.; Wright, Scott; Wright, Warren S. (2009). *Calculus: Early Transcendentals* (3 ed.). Jones & Bartlett Learning. p. xxvii. ISBN 0-7637-5995-3., Extract of page 27

[5] MathPages — Archimedes on Spheres & Cylinders

[6] Boyer, Carl B. (1991). "Archimedes of Syracuse". *A History of Mathematics* (2nd ed.). Wiley. p. 127. ISBN 0-471-54397-7. Greek mathematics sometimes has been described as essentially static, with little regard for the notion of variability; but Archimedes, in his study of the spiral, seems to have found the tangent to a curve through kinematic considerations akin to differential calculus. Thinking of a point on the spiral $r = a\theta$ as subjected to a double motion — a uniform radial motion away from the origin of coordinates and a circular motion about the origin — he seems to have found (through the parallelogram of velocities) the direction of motion (hence of the tangent to the curve) by noting the resultant of the two component motions.

This appears to be the first instance in which a tangent was found to a curve other than a circle.

Archimedes' study of the spiral, a curve that he ascribed to his friend Conon of Alexandria, was part of the Greek search for the solution of the three famous problems.

[7] Katz, V. J. 1995. "Ideas of Calculus in Islam and India." *Mathematics Magazine* (Mathematical Association of America), 68(3):163-174.

[8] Indian mathematics

[9] Boyer, Carl B. (1959). "III. Medieval Contributions". *A History of the Calculus and Its Conceptual Development*. Dover. pp. 79–89. ISBN 978-0-486-60509-8.

[10] Pellegrino, Dana. "Pierre de Fermat". Retrieved 2008-02-24.

[11] Simmons, George F. (2007). *Calculus Gems: Brief Lives and Memorable Mathematics*. Mathematical Association of America. p. 98. ISBN 0-88385-561-5.

[12] Paradís, Jaume; Pla, Josep; Viader, Pelagrí. "Fermat's Treatise On Quadrature: A New Reading" (PDF). Retrieved 2008-02-24.

[13] *The geometrical lectures of Isaac Barrow, translated, with notes and proofs, and a discussion on the advance made therein on the work of his predecessors in the infinitesimal calculus*. Chicago: Open Court. 1916.

[14] Johnston, William; McAllister, Alex (2009). *A Transition to Advanced Mathematics: A Survey Course*. Oxford University Press US. p. 333. ISBN 0-19-531076-4., Chapter 4, p. 333

[15] Reyes 2004, p. 160

[16] Such as Kepler, Descartes, Fermat, Pascal and Wallis. Calinger 1999, p. 556

[17] Foremost among these was Barrow who had created formulas for specific cases and Fermat who created a similar definition for the derivative. For more information; Boyer 184

[18] Calinger 1999, p. 610

[19] Newton, Isaac. "Waste Book". Retrieved 10 January 2012.

[20] Eves, Howard. *An introduction to the history of mathematics, 6th edition*. p. 400.

[21] *Principia*, Florian Cajori 8

[22] Deleuae, Gilles. "DELEUZE / LEIBNIZ Cours Vincennes - 22/04/1980". Retrieved 30 April 2013.

[23] The use of prime to denote the derivative, $f'(x)$, is due to Lagrange.

[24] Katz, Victor J. (1995). "Ideas of Calculus in Islam and India". *Mathematics Magazine* **68** (3): 163–174. doi:10.2307/2691411. JSTOR 2691411. [165–9, 173–4]

3.11 Further reading

- Roero, C.S. (2005). "Gottfried Wilhelm Leibniz, first three papers on the calculus (1684, 1686, 1693)". In Grattan-Guinness, I. *Landmark writings in Western mathematics 1640–1940*. Elsevier. pp. 46–58. ISBN 978-0-444-50871-3.

- Roero, C.S. (1983). "Jakob Bernoulli, attentive student of the work of Archimedes: marginal notes to the edition of Barrow". *Boll. Storia Sci. Mat.* **3** (1): 77–125.

- Boyer, Carl. The History of Calculus. New York: Dover Publications, 1949

- Calinger, Ronald (1999). *A Contextual History of Mathematics*. Toronto: Prentice-Hall. ISBN 0-02-318285-7.

- Reyes, Mitchell (2004). "The Rhetoric in Mathematics: Newton, Leibniz, the Calculus, and the Rhetorical Force of the Infinitesimal". *Quarterly Journal of Speech* **90**: 159–184.

- Grattan-Guinness, Ivor. *The Rainbow of Mathematics: A History of the Mathematical Sciences*, Chapters 5 and 6, W. W. Norton & Company, 2000.

3.12   External links

- A history of the calculus in The MacTutor History of Mathematics archive, 1996.

- Earliest Known Uses of Some of the Words of Mathematics: Calculus & Analysis

- Newton Papers, Cambridge University Digital Library

- (English) (Arabic) The Excursion of Calculus, 1772

Isaac Newton

Gottfried Leibniz.

Chapter 4

History of combinatorics

The **history of combinatorics** is an area of study within the history of mathematics. Its focus ranges from antiquity to modern times.

4.1   Earliest records

The earliest known connection to combinatorics comes from the Rhind papyrus, problem 79, for the implementation of a geometric series.

A Jain text, the Bhagabati Sutra, had the first mention of a combinatorics problem; it asked how many ways one could take six tastes one, two, or three tastes at a time. The Bhagabati Sutra was written around 300 BC, and was the first book to mention the choose function.[1] The next ideas of Combinatorics came from Pingala, who was interested in prosody. Specifically, he wanted to know how many ways a six-syllable meter could be made from short and long notes. He wrote this problem in the Chanda sutra (also Chandahsutra) in the second century BC.[2][3] In addition, he also found the number of meters that had n long notes and k short notes, which is equivalent to finding the binomial coefficients.

The ideas of the Bhagabati were generalised by the Indian mathematician Mahavira in 850 AD, and Pingala's work on prosody was expanded by Bhaskara[1][4] and Hemacandra in 1100 AD. Bhaskara was the first known person to find the generalised choice function, although Brahmagupta may have known earlier.[5] Hemacandra asked how many meters existed of a certain length if a long note was considered to be twice as long as a short note, which is equivalent to finding the Fibonacci numbers.[2]

The ancient Chinese book of divination, I Ching, is about what different hexagrams mean, and to do this one needs to know how many possible hexagrams there were. Since each hexagram is a permutation with repetitions of six lines, where each line can be one of two states, solid or dashed, combinatorics yields the result that there are $2^6 = 64$ hexagrams. A monk also may have counted the number of configurations to a game similar to Go around 700 AD.[6] Although China had relatively few advancements in enumerative combinatorics, they solved a combinatorial design problem, the magic square, around 100 AD.[5]

In Greece, Plutarch wrote that the Xenocrates discovered the number of different syllables possible in the Greek language. This, however, is unlikely because this is one of the few mentions of Combinatorics in Greece. The number they found, 1.002×10^{12} also seems too round to be more than a guess.[6][7]

Magic squares remained an interest of China, and they began to generalise their original 3×3 square between 900 and 1300 AD. China corresponded with the Middle East about this problem in the 13th century.[5] The Middle East also learned about binomial coefficients from Indian work, and found the connection to polynomial expansion.[8]

Abū Bakr ibn Muḥammad ibn al Ḥusayn Al-Karaji (c.953-1029) wrote on the binomial theorem and Pascal's triangle. In a now lost work known only from subsequent quotation by al-Samaw'al Al-Karaji introduced the idea of argument by mathematical induction.

A hexagram

The philosopher and astronomer Rabbi Abraham ibn Ezra (c. 1140) counted the permutations with repetitions in vocalization of Divine Name.[9] He also established the symmetry of binomial coefficients, while a closed formula was obtained later by the talmudist and mathematician Levi ben Gerson (better known as Gersonides), in 1321.[10] The arithmetical triangle— a graphical diagram showing relationships among the binomial coefficients— was presented by mathematicians in treatises dating as far back as the 10th century, and would eventually become known as Pascal's triangle. Later, in Medieval England, campanology provided examples of what is now known as Hamiltonian cycles in certain Cayley graphs on permutations.[11]

4.2   Combinatorics in the West

Combinatorics came to Europe in the 13th century through two mathematicians, Leonardo Fibonacci and Jordanus de Nemore. Fibonacci's Liber Abaci introduced many of the Arabian and Indian ideas to Europe, including that of the

Fibonacci numbers.[12][13] Jordanus was the first person to arrange the binomial coefficients in a triangle, as he did in proposition 70 of *De Arithmetica*. This was also done in the Middle East in 1265, and China around 1300.[5] Today, this triangle is known as Pascal's triangle.

Pascal's contribution to the triangle that bears his name comes from his work on formal proofs about it, in addition to his connection between it and probability.[5] Together with Leibniz and his ideas about partitions in the 17th century,[14] they are considered the founders of modern combinatorics.[15]

Both Pascal and Leibniz understood that algebra and combinatorics corresponded (aka, binomial expansion was equivalent to the choice function). This was expanded by De Moivre, who found the expansion of a multinomial.[16] De Moivre also found the formula for derangements using the principle of inclusion-exclusion, a method different from Nikolaus Bernoulli, who had found them previously.[5] He managed to approximate the binomial coefficients and factorial. Finally, he found a closed form for the Fibonacci numbers by inventing generating functions.[17][18]

In the 18th century, Euler worked on problems of combinatorics. In addition to working on several problems of probability which link to combinatorics, he worked on the knights tour, Graeco-Latin square, Eulerian numbers, and others. He also invented graph theory by solving the Seven Bridges of Königsberg problem, which also led to the formation of topology. Finally, he broke ground with partitions by the use of generating functions.[19]

4.3 Notes

[1] "India". Retrieved 2008-03-05.

[2] Hall, Rachel (2005-02-16). "Math for Poets and Drummers-The Mathematics of Meter" (PDF). Retrieved 2008-03-05.

[3] Kulkarni, Amba. "Recursion and Combinatorial Mathematics in Chandashāstra". arXiv:math/0703658.

[4] Bhaskara. "The Lilavati of Bhaskara". Brown University. Archived from the original on 2008-03-25. Retrieved 2008-03-06.

[5] Biggs, Norman; Keith Lloyd; Robin Wilson (1995). "44". In Ronald Grahm, Martin Grötschel, László Lovász. *Handbook of Combinatorics* (GOOGLE BOOK). MIT Press. pp. 2163–2188. ISBN 0-262-57172-2. Retrieved 2008-03-08.

[6] Dieudonné, J. "The Rhind/Ahmes Papyrus - Mathematics and the Liberal Arts". *Historia Math*. Truman State University. Retrieved 2008-03-06.

[7] Gow, James (1968). *A Short History of Greek Mathematics*. AMS Bookstore. p. 71. ISBN 0-8284-0218-3.

[8] "Middle East". Retrieved 2008-03-08.

[9] The short commentary on Exodus 3:13

[10] History of Combinatorics, chapter in a textbook.

[11] Arthur T. White, "Ringing the Cosets," *Amer. Math. Monthly* **94** (1987), no. 8, 721-746; Arthur T. White, "Fabian Stedman: The First Group Theorist?," *Amer. Math. Monthly* **103** (1996), no. 9, 771-778.

[12] Devlin, Keith (October 2002). "The 800th birthday of the book that brought numbers to the west". *Devlin's Angle*. Retrieved 2008-03-08.

[13] "Fibonacci Sequence- History". Net Industries. 2008. Retrieved 2008-03-08.

[14] Leibniz habilitation thesis *De Arte Combinatoria* was published as a book in 1666 and reprinted later

[15] Dickson, Leonard (2005) [1919]. "Chapter III". *Diophantine Analysis*. History of the Theory of Numbers. Mineola, New York: Dover Publications, Inc. p. 101. ISBN 0-486-44233-0.

[16] Hodgson, James; William Derham; Richard Mead (1708). *Miscellanea Curiosa* (GOOGLE BOOK). Volume II. pp. 183–191. Retrieved 2008-03-08.

[17] O'Connor, John; Edmund Robertson (June 2004). "Abraham de Moivre". *The MacTutor History of Mathematics archive*. Retrieved 2008-03-09.

[18] Pang, Jong-Shi; Olvi Mangasarian (1999). "10.6 Generating Function". In Jong-Shi Pang. *Computational Optimisation* (GOOGLE BOOK). Volume 1. Netherlands: Kluwer Academic Publishers. pp. 182–183. ISBN 0-7923-8480-6. Retrieved 2008-03-09.

[19] "Combinatorics and probability". Retrieved 2008-03-08.

4.4   References

- N.L. Biggs, The roots of combinatorics, *Historia Mathematica* 6 (1979), 109-136.

- Katz, Victor J. (1998). *A History of Mathematics: An Introduction*, 2nd Edition. Addison-Wesley Education Publishers. ISBN 0-321-01618-1.

- O'Connor, John J. and Robertson, Edmund F. (1999–2004). *MacTutor History of Mathematics archive*. St Andrews University.

- Rashed, R. (1994). *The development of Arabic mathematics: between arithmetic and algebra*. London.

- Wilson, R. and Watkins, J. (2013). *Combinatorics: Ancient & Modern*. Oxford.

Chapter 5

History of geometry

Geometry (from the Ancient Greek: γεωμετρία; *geo*- "earth", *-metron* "measurement") arose as the field of knowledge dealing with spatial relationships. Geometry was one of the two fields of pre-modern mathematics, the other being the study of numbers (arithmetic).

Classic geometry was focused in compass and straightedge constructions. Geometry was revolutionized by Euclid, who introduced mathematical rigor and the axiomatic method still in use today. His book, *The Elements* is widely considered the most influential textbook of all time, and was known to all educated people in the West until the middle of the 20th century.[1]

In modern times, geometric concepts have been generalized to a high level of abstraction and complexity, and have been subjected to the methods of calculus and abstract algebra, so that many modern branches of the field are barely recognizable as the descendants of early geometry. (See Areas of mathematics and Algebraic geometry.)

5.1 Early geometry

The earliest recorded beginnings of geometry can be traced to early peoples, who discovered obtuse triangles in the ancient Indus Valley (see Harappan Mathematics), and ancient Babylonia (see Babylonian mathematics) from around 3000 BC. Early geometry was a collection of empirically discovered principles concerning lengths, angles, areas, and volumes, which were developed to meet some practical need in surveying, construction, astronomy, and various crafts. Among these were some surprisingly sophisticated principles, and a modern mathematician might be hard put to derive some of them without the use of calculus. For example, both the Egyptians and the Babylonians were aware of versions of the Pythagorean theorem about 1500 years before Pythagoras; the Egyptians had a correct formula for the volume of a frustum of a square pyramid;

5.1.1 Egyptian geometry

Main article: Egyptian geometry

The ancient Egyptians knew that they could approximate the area of a circle as follows:[2]

Area of Circle ≈ [(Diameter) x 8/9]^2.[2]

Problem 30 of the Ahmes papyrus uses these methods to calculate the area of a circle, according to a rule that the area is equal to the square of 8/9 of the circle's diameter. This assumes that π is 4×(8/9)^2 (or 3.160493...), with an error of slightly over 0.63 percent. This value was slightly less accurate than the calculations of the Babylonians (25/8 = 3.125,

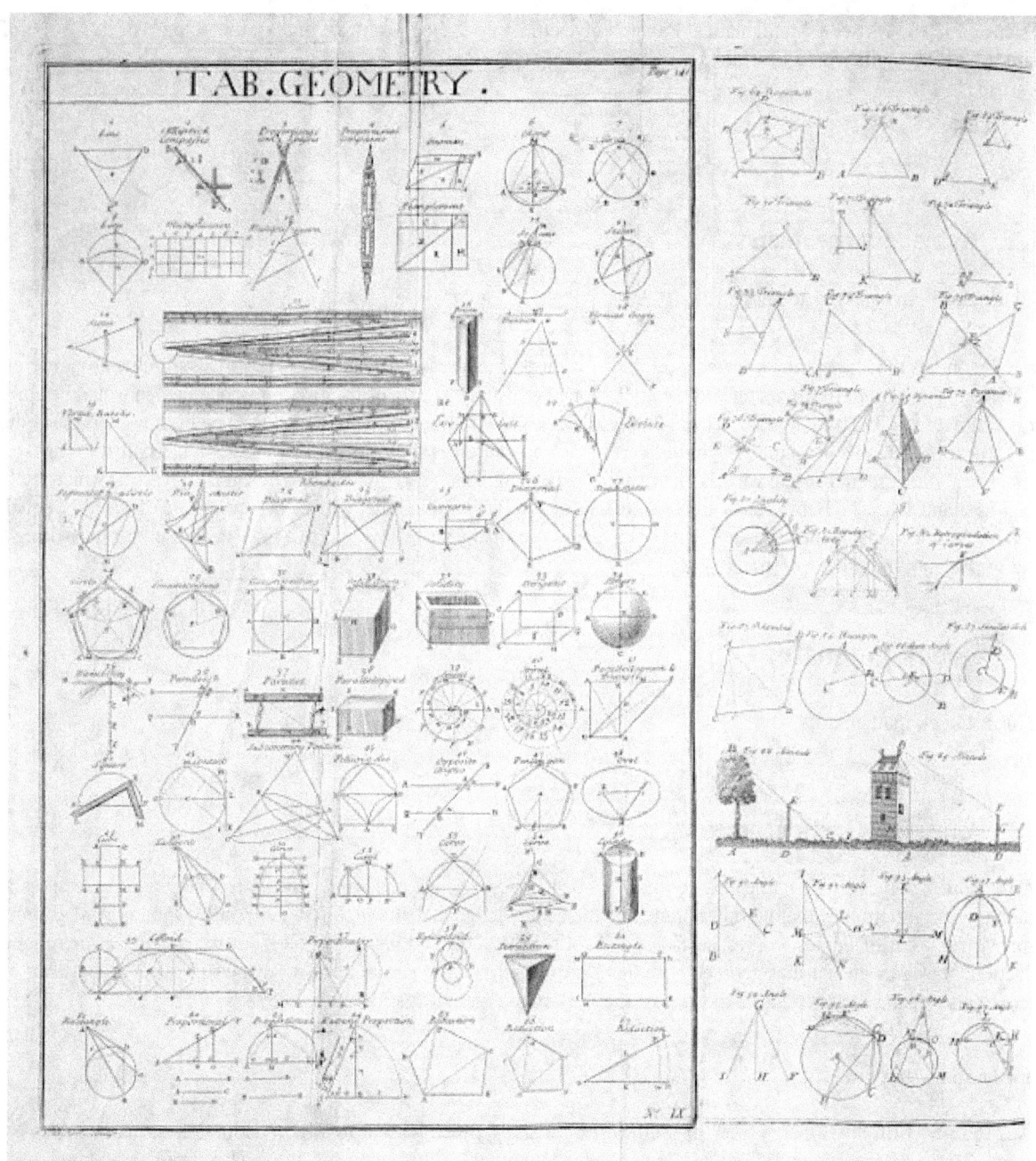

Part of the "Tab.Geometry." (Table of Geometry) from the 1728 Cyclopaedia.

within 0.53 percent), but was not otherwise surpassed until Archimedes' approximation of 211875/67441 = 3.14163, which had an error of just over 1 in 10,000.

Interestingly, Ahmes knew of the modern 22/7 as an approximation for pi, and used it to split a hekat, hekat x 22/x x 7/22 = hekat; however, Ahmes continued to use the traditional 256/81 value for pi for computing his hekat volume found in a cylinder.

Problem 48 involved using a square with side 9 units. This square was cut into a 3x3 grid. The diagonal of the corner squares were used to make an irregular octagon with an area of 63 units. This gave a second value for π of 3.111...

The two problems together indicate a range of values for Pi between 3.11 and 3.16.

Problem 14 in the Moscow Mathematical Papyrus gives the only ancient example finding the volume of a frustum of a pyramid, describing the correct formula:

$$V = \frac{1}{3}h(x_1^2 + x_1 x_2 + x_2^2).$$

5.1.2 Babylonian geometry

Main article: Babylonian mathematics

The Babylonians may have known the general rules for measuring areas and volumes. They measured the circumference of a circle as three times the diameter and the area as one-twelfth the square of the circumference, which would be correct if π is estimated as 3. The volume of a cylinder was taken as the product of the base and the height, however, the volume of the frustum of a cone or a square pyramid was incorrectly taken as the product of the height and half the sum of the bases. The Pythagorean theorem was also known to the Babylonians. Also, there was a recent discovery in which a tablet used π as 3 and 1/8. The Babylonians are also known for the Babylonian mile, which was a measure of distance equal to about seven miles today. This measurement for distances eventually was converted to a time-mile used for measuring the travel of the Sun, therefore, representing time.[3]

5.2 Greek geometry

See also: Greek mathematics

5.2.1 Classical Greek geometry

For the ancient Greek mathematicians, geometry was the crown jewel of their sciences, reaching a completeness and perfection of methodology that no other branch of their knowledge had attained. They expanded the range of geometry to many new kinds of figures, curves, surfaces, and solids; they changed its methodology from trial-and-error to logical deduction; they recognized that geometry studies "eternal forms", or abstractions, of which physical objects are only approximations; and they developed the idea of the "axiomatic method", still in use today.

Thales and Pythagoras

Thales (635-543 BC) of Miletus (now in southwestern Turkey), was the first to whom deduction in mathematics is attributed. There are five geometric propositions for which he wrote deductive proofs, though his proofs have not survived. Pythagoras (582-496 BC) of Ionia, and later, Italy, then colonized by Greeks, may have been a student of Thales, and traveled to Babylon and Egypt. The theorem that bears his name may not have been his discovery, but he was probably one of the first to give a deductive proof of it. He gathered a group of students around him to study mathematics, music, and philosophy, and together they discovered most of what high school students learn today in their geometry courses. In addition, they made the profound discovery of incommensurable lengths and irrational numbers.

Plato

Plato (427-347 BC) is a philosopher that is highly esteemed by the Greeks. There is a story that he had inscribed above the entrance to his famous school, "Let none ignorant of geometry enter here." However, the story is considered to be untrue.[4] Though he was not a mathematician himself, his views on mathematics had great influence. Mathematicians thus accepted his belief that geometry should use no tools but compass and straightedge – never measuring instruments such as a marked ruler or a protractor, because these were a workman's tools, not worthy of a scholar. This dictum led to a

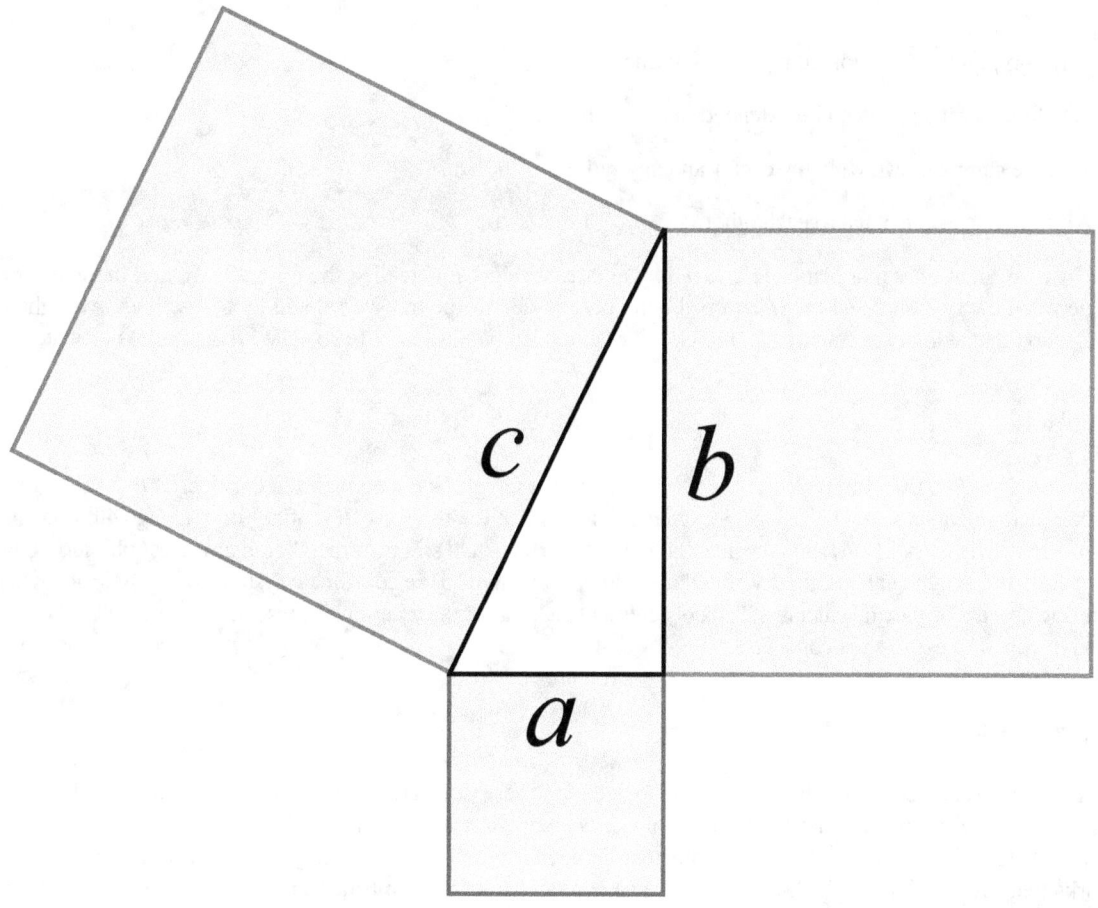

Pythagorean theorem: $a^2 + b^2 = c^2$

deep study of possible compass and straightedge constructions, and three classic construction problems: how to use these tools to trisect an angle, to construct a cube twice the volume of a given cube, and to construct a square equal in area to a given circle. The proofs of the impossibility of these constructions, finally achieved in the 19th century, led to important principles regarding the deep structure of the real number system. Aristotle (384-322 BC), Plato's greatest pupil, wrote a treatise on methods of reasoning used in deductive proofs (see Logic) which was not substantially improved upon until the 19th century.

5.2.2 Hellenistic geometry

Euclid

Euclid (c. 325-265 BC), of Alexandria, probably a student of one of Plato's students, wrote a treatise in 13 books (chapters), titled *The Elements of Geometry*, in which he presented geometry in an ideal axiomatic form, which came to be known as Euclidean geometry. The treatise is not a compendium of all that the Hellenistic mathematicians knew at the time about geometry; Euclid himself wrote eight more advanced books on geometry. We know from other references that Euclid's was not the first elementary geometry textbook, but it was so much superior that the others fell into disuse and were lost. He was brought to the university at Alexandria by Ptolemy I, King of Egypt.

The Elements began with definitions of terms, fundamental geometric principles (called *axioms* or *postulates*), and general quantitative principles (called *common notions*) from which all the rest of geometry could be logically deduced. Following are his five axioms, somewhat paraphrased to make the English easier to read.

1. Any two points can be joined by a straight line.

2. Any finite straight line can be extended in a straight line.

3. A circle can be drawn with any center and any radius.

4. All right angles are equal to each other.

5. If two straight lines in a plane are crossed by another straight line (called the transversal), and the interior angles between the two lines and the transversal lying on one side of the transversal add up to less than two right angles, then on that side of the transversal, the two lines extended will intersect (also called the parallel postulate).

Archimedes

Archimedes (287-212 BC), of Syracuse, Sicily, when it was a Greek city-state, is often considered to be the greatest of the Greek mathematicians, and occasionally even named as one of the three greatest of all time (along with Isaac Newton and Carl Friedrich Gauss). Had he not been a mathematician, he would still be remembered as a great physicist, engineer, and inventor. In his mathematics, he developed methods very similar to the coordinate systems of analytic geometry, and the limiting process of integral calculus. The only element lacking for the creation of these fields was an efficient algebraic notation in which to express his concepts.

After Archimedes

After Archimedes, Hellenistic mathematics began to decline. There were a few minor stars yet to come, but the golden age of geometry was over. Proclus (410-485), author of *Commentary on the First Book of Euclid*, was one of the last important players in Hellenistic geometry. He was a competent geometer, but more importantly, he was a superb commentator on the works that preceded him. Much of that work did not survive to modern times, and is known to us only through his commentary. The Roman Republic and Empire that succeeded and absorbed the Greek city-states produced excellent engineers, but no mathematicians of note.

The great Library of Alexandria was later burned. There is a growing consensus among historians that the Library of Alexandria likely suffered from several destructive events, but that the destruction of Alexandria's pagan temples in the late 4th century was probably the most severe and final one. The evidence for that destruction is the most definitive and secure. Caesar's invasion may well have led to the loss of some 40,000-70,000 scrolls in a warehouse adjacent to the port (as Luciano Canfora argues, they were likely copies produced by the Library intended for export), but it is unlikely to have affected the Library or Museum, given that there is ample evidence that both existed later.

Civil wars, decreasing investments in maintenance and acquisition of new scrolls and generally declining interest in non-religious pursuits likely contributed to a reduction in the body of material available in the Library, especially in the 4th century. The Serapeum was certainly destroyed by Theophilus in 391, and the Museum and Library may have fallen victim to the same campaign.

5.3 Indian geometry

See also: Indian mathematics

5.3.1 Vedic period

The *Satapatha Brahmana* (ninth century BC) contains rules for ritual geometric constructions that are similar to the *Sulba Sutras*.[5]

The *Śulba Sūtras* (literally, "Aphorisms of the Chords" in Vedic Sanskrit) (c. 700-400 BC) list rules for the construction of fire altars for rituals.[6] Most mathematical problems considered in the *Śulba Sūtras* spring from "a single theological requirement,"[7] that of constructing fire altars which have different shapes but occupy the same area. The altars were required to be constructed of five layers of burnt brick, with the further condition that each layer consist of 200 bricks and that no two adjacent layers have congruent arrangements of bricks.[7]

According to (Hayashi 2005, p. 363), the *Śulba Sūtras* contain "the earliest extant verbal expression of the Pythagorean Theorem in the world, although it had already been known to the Old Babylonians."

> The diagonal rope (*akṣṇayā-rajju*) of an oblong (rectangle) produces both which the flank (*pārśvamāni*) and the horizontal (*tiryaṇmānī*) <ropes> produce separately."[8]

Since the statement is a *sūtra*, it is necessarily compressed and what the ropes *produce* is not elaborated on, but the context clearly implies the square areas constructed on their lengths, and would have been explained so by the teacher to the student.[8]

They contain lists of Pythagorean triples,[9] which are particular cases of Diophantine equations.[10] They also contain statements (that with hindsight we know to be approximate) about squaring the circle and "circling the square."[11]

Baudhayana (c. eighth century BC) composed the *Baudhayana Sulba Sutra*, the best-known *Sulba Sutra*, which contains examples of simple Pythagorean triples, such as: $(3, 4, 5)$, $(5, 12, 13)$, $(8, 15, 17)$, $(7, 24, 25)$, and $(12, 35, 37)$ [12] as well as a statement of the Pythagorean theorem for the sides of a square: "The rope which is stretched across the diagonal of a square produces an area double the size of the original square."[12] It also contains the general statement of the Pythagorean theorem (for the sides of a rectangle): "The rope stretched along the length of the diagonal of a rectangle makes an area which the vertical and horizontal sides make together."[12]

According to mathematician S. G. Dani, the Babylonian cuneiform tablet Plimpton 322 written c. 1850 BC[13] "contains fifteen Pythagorean triples with quite large entries, including (13500, 12709, 18541) which is a primitive triple,[14] indicating, in particular, that there was sophisticated understanding on the topic" in Mesopotamia in 1850 BC. "Since these tablets predate the Sulbasutras period by several centuries, taking into account the contextual appearance of some of the triples, it is reasonable to expect that similar understanding would have been there in India."[15] Dani goes on to say:

> "As the main objective of the *Sulvasutras* was to describe the constructions of altars and the geometric principles involved in them, the subject of Pythagorean triples, even if it had been well understood may still not have featured in the *Sulvasutras*. The occurrence of the triples in the *Sulvasutras* is comparable to mathematics that one may encounter in an introductory book on architecture or another similar applied area, and would not correspond directly to the overall knowledge on the topic at that time. Since, unfortunately, no other contemporaneous sources have been found it may never be possible to settle this issue satisfactorily."[15]

In all, three *Sulba Sutras* were composed. The remaining two, the *Manava Sulba Sutra* composed by Manava (fl. 750-650 BC) and the *Apastamba Sulba Sutra*, composed by Apastamba (c. 600 BC), contained results similar to the *Baudhayana Sulba Sutra*.

5.3.2   Classical period

In the Bakhshali manuscript, there is a handful of geometric problems (including problems about volumes of irregular solids). The Bakhshali manuscript also "employs a decimal place value system with a dot for zero."[16] Aryabhata's *Aryabhatiya* (499) includes the computation of areas and volumes.

Brahmagupta wrote his astronomical work *Brāhma Sphuṭa Siddhānta* in 628. Chapter 12, containing 66 Sanskrit verses, was divided into two sections: "basic operations" (including cube roots, fractions, ratio and proportion, and barter) and "practical mathematics" (including mixture, mathematical series, plane figures, stacking bricks, sawing of timber, and piling of grain).[17] In the latter section, he stated his famous theorem on the diagonals of a cyclic quadrilateral:[17]

Brahmagupta's theorem: If a cyclic quadrilateral has diagonals that are perpendicular to each other, then the perpendicular line drawn from the point of intersection of the diagonals to any side of the quadrilateral always bisects the opposite side.

Chapter 12 also included a formula for the area of a cyclic quadrilateral (a generalization of Heron's formula), as well as a complete description of rational triangles (*i.e.* triangles with rational sides and rational areas).

Brahmagupta's formula: The area, A, of a cyclic quadrilateral with sides of lengths a, b, c, d, respectively, is given by

$$A = \sqrt{(s-a)(s-b)(s-c)(s-d)}$$

where s, the semiperimeter, given by: $s = \frac{a+b+c+d}{2}$.

Brahmagupta's Theorem on rational triangles: A triangle with rational sides a, b, c and rational area is of the form:

$$a = \frac{u^2}{v} + v, \quad b = \frac{u^2}{w} + w, \quad c = \frac{u^2}{v} + \frac{u^2}{w} - (v+w)$$

for some rational numbers u, v, and w.[18]

5.4 Chinese geometry

See also: Chinese mathematics

The first definitive work (or at least oldest existent) on geometry in China was the *Mo Jing*, the Mohist canon of the early philosopher Mozi (470-390 BC). It was compiled years after his death by his followers around the year 330 BC.[19] Although the *Mo Jing* is the oldest existent book on geometry in China, there is the possibility that even older written material existed. However, due to the infamous Burning of the Books in a political maneuver by the Qin Dynasty ruler Qin Shihuang (r. 221-210 BC), multitudes of written literature created before his time were purged. In addition, the *Mo Jing* presents geometrical concepts in mathematics that are perhaps too advanced not to have had a previous geometrical base or mathematic background to work upon.

The *Mo Jing* described various aspects of many fields associated with physical science, and provided a small wealth of information on mathematics as well. It provided an 'atomic' definition of the geometric point, stating that a line is separated into parts, and the part which has no remaining parts (i.e. cannot be divided into smaller parts) and thus forms the extreme end of a line is a point.[19] Much like Euclid's first and third definitions and Plato's 'beginning of a line', the *Mo Jing* stated that "a point may stand at the end (of a line) or at its beginning like a head-presentation in childbirth. (As to its invisibility) there is nothing similar to it."[20] Similar to the atomists of Democritus, the *Mo Jing* stated that a point is the smallest unit, and cannot be cut in half, since 'nothing' cannot be halved.[20] It stated that two lines of equal length will always finish at the same place,[20] while providing definitions for the *comparison of lengths* and for *parallels*,[21] along with principles of space and bounded space.[22] It also described the fact that planes without the quality of thickness cannot be piled up since they cannot mutually touch.[23] The book provided definitions for circumference, diameter, and radius, along with the definition of volume.[24]

The Han Dynasty (202 BC-220 AD) period of China witnessed a new flourishing of mathematics. One of the oldest Chinese mathematical texts to present geometric progressions was the *Suàn shù shū* of 186 BC, during the Western Han era. The mathematician, inventor, and astronomer Zhang Heng (78-139 AD) used geometrical formulas to solve mathematical problems. Although rough estimates for pi (π) were given in the *Zhou Li* (compiled in the 2nd century BC),[25] it was Zhang Heng who was the first to make a concerted effort at creating a more accurate formula for pi. Zhang Heng approximated pi as 730/232 (or approx 3.1466), although he used another formula of pi in finding a spherical volume, using the square root of 10 (or approx 3.162) instead. Zu Chongzhi (429-500 AD) improved the accuracy of the approximation of pi to between 3.1415926 and 3.1415927, with $^{355}/_{113}$ (🔲🔲, Milü, detailed approximation) and $^{22}/_7$ (🔲🔲, Yuelü, rough approximation) being the other notable approximation.[26] In comparison to later works, the formula for pi given by the French mathematician Franciscus Vieta (1540-1603) fell halfway between Zu's approximations.

5.4.1 *The Nine Chapters on the Mathematical Art*

The Nine Chapters on the Mathematical Art, the title of which first appeared by 179 AD on a bronze inscription, was edited and commented on by the 3rd century mathematician Liu Hui from the Kingdom of Cao Wei. This book included many problems where geometry was applied, such as finding surface areas for squares and circles, the volumes of solids in various three-dimensional shapes, and included the use of the Pythagorean theorem. The book provided illustrated proof for the Pythagorean theorem,[27] contained a written dialogue between of the earlier Duke of Zhou and Shang Gao on the properties of the right angle triangle and the Pythagorean theorem, while also referring to the astronomical gnomon, the circle and square, as well as measurements of heights and distances.[28] The editor Liu Hui listed pi as 3.141014 by using a 192 sided polygon, and then calculated pi as 3.14159 using a 3072 sided polygon. This was more accurate than Liu Hui's contemporary Wang Fan, a mathematician and astronomer from Eastern Wu, would render pi as 3.1555 by using $^{142}/_{45}$.[29] Liu Hui also wrote of mathematical surveying to calculate distance measurements of depth, height, width, and surface area. In terms of solid geometry, he figured out that a wedge with rectangular base and both sides sloping could be broken down into a pyramid and a tetrahedral wedge.[30] He also figured out that a wedge with trapezoid base and both sides sloping could be made to give two tetrahedral wedges separated by a pyramid.[30] Furthermore, Liu Hui described Cavalieri's principle on volume, as well as Gaussian elimination. From the *Nine Chapters*, it listed the following geometrical formulas that were known by the time of the Former Han Dynasty (202 BCE–9 CE).

Areas for the[31]

Volumes for the[30]

Continuing the geometrical legacy of ancient China, there were many later figures to come, including the famed astronomer and mathematician Shen Kuo (1031-1095 CE), Yang Hui (1238-1298) who discovered Pascal's Triangle, Xu Guangqi (1562-1633), and many others.

5.5   Islamic geometry

See also: Islamic mathematics

Although the Islamic mathematicians are most famed for their work on algebra, number theory and number systems, they also made considerable contributions to geometry, trigonometry and mathematical astronomy, and were responsible for the development of algebraic geometry. Geometrical magnitudes were treated as "algebraic objects" by most Islamic mathematicians however.

Al-Mahani (born 820) conceived the idea of reducing geometrical problems such as duplicating the cube to problems in algebra. Al-Karaji (born 953) completely freed algebra from geometrical operations and replaced them with the arithmetical type of operations which are at the core of algebra today.

5.5.1   Thabit family and other early geometers

Thābit ibn Qurra (known as Thebit in Latin) (born 836) contributed to a number of areas in mathematics, where he played an important role in preparing the way for such important mathematical discoveries as the extension of the concept of number to (positive) real numbers, integral calculus, theorems in spherical trigonometry, analytic geometry, and non-Euclidean geometry. In astronomy Thabit was one of the first reformers of the Ptolemaic system, and in mechanics he was a founder of statics. An important geometrical aspect of Thabit's work was his book on the composition of ratios. In this book, Thabit deals with arithmetical operations applied to ratios of geometrical quantities. The Greeks had dealt with geometric quantities but had not thought of them in the same way as numbers to which the usual rules of arithmetic could be applied. By introducing arithmetical operations on quantities previously regarded as geometric and non-numerical, Thabit started a trend which led eventually to the generalisation of the number concept.

In some respects, Thabit is critical of the ideas of Plato and Aristotle, particularly regarding motion. It would seem that here his ideas are based on an acceptance of using arguments concerning motion in his geometrical arguments. Another important contribution Thabit made to geometry was his generalization of the Pythagorean theorem, which he extended from special right triangles to all triangles in general, along with a general proof.[32]

Ibrahim ibn Sinan ibn Thabit (born 908), who introduced a method of integration more general than that of Archimedes, and al-Quhi (born 940) were leading figures in a revival and continuation of Greek higher geometry in the Islamic world. These mathematicians, and in particular Ibn al-Haytham, studied optics and investigated the optical properties of mirrors made from conic sections.

Astronomy, time-keeping and geography provided other motivations for geometrical and trigonometrical research. For example Ibrahim ibn Sinan and his grandfather Thabit ibn Qurra both studied curves required in the construction of sundials. Abu'l-Wafa and Abu Nasr Mansur both applied spherical geometry to astronomy.

5.5.2 Geometric architecture

Recent discoveries have shown that geometrical quasicrystal patterns were first employed in the girih tiles found in medieval Islamic architecture dating back over five centuries ago. A 2007 paper in the journal *Science* suggesting that girih tilings possessed properties consistent with self-similar fractal quasicrystalline tilings such as the Penrose tilings, predating them by five centuries.[33][34]

5.6 Modern geometry

5.6.1 The 17th century

When Europe began to emerge from its Dark Ages, the Hellenistic and Islamic texts on geometry found in Islamic libraries were translated from Arabic into Latin. The rigorous deductive methods of geometry found in Euclid's *Elements of Geometry* were relearned, and further development of geometry in the styles of both Euclid (Euclidean geometry) and Khayyam (algebraic geometry) continued, resulting in an abundance of new theorems and concepts, many of them very profound and elegant.

In the early 17th century, there were two important developments in geometry. The first and most important was the creation of analytic geometry, or geometry with coordinates and equations, by René Descartes (1596–1650) and Pierre de Fermat (1601–1665). This was a necessary precursor to the development of calculus and a precise quantitative science of physics. The second geometric development of this period was the systematic study of projective geometry by Girard Desargues (1591–1661). Projective geometry is the study of geometry without measurement, just the study of how points align with each other. There had been some early work in this area by Hellenistic geometers, notably Pappus (c. 340). The greatest flowering of the field occurred with Jean-Victor Poncelet (1788–1867).

In the late 17th century, calculus was developed independently and almost simultaneously by Isaac Newton (1642–1727) and Gottfried Wilhelm Leibniz (1646–1716). This was the beginning of a new field of mathematics now called analysis. Though not itself a branch of geometry, it is applicable to geometry, and it solved two families of problems that had long been almost intractable: finding tangent lines to odd curves, and finding areas enclosed by those curves. The methods of calculus reduced these problems mostly to straightforward matters of computation.

5.6.2 The 18th and 19th centuries

Non-Euclidean geometry

The very old problem of proving Euclid's Fifth Postulate, the "Parallel Postulate", from his first four postulates had never been forgotten. Beginning not long after Euclid, many attempted demonstrations were given, but all were later found to be faulty, through allowing into the reasoning some principle which itself had not been proved from the first four postulates. Though Omar Khayyám was also unsuccessful in proving the parallel postulate, his criticisms of Euclid's theories of parallels and his proof of properties of figures in non-Euclidean geometries contributed to the eventual development of non-Euclidean geometry. By 1700 a great deal had been discovered about what can be proved from the first four, and what the pitfalls were in attempting to prove the fifth. Saccheri, Lambert, and Legendre each did excellent work on the problem in the 18th century, but still fell short of success. In the early 19th century, Gauss, Johann Bolyai, and Lobatchewsky, each independently, took a different approach. Beginning to suspect that it was impossible to prove the Parallel Postulate, they set out to develop a self-consistent geometry in which that postulate was false. In this they were successful, thus creating the first non-Euclidean geometry. By 1854, Bernhard Riemann, a student of Gauss, had applied methods of calculus in a ground-breaking study of the intrinsic (self-contained) geometry of all smooth surfaces, and thereby found a different non-Euclidean geometry. This work of Riemann later became fundamental for Einstein's theory of relativity.

It remained to be proved mathematically that the non-Euclidean geometry was just as self-consistent as Euclidean geometry, and this was first accomplished by Beltrami in 1868. With this, non-Euclidean geometry was established on an equal mathematical footing with Euclidean geometry.

While it was now known that different geometric theories were mathematically possible, the question remained, "Which one of these theories is correct for our physical space?" The mathematical work revealed that this question must be answered by physical experimentation, not mathematical reasoning, and uncovered the reason why the experimentation must involve immense (interstellar, not earth-bound) distances. With the development of relativity theory in physics, this question became vastly more complicated.

Introduction of mathematical rigor

All the work related to the Parallel Postulate revealed that it was quite difficult for a geometer to separate his logical reasoning from his intuitive understanding of physical space, and, moreover, revealed the critical importance of doing so. Careful examination had uncovered some logical inadequacies in Euclid's reasoning, and some unstated geometric principles to which Euclid sometimes appealed. This critique paralleled the crisis occurring in calculus and analysis regarding the meaning of infinite processes such as convergence and continuity. In geometry, there was a clear need for a new set of axioms, which would be complete, and which in no way relied on pictures we draw or on our intuition of space. Such axioms, now known as Hilbert's axioms, were given by David Hilbert in 1894 in his dissertation *Grundlagen der Geometrie* (*Foundations of Geometry*). Some other complete sets of axioms had been given a few years earlier, but did not match Hilbert's in economy, elegance, and similarity to Euclid's axioms.

Analysis situs, or topology

In the mid-18th century, it became apparent that certain progressions of mathematical reasoning recurred when similar ideas were studied on the number line, in two dimensions, and in three dimensions. Thus the general concept of a metric space was created so that the reasoning could be done in more generality, and then applied to special cases. This method of studying calculus- and analysis-related concepts came to be known as analysis situs, and later as topology. The important topics in this field were properties of more general figures, such as connectedness and boundaries, rather than properties like straightness, and precise equality of length and angle measurements, which had been the focus of Euclidean and non-Euclidean geometry. Topology soon became a separate field of major importance, rather than a sub-field of geometry or analysis.

5.6.3   The 20th century

Developments in algebraic geometry included the study of curves and surfaces over finite fields as demonstrated by the works of among others André Weil, Alexander Grothendieck, and Jean-Pierre Serre as well as over the real or complex numbers. Finite geometry itself, the study of spaces with only finitely many points, found applications in coding theory and cryptography. With the advent of the computer, new disciplines such as computational geometry or digital geometry deal with geometric algorithms, discrete representations of geometric data, and so forth.

5.7   Timeline

Main article: Timeline of geometry

5.8   See also

- *Flatland*, a book by "A. Square" about two– and three-dimensional space, to understand the concept of four dimensions

- History of mathematics

- Important publications in geometry

- Interactive geometry software

- List of geometry topics

5.9 Notes

[1] Howard Eves, *An Introduction to the History of Mathematics*, Saunders: 1990 (ISBN 0-03-029558-0), p. 141: "No work, except The Bible, has been more widely used...."

[2] Ray C. Jurgensen, Alfred J. Donnelly, and Mary P. Dolciani. Editorial Advisors Andrew M. Gleason, Albert E. Meder, Jr. *Modern School Mathematics: Geometry* (Student's Edition). Houghton Mifflin Company, Boston, 1972, p. 52. ISBN 0-395-13102-2. Teachers Edition ISBN 0-395-13103-0.

[3] Eves, Chapter 2.

[4] Cherowitzo, Bill. "What precisely was written over the door of Plato's Academy?" (PDF). *http://www.math.ucdenver.edu/"*. *Retrieved 8 April 2015.*

[5] A. Seidenberg, 1978. The origin of mathematics. Archive for the history of Exact Sciences, vol 18.

[6] (Staal 1999)

[7] (Hayashi 2003, p. 118)

[8] (Hayashi 2005, p. 363)

[9] Pythagorean triples are triples of integers (a, b, c) with the property: $a^2 + b^2 = c^2$. Thus, $3^2 + 4^2 = 5^2$, $8^2 + 15^2 = 17^2$, $12^2 + 35^2 = 37^2$ etc.

[10] (Cooke 2005, p. 198): "The arithmetic content of the *Śulva Sūtras* consists of rules for finding Pythagorean triples such as (3, 4, 5), (5, 12, 13), (8, 15, 17), and (12, 35, 37). It is not certain what practical use these arithmetic rules had. The best conjecture is that they were part of religious ritual. A Hindu home was required to have three fires burning at three different altars. The three altars were to be of different shapes, but all three were to have the same area. These conditions led to certain "Diophantine" problems, a particular case of which is the generation of Pythagorean triples, so as to make one square integer equal to the sum of two others."

[11] (Cooke 2005, pp. 199–200): "The requirement of three altars of equal areas but different shapes would explain the interest in transformation of areas. Among other transformation of area problems the Hindus considered in particular the problem of squaring the circle. The *Bodhayana Sutra* states the converse problem of constructing a circle equal to a given square. The following approximate construction is given as the solution.... this result is only approximate. The authors, however, made no distinction between the two results. In terms that we can appreciate, this construction gives a value for π of 18 $(3 - 2\sqrt{2})$, which is about 3.088."

[12] (Joseph 2000, p. 229)

[13] Mathematics Department, University of British Columbia, *The Babylonian tabled Plimpton 322*.

[14] Three positive integers (a, b, c) form a *primitive* Pythagorean triple if $c^2 = a^2 + b^2$ and if the highest common factor of a, b, c is 1. In the particular Plimpton322 example, this means that $13500^2 + 12709^2 = 18541^2$ and that the three numbers do not have any common factors. However some scholars have disputed the Pythagorean interpretation of this tablet; see Plimpton 322 for details.

[15] (Dani 2003)

[16] (Hayashi 2005, p. 371)

[17] (Hayashi 2003, pp. 121–122)

[18] (Stillwell 2004, p. 77)

[19] Needham, Volume 3, 91.

[20] Needham, Volume 3, 92.

[21] Needham, Volume 3, 92-93.

[22] Needham, Volume 3, 93.

[23] Needham, Volume 3, 93-94.

[24] Needham, Volume 3, 94.

[25] Needham, Volume 3, 99.

[26] Needham, Volume 3, 101.

[27] Needham, Volume 3, 22.

[28] Needham, Volume 3, 21.

[29] Needham, Volume 3, 100.

[30] Needham, Volume 3, 98–99.

[31] Needham, Volume 3, 98.

[32] Sayili, Aydin (1960). "Thabit ibn Qurra's Generalization of the Pythagorean Theorem". *Isis* **51** (1): 35–37. doi:10.1086/348837.

[33] Peter J. Lu and Paul J. Steinhardt (2007), "Decagonal and Quasi-crystalline Tilings in Medieval Islamic Architecture" (PDF), *Science* **315** (5815): 1106–1110, Bibcode:2007Sci...315.1106L, doi:10.1126/science.1135491, PMID 17322056.

[34] Supplemental figures

5.10   References

- Cooke, Roger (2005), *The History of Mathematics:*, New York: Wiley-Interscience, 632 pages, ISBN 0-471-44459-6

- Dani, S. G. (July 25, 2003), "Pythogorean Triples in the Sulvasutras" (PDF), *Current Science* **85** (2): 219–224

- Eder, Michelle (2000), *Views of Euclid's Parallel Postulate in Ancient Greece and in Medieval Islam*, Rutgers University, retrieved 2008-01-23

- Hayashi, Takao (2003), "Indian Mathematics", in Grattan-Guinness, Ivor, *Companion Encyclopedia of the History and Philosophy of the Mathematical Sciences* **1**, Baltimore, MD: The Johns Hopkins University Press, 976 pages, pp. 118–130, ISBN 0-8018-7396-7

- Hayashi, Takao (2005), "Indian Mathematics", in Flood, Gavin, *The Blackwell Companion to Hinduism*, Oxford: Basil Blackwell, 616 pages, pp. 360–375, ISBN 978-1-4051-3251-0

- Joseph, G. G. (2000), *The Crest of the Peacock: The Non-European Roots of Mathematics*, Princeton, NJ: Princeton University Press, 416 pages, ISBN 0-691-00659-8

- Katz, Victor J. (1998), *History of Mathematics: An Introduction*, Addison-Wesley, ISBN 0-321-01618-1, OCLC 38199387 60154481

- Needham, Joseph (1986), *Science and Civilization in China: Volume 3, Mathematics and the Sciences of the Heavens and the Earth*, Taipei: Caves Books Ltd

- Rozenfeld, Boris A. (1988), *A History of Non-Euclidean Geometry: Evolution of the Concept of a Geometric Space*, Springer Science+Business Media, ISBN 0-387-96458-4, OCLC 15550634 230166667 230980046 77693662

- Smith, John D. (1992), "The Remarkable Ibn al-Haytham", *The Mathematical Gazette* (Mathematical Association) **76** (475): 189–198, doi:10.2307/3620392, JSTOR 3620392

- Staal, Frits (1999), "Greek and Vedic Geometry", *Journal of Indian Philosophy,* **27** (1–2): 105–127, doi:10.1023/A:1004364417

- Stillwell, John (2004), *Berlin and New York: Mathematics and its History* (2 ed.), Springer, 568 pages, ISBN 0-387-95336-1

5.11 External links

- Islamic Geometry

- Geometry in the 19th Century at the Stanford Encyclopedia of Philosophy

- Arabic mathematics : forgotten brilliance?

Woman teaching geometry. *Illustration at the beginning of a medieval translation of Euclid's Elements, (c. 1310)*

Geometry was connected to the divine for most medieval scholars. The compass in this 13th-century manuscript is a symbol of God's act of Creation.

Rigveda *manuscript in Devanagari.*

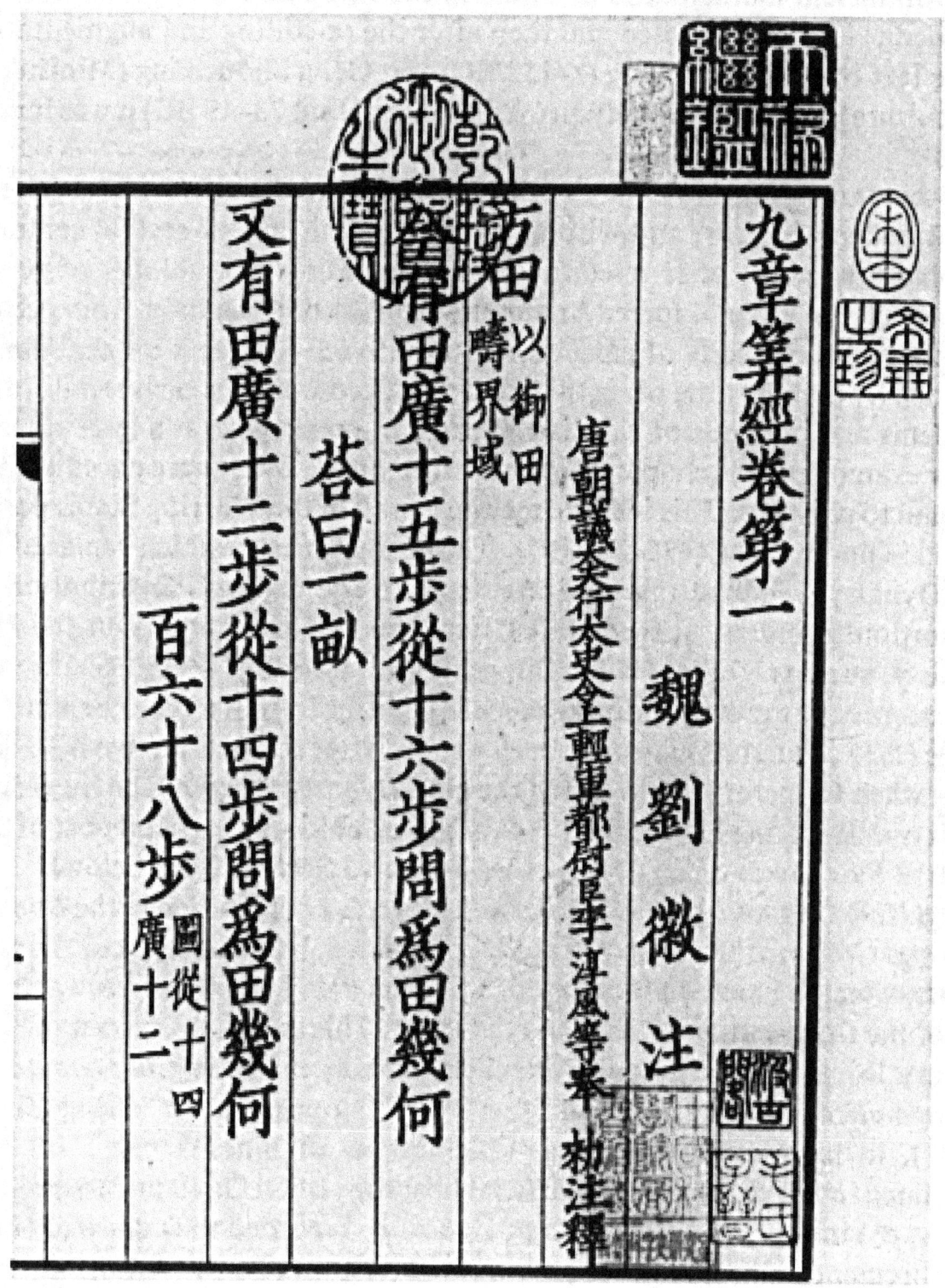

The Nine Chapters on the Mathematical Art, *first compiled in 179 AD, with added commentary in the 3rd century by Liu Hui.*

The Sea Island Mathematical Manual, *Liu Hui, 3rd century.*

An engraving by Albrecht Dürer featuring Mashallah, from the title page of the De scientia motus orbis *(Latin version with engraving, 1504). As in many medieval illustrations, the compass here is an icon of religion as well as science, in reference to God as the architect of creation*

DISCOURS
DE LA METHODE
Pour bien conduire sa raison, & chercher

la verité dans les sciences.

Plus

LA DIOPTRIQVE.
LES METEORES.
Et
LA GEOMETRIE.

Qui sont des essais de cete Methode.

a Leyde

De l'Imprimerie de Ian Maire.

cIↄ Iↄc xxxvii.

Avec Privilege.

Discourse on Method by René Descartes

William Blake's "Newton" is a demonstration of his opposition to the 'single-vision' of scientific materialism; here, Isaac Newton is shown as 'divine geometer' (1795)

Chapter 6

History of logic

The **history of logic** is the study of the development of the science of valid inference (logic). Formal logic was developed in ancient times in China, India, and Greece. Greek logic, particularly Aristotelian logic, found wide application and acceptance in science and mathematics.

Aristotle's logic was further developed by Islamic and Christian philosophers in the Middle Ages, reaching a high point in the mid-fourteenth century. The period between the fourteenth century and the beginning of the nineteenth century was largely one of decline and neglect, and is regarded as barren by at least one historian of logic.[1]

Logic was revived in the mid-nineteenth century, at the beginning of a revolutionary period when the subject developed into a rigorous and formalistic discipline whose exemplar was the exact method of proof used in mathematics. The development of the modern "symbolic" or "mathematical" logic during this period is the most significant in the two-thousand-year history of logic, and is arguably one of the most important and remarkable events in human intellectual history.[2]

Progress in mathematical logic in the first few decades of the twentieth century, particularly arising from the work of Gödel and Tarski, had a significant impact on analytic philosophy and philosophical logic, particularly from the 1950s onwards, in subjects such as modal logic, temporal logic, deontic logic, and relevance logic.

6.1   Prehistory of logic

Valid reasoning has been employed in all periods of human history. However, logic studies the *principles* of valid reasoning, inference and demonstration. It is probable that the idea of demonstrating a conclusion first arose in connection with geometry, which originally meant the same as "land measurement".[3] In particular, the ancient Egyptians had empirically discovered some truths of geometry, such as the formula for the volume of a truncated pyramid.[4]

Another origin can be seen in Babylonia. Esagil-kin-apli's medical *Diagnostic Handbook* in the 11th century BC was based on a logical set of axioms and assumptions,[5] while Babylonian astronomers in the 8th and 7th centuries BC employed an internal logic within their predictive planetary systems, an important contribution to the philosophy of science.[6]

6.2   Logic in Greek philosophy

6.2.1   Before Plato

While the ancient Egyptians empirically discovered some truths of geometry, the great achievement of the ancient Greeks was to replace empirical methods by demonstrative science. The systematic study of this seems to have begun with the school of Pythagoras in the late sixth century BC.[4] The three basic principles of geometry are as follows:

- Certain propositions must be accepted as true without demonstration; such a proposition is known as an axiom of geometry.

- Every proposition that is not an axiom of geometry must be demonstrated as following from the axioms of geometry; such a demonstration is known as a proof or a "derivation" of the proposition.

- The proof must be *formal*; that is, the derivation of the proposition must be independent of the particular subject matter in question.[4]

Fragments of early proofs are preserved in the works of Plato and Aristotle,[7] and the idea of a deductive system was probably known in the Pythagorean school and the Platonic Academy.[4]

Separately from geometry, the idea of a standard argument pattern is found in the method of proof known as *reductio ad absurdum*, which was used by Zeno of Elea, a pre-Socratic philosopher of the fifth century BC. This is the technique of drawing an obviously false (that is, "absurd") conclusion from an assumption, thus demonstrating that the assumption is false.[8] Plato's Parmenides portrays Zeno as claiming to have written a book defending the monism of Parmenides by demonstrating the absurd consequence of assuming that there is plurality. Other philosophers who practised such *dialectic* reasoning were the "minor Socratics", including Euclid of Megara, who were probably followers of Parmenides and Zeno. The members of this school were called "dialecticians" (from a Greek word meaning "to discuss").

Further evidence that pre-Aristotelian thinkers were concerned with the principles of reasoning is found in the fragment called *dissoi logoi*, probably written at the beginning of the fourth century BC. This is part of a protracted debate about truth and falsity.[9]

In the case of the classical Greek city-states, interest in argumentation was also stimulated by the activities of the Rhetoricians or Orators and the Sophists, who used arguments to defend or attack a thesis, both in legal and political contexts.[10]

6.2.2 Plato's logic

None of the surviving works of the great fourth-century philosopher Plato (428–347 BC) include any formal logic,[11] but they include important contributions to the field of philosophical logic. Plato raises three questions:

- What is it that can properly be called true or false?

- What is the nature of the connection between the assumptions of a valid argument and its conclusion?

- What is the nature of definition?

The first question arises in the dialogue *Theaetetus*, where Plato identifies thought or opinion with talk or discourse (*logos*).[12] The second question is a result of Plato's theory of Forms. Forms are not things in the ordinary sense, nor strictly ideas in the mind, but they correspond to what philosophers later called universals, namely an abstract entity common to each set of things that have the same name. In both the *Republic* and the *Sophist*, Plato suggests that the necessary connection between the assumptions of a valid argument and its conclusion corresponds to a necessary connection between "forms".[13] The third question is about definition. Many of Plato's dialogues concern the search for a definition of some important concept (justice, truth, the Good), and it is likely that Plato was impressed by the importance of definition in mathematics.[14] What underlies every definition is a Platonic Form, the common nature present in different particular things. Thus, a definition reflects the ultimate object of understanding, and is the foundation of all valid inference. This had a great influence on Aristotle, in particular Aristotle's notion of the essence of a thing.[15]

6.2.3 Aristotle's logic

Main article: Organon

The logic of Aristotle, and particularly his theory of the syllogism, has had an enormous influence in Western thought.[16] His logical works, called the *Organon*, are the earliest formal study of logic that have come down to modern times. Though it is difficult to determine the dates, the probable order of writing of Aristotle's logical works is:

Plato's academy

- *The Categories*, a study of the ten kinds of primitive term.

- *The Topics* (with an appendix called *On Sophistical Refutations*), a discussion of dialectics.

- *On Interpretation*, an analysis of simple categorical propositions into simple terms, negation, and signs of quantity. It also contains a comprehensive treatment of the notions of opposition and conversion; chapter 7 is at the origin of the square of opposition (or logical square); chapter 9 contains the beginning of modal logic.

- *The Prior Analytics*, a formal analysis of what makes a syllogism (a valid argument, according to Aristotle).

- *The Posterior Analytics*, a study of scientific demonstration, containing Aristotle's mature views on logic.

These works are of outstanding importance in the history of logic. Aristotle was the first logician to attempt a systematic analysis of logical syntax, of noun (or *term*), and of verb. In the *Categories*, he attempts to discern all the possible things to which a term can refer; this idea underpins his philosophical work *Metaphysics*, which itself had a profound influence

Aristotle's logic was still influential in the Renaissance

on Western thought. He was the first to deal with the principles of contradiction and excluded middle in a systematic way. He was the first *formal logician*, in that he demonstrated the principles of reasoning by employing variables to show the underlying logical form of an argument. He was looking for relations of dependence which characterise necessary inference, and distinguished the validity of these relations, from the truth of the premises (the soundness of the argument). The *Prior Analytics* contains his exposition of the "syllogism", where three important principles are applied for the first time in history: the use of variables, a purely formal treatment, and the use of an axiomatic system. He also developed a theory of non-formal logic (*i.e.*, the theory of fallacies), which is presented in *Topics* and *Sophistical Refutations*.[17]

6.2.4 Stoic logic

The other great school of Greek logic is that of the Stoics.[18] Stoic logic traces its roots back to the late 5th century BC philosopher Euclid of Megara, a pupil of Socrates and slightly older contemporary of Plato. His pupils and successors were called "Megarians", or "Eristics", and later the "Dialecticians". The two most important dialecticians of the Megarian school were Diodorus Cronus and Philo, who were active in the late 4th century BC. The Stoics adopted the Megarian logic and systemized it. The most important member of the school was Chrysippus (c. 278–c. 206 BC), who was its third head, and who formalized much of Stoic doctrine. He is supposed to have written over 700 works, including at least 300 on logic, almost none of which survive.[19][20] Unlike with Aristotle, we have no complete works by the Megarians or the early Stoics, and have to rely mostly on accounts (sometimes hostile) by later sources, including prominently Diogenes Laertius, Sextus Empiricus, Galen, Aulus Gellius, Alexander of Aphrodisias, and Cicero.[21]

Three significant contributions of the Stoic school were (i) their account of modality, (ii) their theory of the Material conditional, and (iii) their account of meaning and truth.[22]

Chrysippus of Soli

- *Modality*. According to Aristotle, the Megarians of his day claimed there was no distinction between potentiality and actuality.[23] Diodorus Cronus defined the possible as that which either is or will be, the impossible as what will not be true, and the contingent as that which either is already, or will be false.[24] Diodorus is also famous for what is known as his Master argument, which states that each pair of the following 3 propositions contradicts the third proposition:

- Everything that is past is true and necessary.
- The impossible does not follow from the possible.
- What neither is nor will be is possible.

Diodorus used the plausibility of the first two to prove that nothing is possible if it neither is nor will be true.[25] Chrysippus, by contrast, denied the second premise and said that the impossible could follow from the possible.[26]

- *Conditional statements.* The first logicians to debate conditional statements were Diodorus and his pupil Philo of Megara. Sextus Empiricus refers three times to a debate between Diodorus and Philo. Philo regarded a conditional as true unless it has both a true antecedent and a false consequent. Precisely, let T_0 and T_1 be true statements, and let F_0 and F_1 be false statements; then, according to Philo, each of the following conditionals is a true statement, because it is not the case that the consequent is false while the antecedent is true (it is not the case that a false statement is asserted to follow from a true statement):

  - If T_0, then T_1
  - If F_0, then T_0
  - If F_0, then F_1

  The following conditional does not meet this requirement, and is therefore a false statement according to Philo:

  - If T_0, *then* F_0

  Indeed, Sextus says "According to [Philo], there are three ways in which a conditional may be true, and one in which it may be false."[27] Philo's criterion of truth is what would now be called a truth-functional definition of "if ... then"; it is the definition used in modern logic.

  In contrast, Diodorus allowed the validity of conditionals only when the antecedent clause could never lead to an untrue conclusion.[27][28][29] A century later, the Stoic philosopher Chrysippus attacked the assumptions of both Philo and Diodorus.

- *Meaning and truth.* The most important and striking difference between Megarian-Stoic logic and Aristotelian logic is that Megarian-Stoic logic concerns propositions, not terms, and is thus closer to modern propositional logic.[30] The Stoics distinguished between utterance (*phone*), which may be noise, speech (*lexis*), which is articulate but which be meaningless, and discourse (*logos*), which is meaningful utterance. The most original part of their theory is the idea that what is expressed by a sentence, called a *lekton*, is something real; this corresponds to what is now called a *proposition*. Sextus says that according to the Stoics, three things are linked together: that which signifies, that which is signified, and the object; for example, that which signifies is the word *Dion*, and that which is signified is what Greeks understand but barbarians do not, and the object is Dion himself.[31]

6.3 Logic in Asia

6.3.1 Logic in India

Main article: Indian logic

Logic began independently in ancient India and continued to develop to early modern times without any known influence from Greek logic.[32] Medhatithi Gautama (c. 6th century BC) founded the *anviksiki* school of logic.[33] The *Mahabharata* (12.173.45), around the 5th century BC, refers to the *anviksiki* and *tarka* schools of logic. Pāṇini (c. 5th century BC) developed a form of logic (to which Boolean logic has some similarities) for his formulation of Sanskrit grammar. Logic is described by Chanakya (c. 350-283 BC) in his *Arthashastra* as an independent field of inquiry.[34]

Two of the six Indian schools of thought deal with logic: Nyaya and Vaisheshika. The Nyaya Sutras of Aksapada Gautama (c. 2nd century AD) constitute the core texts of the Nyaya school, one of the six orthodox schools of Hindu philosophy. This realist school developed a rigid five-member schema of inference involving an initial premise, a reason, an example, an application, and a conclusion.[35] The idealist Buddhist philosophy became the chief opponent to the Naiyayikas. Nagarjuna (c. 150-250 AD), the founder of the Madhyamika ("Middle Way") developed an analysis known as the catuṣkoṭi (Sanskrit), a "four-cornered" system of argumentation that involves the systematic examination and rejection of each of the 4 possibilities of a proposition, P:

1. P; that is, being.

2. not P; that is, not being.

3. P and not P; that is, being and not being.

4. not (P or not P); that is, neither being nor not being.

It is interesting to note that under propositional logic, De Morgan's laws imply that this is equivalent to the third case (P and not P), and is therefore superfluous; there are actually only 3 cases to consider.

However, Dignaga (c 480-540 AD) is sometimes said to have developed a formal syllogism,[36] and it was through him and his successor, Dharmakirti, that Buddhist logic reached its height; it is contested whether their analysis actually constitutes a formal syllogistic system. In particular, their analysis centered on the definition of an inference-warranting relation, "vyapti", also known as invariable concomitance or pervasion.[37] To this end, a doctrine known as "apoha" or differentiation was developed.[38] This involved what might be called inclusion and exclusion of defining properties.

The difficulties involved in this enterprise, in part, stimulated the neo-scholastic school of Navya-Nyāya, which developed a formal analysis of inference in the sixteenth century. This later school began around eastern India and Bengal, and developed theories resembling modern logic, such as Gottlob Frege's "distinction between sense and reference of proper names" and his "definition of number," as well as the Navya-Nyaya theory of "restrictive conditions for universals" anticipating some of the developments in modern set theory.[39] Since 1824, Indian logic attracted the attention of many Western scholars, and has had an influence on important 19th-century logicians such as Charles Babbage, Augustus De Morgan, and particularly George Boole, as confirmed by his wife Mary Everest Boole, who wrote in 1901 an "open letter to Dr Bose", which was titled "Indian Thought and Western Science in the Nineteenth Century" and stated:[40][41] "Think what must have been the effect of the intense Hinduizing of three such men as Babbage, De Morgan and George Boole on the mathematical atmosphere of 1830-1865".

Dignāga's famous "wheel of reason" (*Hetucakra*) is a method of indicating when one thing (such as smoke) can be taken as an invariable sign of another thing (like fire), but the inference is often inductive and based on past observation. Matilal remarks that Dignāga's analysis is much like John Stuart Mill's Joint Method of Agreement and Difference, which is inductive.[42]

In addition, the traditional five-member Indian syllogism, though deductively valid, has repetitions that are unnecessary to its logical validity. As a result, some commentators see the traditional Indian syllogism as a rhetorical form that is entirely natural in many cultures of the world, and yet not as a logical form—not in the sense that all logically unnecessary elements have been omitted for the sake of analysis.

6.3.2   Logic in China

Main article: Logic in China

In China, a contemporary of Confucius, Mozi, "Master Mo", is credited with founding the Mohist school, whose canons dealt with issues relating to valid inference and the conditions of correct conclusions. In particular, one of the schools that grew out of Mohism, the Logicians, are credited by some scholars for their early investigation of formal logic. Due to the harsh rule of Legalism in the subsequent Qin Dynasty, this line of investigation disappeared in China until the introduction of Indian philosophy by Buddhists.

6.4 Medieval logic

6.4.1 Logic in the Middle East

Main article: Logic in Islamic philosophy
See also: Avicennian logic

The works of Al-Kindi, Al-Farabi, Avicenna, Al-Ghazali, Averroes and other Muslim logicians were based on Aris-

A text by Avicenna, founder of Avicennian logic

totelian logic and were important in communicating the ideas of the ancient world to the medieval West.[43] Al-Farabi

(Alfarabi) (873–950) was an Aristotelian logician who discussed the topics of future contingents, the number and relation of the categories, the relation between logic and grammar, and non-Aristotelian forms of inference.[44] Al-Farabi also considered the theories of conditional syllogisms and analogical inference, which were part of the Stoic tradition of logic rather than the Aristotelian.[45]

Ibn Sina (Avicenna) (980–1037) was the founder of Avicennian logic, which replaced Aristotelian logic as the dominant system of logic in the Islamic world,[46] and also had an important influence on Western medieval writers such as Albertus Magnus.[47] Avicenna wrote on the hypothetical syllogism[48] and on the propositional calculus, which were both part of the Stoic logical tradition.[49] He developed an original "temporally modalized" syllogistic theory, involving temporal logic and modal logic.[44] He also made use of inductive logic, such as the methods of agreement, difference, and concomitant variation which are critical to the scientific method.[48] One of Avicenna's ideas had a particularly important influence on Western logicians such as William of Ockham: Avicenna's word for a meaning or notion (*ma'na*), was translated by the scholastic logicians as the Latin *intentio*; in medieval logic and epistemology, this is a sign in the mind that naturally represents a thing.[50] This was crucial to the development of Ockham's conceptualism: A universal term (*e.g.,* "man") does not signify a thing existing in reality, but rather a sign in the mind (*intentio in intellectu*) which represents many things in reality; Ockham cites Avicenna's commentary on *Metaphysics* V in support of this view.[51]

Fakhr al-Din al-Razi (b. 1149) criticised Aristotle's "first figure" and formulated an early system of inductive logic, foreshadowing the system of inductive logic developed by John Stuart Mill (1806–1873).[52] Al-Razi's work was seen by later Islamic scholars as marking a new direction for Islamic logic, towards a Post-Avicennian logic. This was further elaborated by his student Afdaladdîn al-Khûnajî (d. 1249), who developed a form of logic revolving around the subject matter of conceptions and assents. In response to this tradition, Nasir al-Din al-Tusi (1201–1274) began a tradition of Neo-Avicennian logic which remained faithful to Avicenna's work and existed as an alternative to the more dominant Post-Avicennian school over the following centuries.[53]

The Illuminationist school was founded by Shahab al-Din Suhrawardi (1155–1191), who developed the idea of "decisive necessity", which refers to the reduction of all modalities (necessity, possibility, contingency and impossibility) to the single mode of necessity.[54] Ibn al-Nafis (1213–1288) wrote a book on Avicennian logic, which was a commentary of Avicenna's *Al-Isharat* (*The Signs*) and *Al-Hidayah* (*The Guidance*).[55] Ibn Taymiyyah (1263–1328), wrote the *Ar-Radd 'ala al-Mantiqiyyin*, where he argued against the usefulness, though not the validity, of the syllogism[56] and in favour of inductive reasoning.[52] Ibn Taymiyyah also argued against the certainty of syllogistic arguments and in favour of analogy; his argument is that concepts founded on induction are themselves not certain but only probable, and thus a syllogism based on such concepts is no more certain than an argument based on analogy. He further claimed that induction itself is founded on a process of analogy. His model of analogical reasoning was based on that of juridical arguments.[57][58] This model of analogy has been used in the recent work of John F. Sowa.[58]

The *Sharh al-takmil fi'l-mantiq* written by Muhammad ibn Fayd Allah ibn Muhammad Amin al-Sharwani in the 15th century is the last major Arabic work on logic that has been studied.[59] However, "thousands upon thousands of pages" on logic were written between the 14th and 19th centuries, though only a fraction of the texts written during this period have been studied by historians, hence little is known about the original work on Islamic logic produced during this later period.[53]

6.4.2   Logic in medieval Europe

Main article: Term logic

"Medieval logic" (also known as "Scholastic logic") generally means the form of Aristotelian logic developed in medieval Europe throughout roughly the period 1200–1600.[60] For centuries after Stoic logic had been formulated, it was the dominant system of logic in the classical world. When the study of logic resumed after the Dark Ages, the main source was the work of the Christian philosopher Boethius, who was familiar with some of Aristotle's logic, but almost none of the work of the Stoics.[61] Until the twelfth century, the only works of Aristotle available in the West were the *Categories*, *On Interpretation*, and Boethius's translation of the Isagoge of Porphyry (a commentary on the Categories). These works were known as the "Old Logic" (*Logica Vetus* or *Ars Vetus*). An important work in this tradition was the *Logica Ingredientibus* of Peter Abelard (1079–1142). His direct influence was small,[62] but his influence through pupils such as John of Salisbury was great, and his method of applying rigorous logical analysis to theology shaped the way that theological criticism developed in the period that followed.[63]

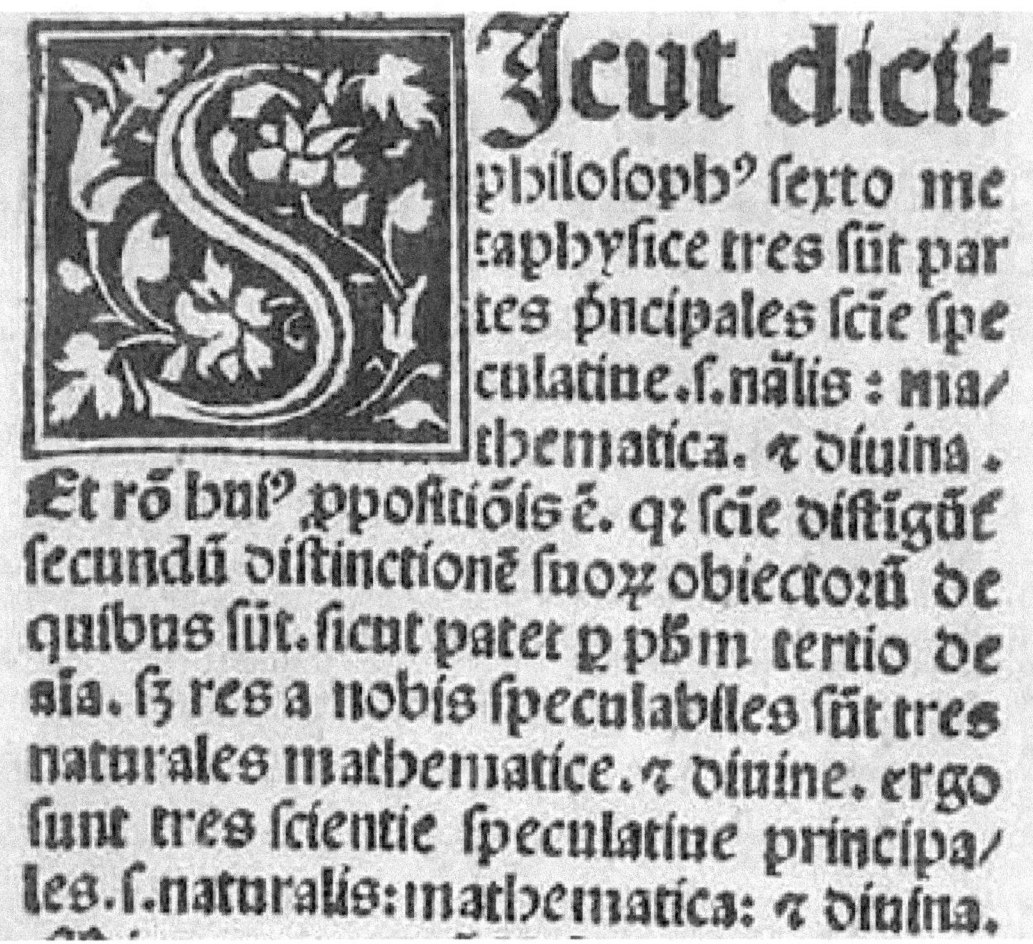

Brito's questions on the Old Logic

By the early thirteenth century, the remaining works of Aristotle's *Organon* (including the *Prior Analytics*, *Posterior Analytics*, and the *Sophistical Refutations*) had been recovered in the West and were revived by Saint Thomas Aquinas.[64] Logical work until then was mostly paraphrasis or commentary on the work of Aristotle.[65] The period from the middle of the thirteenth to the middle of the fourteenth century was one of significant developments in logic, particularly in three areas which were original, with little foundation in the Aristotelian tradition that came before. These were:[66]

- The theory of supposition. Supposition theory deals with the way that predicates (*e.g.,* 'man') range over a domain of individuals (*e.g.,* all men).[67] In the proposition 'every man is an animal', does the term 'man' range over or 'supposit for' men existing just in the present, or does the range include past and future men? Can a term supposit for a non-existing individual? Some medievalists have argued that this idea is a precursor of modern first-order logic.[68] "The theory of supposition with the associated theories of *copulatio* (sign-capacity of adjectival terms), *ampliatio* (widening of referential domain), and *distributio* constitute one of the most original achievements of Western medieval logic".[69]

- The theory of syncategoremata. Syncategoremata are terms which are necessary for logic, but which, unlike *categorematic* terms, do not signify on their own behalf, but 'co-signify' with other words. Examples of syncategoremata are 'and', 'not', 'every', 'if', and so on.

- The theory of consequences. A consequence is a hypothetical, conditional proposition: two propositions joined by the terms 'if ... then'. For example 'if a man runs, then God exists' (*Si homo currit, Deus est*).[70] A fully developed theory of consequences is given in Book III of William of Ockham's work Summa Logicae. There, Ockham distinguishes between 'material' and 'formal' consequences, which are roughly equivalent to the modern material implication and logical implication respectively. Similar accounts are given by Jean Buridan and Albert of Saxony.

The last great works in this tradition are the *Logic* of John Poinsot (1589–1644, known as John of St Thomas), the *Metaphysical Disputations* of Francisco Suarez (1548–1617), and the *Logica Demonstrativa* of Giovanni Girolamo Saccheri (1667–1733).

6.5   Traditional logic

6.5.1   The textbook tradition

Traditional logic generally means the textbook tradition that begins with Antoine Arnauld's and Pierre Nicole's *Logic, or the Art of Thinking*, better known as the *Port-Royal Logic*.[71] Published in 1662, it was the most influential work on logic in England until the nineteenth century.[72] The book presents a loosely Cartesian doctrine (that the proposition is a combining of ideas rather than terms, for example) within a framework that is broadly derived from Aristotelian and medieval term logic. Between 1664 and 1700, there were eight editions, and the book had considerable influence after that.[72] The account of propositions that Locke gives in the *Essay* is essentially that of Port-Royal: "Verbal propositions, which are words, [are] the signs of our ideas, put together or separated in affirmative or negative sentences. So that proposition consists in the putting together or separating these signs, according as the things which they stand for agree or disagree." (Locke, *An Essay Concerning Human Understanding*, IV. 5. 6)

Another influential work was the *Novum Organum* by Francis Bacon, published in 1620. The title translates as "new instrument". This is a reference to Aristotle's work known as the *Organon*. In this work, Bacon rejects the syllogistic method of Aristotle in favor of an alternative procedure "which by slow and faithful toil gathers information from things and brings it into understanding".[73] This method is known as inductive reasoning, a method which starts from empirical observation and proceeds to lower axioms or propositions; from these lower axioms, more general ones can be induced. For example, in finding the cause of a *phenomenal nature* such as heat, 3 lists should be constructed:

- The presence list: a list of every situation where heat is found.

- The absence list: a list of every situation that is similar to at least one of those of the presence list, except for the lack of heat.

- The variability list: a list of every situation where heat can vary.

Then, the *form nature* (or cause) of heat may be defined as that which is common to every situation of the presence list, and which is lacking from every situation of the absence list, and which varies by degree in every situation of the variability list.

Other works in the textbook tradition include Isaac Watts's *Logick: Or, the Right Use of Reason* (1725), Richard Whately's *Logic* (1826), and John Stuart Mill's *A System of Logic* (1843). Although the latter was one of the last great works in the tradition, Mill's view that the foundations of logic lie in introspection[74] influenced the view that logic is best understood as a branch of psychology, a view which dominated the next fifty years of its development, especially in Germany.[75]

6.5.2   Logic in Hegel's philosophy

G.W.F. Hegel indicated the importance of logic to his philosophical system when he condensed his extensive *Science of Logic* into a shorter work published in 1817 as the first volume of his *Encyclopaedia of the Philosophical Sciences*. The "Shorter" or "Encyclopaedia" *Logic*, as it is often known, lays out a series of transitions which leads from the most empty and abstract of categories—Hegel begins with "Pure Being" and "Pure Nothing"—to the "Absolute, the category which contains and resolves all the categories which preceded it. Despite the title, Hegel's *Logic* is not really a contribution to the science of valid inference. Rather than deriving conclusions about concepts through valid inference from premises, Hegel seeks to show that thinking about one concept compels thinking about another concept (one cannot, he argues, possess the concept of "Quality" without the concept of "Quantity"); this compulsion is, supposedly, not a matter of individual psychology, because it arises almost organically from the content of the concepts themselves. His purpose is to show the rational structure of the "Absolute"—indeed of rationality itself. The method by which thought is driven from one concept to its contrary, and then to further concepts, is known as the Hegelian dialectic.

Although Hegel's *Logic* has had little impact on mainstream logical studies, its influence can be seen elsewhere:

- Carl von Prantl's *Geschichte der Logik in Abendland* (1855–1867).[76]

- The work of the British Idealists, such as F.H. Bradley's *Principles of Logic* (1883).

- The economic, political, and philosophical studies of Karl Marx, and in the various schools of Marxism.

6.5.3   Logic and psychology

Between the work of Mill and Frege stretched half a century during which logic was widely treated as a descriptive science, an empirical study of the structure of reasoning, and thus essentially as a branch of psychology.[77] The German psychologist Wilhelm Wundt, for example, discussed deriving "the logical from the psychological laws of thought", emphasizing that "psychological thinking is always the more comprehensive form of thinking."[78] This view was widespread among German philosophers of the period:

- Theodor Lipps described logic as "a specific discipline of psychology".[79]

- Christoph von Sigwart understood logical necessity as grounded in the individual's compulsion to think in a certain way.[80]

- Benno Erdmann argued that "logical laws only hold within the limits of our thinking".[81]

Such was the dominant view of logic in the years following Mill's work.[82] This psychological approach to logic was rejected by Gottlob Frege. It was also subjected to an extended and destructive critique by Edmund Husserl in the first volume of his *Logical Investigations* (1900), an assault which has been described as "overwhelming".[83] Husserl argued forcefully that grounding logic in psychological observations implied that all logical truths remained unproven, and that skepticism and relativism were unavoidable consequences.

Such criticisms did not immediately extirpate what is called "psychologism". For example, the American philosopher Josiah Royce, while acknowledging the force of Husserl's critique, remained "unable to doubt" that progress in psychology would be accompanied by progress in logic, and vice versa.[84]

6.6   Rise of modern logic

The period between the fourteenth century and the beginning of the nineteenth century had been largely one of decline and neglect, and is generally regarded as barren by historians of logic.[1] The revival of logic occurred in the mid-nineteenth century, at the beginning of a revolutionary period where the subject developed into a rigorous and formalistic discipline whose exemplar was the exact method of proof used in mathematics. The development of the modern "symbolic" or "mathematical" logic during this period is the most significant in the 2000-year history of logic, and is arguably one of the most important and remarkable events in human intellectual history.[2]

A number of features distinguish modern logic from the old Aristotelian or traditional logic, the most important of which are as follows:[85] Modern logic is fundamentally a *calculus* whose rules of operation are determined only by the *shape* and not by the *meaning* of the symbols it employs, as in mathematics. Many logicians were impressed by the "success" of mathematics, in that there had been no prolonged dispute about any truly mathematical result. C.S. Peirce noted[86] that even though a mistake in the evaluation of a definite integral by Laplace led to an error concerning the moon's orbit that persisted for nearly 50 years, the mistake, once spotted, was corrected without any serious dispute. Peirce contrasted this with the disputation and uncertainty surrounding traditional logic, and especially reasoning in metaphysics. He argued that a truly "exact" logic would depend upon mathematical, i.e., "diagrammatic" or "iconic" thought. "Those who follow such methods will ... escape all error except such as will be speedily corrected after it is once suspected". Modern logic is also "constructive" rather than "abstractive"; i.e., rather than abstracting and formalising theorems derived from ordinary language (or from psychological intuitions about validity), it constructs theorems by formal methods, then looks for an interpretation in ordinary language. It is entirely symbolic, meaning that even the logical constants (which the medieval logicians called "syncategoremata") and the categoric terms are expressed in symbols.

6.7   Periods of modern logic

The development of modern logic falls into roughly five periods:[87]

- The **embryonic period** from Leibniz to 1847, when the notion of a logical calculus was discussed and developed, particularly by Leibniz, but no schools were formed, and isolated periodic attempts were abandoned or went unnoticed.

- The **algebraic period** from Boole's Analysis to Schröder's *Vorlesungen*. In this period, there were more practitioners, and a greater continuity of development.

- The **logicist period** from the Begriffsschrift of Frege to the *Principia Mathematica* of Russell and Whitehead. This was dominated by the "logicist school", whose aim was to incorporate the logic of all mathematical and scientific discourse in a single unified system, and which, taking as a fundamental principle that all mathematical truths are logical, did not accept any non-logical terminology. The major logicists were Frege, Russell, and the early Wittgenstein.[88] It culminates with the *Principia*, an important work which includes a thorough examination and attempted solution of the antinomies which had been an obstacle to earlier progress.

- The **metamathematical period** from 1910 to the 1930s, which saw the development of metalogic, in the finitist system of Hilbert, and the non-finitist system of Löwenheim and Skolem, the combination of logic and metalogic in the work of Gödel and Tarski. Gödel's incompleteness theorem of 1931 was one of the greatest achievements in the history of logic. Later in the 1930s, Gödel developed the notion of set-theoretic constructibility.

- The **period after World War II**, when mathematical logic branched into four inter-related but separate areas of research: model theory, proof theory, computability theory, and set theory, and its ideas and methods began to influence philosophy.

6.7.1   Embryonic period

The idea that inference could be represented by a purely mechanical process is found as early as Raymond Llull, who proposed a (somewhat eccentric) method of drawing conclusions by a system of concentric rings. The work of logicians such as the Oxford Calculators[89] led to a method of using letters instead of writing out logical calculations (*calculationes*) in words, a method used, for instance, in the *Logica magna* of Paul of Venice. Three hundred years after Llull, the English philosopher and logician Thomas Hobbes suggested that all logic and reasoning could be reduced to the mathematical operations of addition and subtraction.[90] The same idea is found in the work of Leibniz, who had read both Llull and Hobbes, and who argued that logic can be represented through a combinatorial process or calculus. But, like Llull and Hobbes, he failed to develop a detailed or comprehensive system, and his work on this topic was not published until long after his death. Leibniz says that ordinary languages are subject to "countless ambiguities" and are unsuited for a calculus, whose task is to expose mistakes in inference arising from the forms and structures of words;[91] hence, he proposed to identify an alphabet of human thought comprising fundamental concepts which could be composed to express complex ideas,[92] and create a *calculus ratiocinator* that would make all arguments "as tangible as those of the Mathematicians, so that we can find our error at a glance, and when there are disputes among persons, we can simply say: Let us calculate."[93]

Gergonne (1816) said that reasoning does not have to be about objects about which one has perfectly clear ideas, since algebraic operations can be carried out without having any idea of the meaning of the symbols involved.[94] Bolzano anticipated a fundamental idea of modern proof theory when he defined logical consequence or "deducibility" in terms of variables:[95]

> Hence I say that propostions M, N, O,... are *deducible* from propositions A, B, C, D,... with respect to variable parts i, j,..., if every class of ideas whose substitution for i, j,... makes all of A, B, C, D,... true, also makes all of M, N, O,... true. Occasionally, since it is customary, I shall say that propositions M, N, O,... *follow*, or can be *inferred* or *derived*, from A, B, C, D,.... Propositions A, B, C, D,... I shall call the *premises*, M, N, O,... the *conclusions*.

This is now known as semantic validity.

6.7.2 Algebraic period

Modern logic begins with what is known as the "algebraic school", originating with Boole and including Peirce, Jevons, Schröder and Venn.[96] Their objective was to develop a calculus to formalise reasoning in the area of classes, propositions and probabilities. The school begins with Boole's seminal work *Mathematical Analysis of Logic* which appeared in 1847, although De Morgan (1847) is its immediate precursor.[97] The fundamental idea of Boole's system is that algebraic formulae can be used to express logical relations. This idea occurred to Boole in his teenage years, working as an usher in a private school in Lincoln, Lincolnshire.[98] For example, let x and y stand for classes let the symbol = signify that the classes have the same members, xy stand for the class containing all and only the members of x and y and so on. Boole calls these *elective symbols*, i.e. symbols which select certain objects for consideration.[99] An expression in which elective symbols are used is called an *elective function*, and an equation of which the members are elective functions, is an *elective equation*.[100] The theory of elective functions and their "development" is essentially the modern idea of truth-functions and their expression in disjunctive normal form.[99]

Boole's system admits of two interpretations, in class logic, and propositional logic. Boole distinguished between "primary propositions" which are the subject of syllogistic theory, and "secondary propositions", which are the subject of propositional logic, and showed how under different "interpretations" the same algebraic system could represent both. An example of a primary proposition is "All inhabitants are either Europeans or Asiatics." An example of a secondary proposition is "Either all inhabitants are Europeans or they are all Asiatics."[101] These are easily distinguished in modern propositional calculus, where it is also possible to show that the first follows from the second, but it is a significant disadvantage that there is no way of representing this in the Boolean system.[102]

In his *Symbolic Logic* (1881), John Venn used diagrams of overlapping areas to express Boolean relations between classes or truth-conditions of propositions. In 1869 Jevons realised that Boole's methods could be mechanised, and constructed a "logical machine" which he showed to the Royal Society the following year.[99] In 1885 Allan Marquand proposed an electrical version of the machine that is still extant (picture at the Firestone Library).

The defects in Boole's system (such as the use of the letter *v* for existential propositions) were all remedied by his followers. Jevons published *Pure Logic, or the Logic of Quality apart from Quantity* in 1864, where he suggested a symbol to signify exclusive or, which allowed Boole's system to be greatly simplified.[103] This was usefully exploited by Schröder when he set out theorems in parallel columns in his *Vorlesungen* (1890–1905). Peirce (1880) showed how all the Boolean elective functions could be expressed by the use of a single primitive binary operation, "neither ... nor ..." and equally well "not both ... and ...",[104] however, like many of Peirce's innovations, this remained unknown or unnoticed until Sheffer rediscovered it in 1913.[105] Boole's early work also lacks the idea of the logical sum which originates in Peirce (1867), Schröder (1877) and Jevons (1890),[106] and the concept of inclusion, first suggested by Gergonne (1816) and clearly articulated by Peirce (1870).

The success of Boole's algebraic system suggested that all logic must be capable of algebraic representation, and there were attempts to express a logic of relations in such form, of which the most ambitious was Schröder's monumental *Vorlesungen über die Algebra der Logik* ("Lectures on the Algebra of Logic", vol iii 1895), although the original idea was again anticipated by Peirce.[107]

Boole's unwavering acceptance of Aristotle's logic is emphasized by the historian of logic John Corcoran in an accessible introduction to *Laws of Thought*[108] Corcoran also wrote a point-by-point comparison of *Prior Analytics* and *Laws of Thought*.[109] According to Corcoran, Boole fully accepted and endorsed Aristotle's logic. Boole's goals were "to go under, over, and beyond" Aristotle's logic by 1) providing it with mathematical foundations involving equations, 2) extending the class of problems it could treat — from assessing validity to solving equations — and 3) expanding the range of applications it could handle — e.g. from propositions having only two terms to those having arbitrarily many.

More specifically, Boole agreed with what Aristotle said; Boole's 'disagreements', if they might be called that, concern what Aristotle did not say. First, in the realm of foundations, Boole reduced the four propositional forms of Aristotle's logic to formulas in the form of equations — by itself a revolutionary idea. Second, in the realm of logic's problems, Boole's addition of equation solving to logic — another revolutionary idea — involved Boole's doctrine that Aristotle's rules of inference (the "perfect syllogisms") must be supplemented by rules for equation solving. Third, in the realm of applications, Boole's system could handle multi-term propositions and arguments whereas Aristotle could handle only two-termed subject-predicate propositions and arguments. For example, Aristotle's system could not deduce "No quadrangle that is a square is a rectangle that is a rhombus" from "No square that is a quadrangle is a rhombus that is a rectangle" or from "No rhombus that is a rectangle is a square that is a quadrangle".

6.7.3   Logicist period

After Boole, the next great advances were made by the German mathematician Gottlob Frege. Frege's objective was the program of Logicism, i.e. demonstrating that arithmetic is identical with logic.[110] Frege went much further than any of his predecessors in his rigorous and formal approach to logic, and his calculus or Begriffsschrift is important.[110] Frege also tried to show that the concept of number can be defined by purely logical means, so that (if he was right) logic includes arithmetic and all branches of mathematics that are reducible to arithmetic. He was not the first writer to suggest this. In his pioneering work *Die Grundlagen der Arithmetik* (The Foundations of Arithmetic), sections 15–17, he acknowledges the efforts of Leibniz, J.S. Mill as well as Jevons, citing the latter's claim that "algebra is a highly developed logic, and number but logical discrimination."[111]

Frege's first work, the *Begriffsschrift* ("concept script") is a rigorously axiomatised system of propositional logic, relying on just two connectives (negational and conditional), two rules of inference (*modus ponens* and substitution), and six axioms. Frege referred to the "completeness" of this system, but was unable to prove this.[112] The most significant innovation, however, was his explanation of the quantifier in terms of mathematical functions. Traditional logic regards the sentence "Caesar is a man" as of fundamentally the same form as "all men are mortal." Sentences with a proper name subject were regarded as universal in character, interpretable as "every Caesar is a man".[113] Frege argued that the quantifier expression "all men" does not have the same logical or semantic form as "all men", and that the universal proposition "every A is B" is a complex proposition involving two *functions*, namely ' – is A' and ' – is B' such that whatever satisfies the first, also satisfies the second. In modern notation, this would be expressed as

$$(x)\ Ax \to Bx$$

In English, "for all x, if Ax then Bx". Thus only singular propositions are of subject-predicate form, and they are irreducibly singular, i.e. not reducible to a general proposition. Universal and particular propositions, by contrast, are not of simple subject-predicate form at all. If "all mammals" were the logical subject of the sentence "all mammals are land-dwellers", then to negate the whole sentence we would have to negate the predicate to give "all mammals are *not* land-dwellers". But this is not the case.[114] This functional analysis of ordinary-language sentences later had a great impact on philosophy and linguistics.

This means that in Frege's calculus, Boole's "primary" propositions can be represented in a different way from "secondary" propositions. "All inhabitants are either Europeans or Asiatics" is

$$(x)\ [\ I(x) \to (E(x) \vee A(x))\ ]$$

whereas "All the inhabitants are Europeans or all the inhabitants are Asiatics" is

$$(x)\ (I(x) \to E(x)) \vee (x)\ (I(x) \to A(x))$$

As Frege remarked in a critique of Boole's calculus:

> "The real difference is that I avoid [the Boolean] division into two parts ... and give a homogeneous presentation of the lot. In Boole the two parts run alongside one another, so that one is like the mirror image of the other, but for that very reason stands in no organic relation to it'[115]

As well as providing a unified and comprehensive system of logic, Frege's calculus also resolved the ancient problem of multiple generality. The ambiguity of "every girl kissed a boy" is difficult to express in traditional logic, but Frege's logic captures this through the different scope of the quantifiers. Thus

(x) [girl(x) -> E(y) (boy(y) & kissed(x,y))]

means that to every girl there corresponds some boy (any one will do) who the girl kissed. But

E(x) [boy(x) & (y) (girl(y) -> kissed(y,x))]

means that there is some particular boy whom every girl kissed. Without this device, the project of logicism would have been doubtful or impossible. Using it, Frege provided a definition of the ancestral relation, of the many-to-one relation, and of mathematical induction.[116]

This period overlaps with the work of what is known as the "mathematical school", which included Dedekind, Pasch, Peano, Hilbert, Zermelo, Huntington, Veblen and Heyting. Their objective was the axiomatisation of branches of mathematics like geometry, arithmetic, analysis and set theory.

The logicist project received a near-fatal setback with the discovery of a paradox in 1901 by Bertrand Russell. This proved that the Frege's naive set theory led to a contradiction. Frege's theory is that for any formal criterion, there is a set of all objects that meet the criterion. Russell showed that a set containing exactly the sets that are not members of themselves would contradict its own definition (if it is not a member of itself, it is a member of itself, and if it is a member of itself, it is not).[117] This contradiction is now known as Russell's paradox. One important method of resolving this paradox was proposed by Ernst Zermelo.[118] Zermelo set theory was the first axiomatic set theory. It was developed into the now-canonical Zermelo–Fraenkel set theory (ZF).

The monumental Principia Mathematica, a three-volume work on the foundations of mathematics, written by Russell and Alfred North Whitehead and published 1910–13 also included an attempt to resolve the paradox, by means of an elaborate system of types: a set of elements is of a different type than is each of its elements (set is not the element; one element is not the set) and one cannot speak of the "set of all sets". The *Principia* was an attempt to derive all mathematical truths from a well-defined set of axioms and inference rules in symbolic logic.

6.7.4 Metamathematical period

The names of Gödel and Tarski dominate the 1930s,[119] a crucial period in the development of metamathematics – the study of mathematics using mathematical methods to produce metatheories, or mathematical theories about other mathematical theories. Early investigations into metamathematics had been driven by Hilbert's program. which sought to resolve the ongoing crisis in the foundations of mathematics by grounding all of mathematics to a finite set of axioms, proving its consistency by "finitistic" means and providing a procedure which would decide the truth or falsity of any mathematical statement. Work on metamathematics culminated in the work of Gödel, who in 1929 showed that a given first-order sentence is deducible if and only if it is logically valid – i.e. it is true in every structure for its language. This is known as Gödel's completeness theorem. A year later, he proved two important theorems, which showed Hibert's program to be unattainable in its original form. The first is that no consistent system of axioms whose theorems can be listed by an effective procedure such as an algorithm or computer program is capable of proving all facts about the natural numbers. For any such system, there will always be statements about the natural numbers that are true, but that are unprovable within the system. The second is that if such a system is also capable of proving certain basic facts about the natural numbers, then the system cannot prove the consistency of the system itself. These two results are known as Gödel's incompleteness theorems, or simply *Gödel's Theorem*. Later in the decade, Gödel developed the concept of set-theoretic constructibility, as part of his proof that the axiom of choice and the continuum hypothesis are consistent with Zermelo–Fraenkel set theory.

In proof theory, Gerhard Gentzen developed natural deduction and the sequent calculus. The former attempts to model logical reasoning as it 'naturally' occurs in practice and is most easily applied to intuitionistic logic, while the latter was devised to clarify the derivation of logical proofs in any formal system. Since Gentzen's work, natural deduction and sequent calculi have been widely applied in the fields of proof theory, mathematical logic and computer science. Gentzen also proved normalization and cut-elimination theorems for intuitionistic and classical logic which could be used to reduce logical proofs to a normal form.[120][121]

Alfred Tarski, a pupil of Łukasiewicz, is best known for his definition of truth and logical consequence, and the semantic concept of logical satisfaction. In 1933, he published (in Polish) *The concept of truth in formalized languages*, in which he proposed his semantic theory of truth: a sentence such as "snow is white" is true if and only if snow is white. Tarski's theory separated the metalanguage, which makes the statement about truth, from the object language, which contains the sentence whose truth is being asserted, and gave a correspondence (the T-schema) between phrases in the object language and elements of an interpretation. Tarski's approach to the difficult idea of explaining truth has been enduringly influential in logic and philosophy, especially in the development of model theory.[122] Tarski also produced important work on the methodology of deductive systems, and on fundamental principles such as completeness, decidability, consistency and definability. According to Anita Feferman, Tarski "changed the face of logic in the twentieth century".[123]

Alonzo Church and Alan Turing proposed formal models of computability, giving independent negative solutions to Hilbert's *Entscheidungsproblem* in 1936 and 1937, respectively. The *Entscheidungsproblem* asked for a procedure that, given any formal mathematical statement, would algorithmically determine whether the statement is true. Church and Turing proved there is no such procedure; Turing's paper introduced the halting problem as a key example of a mathematical problem without an algorithmic solution.

Church's system for computation developed into the modern λ-calculus, while the Turing machine became a standard model for a general-purpose computing device. It was soon shown that many other proposed models of computation were equivalent in power to those proposed by Church and Turing. These results led to the Church–Turing thesis that any deterministic algorithm that can be carried out by a human can be carried out by a Turing machine. Church proved additional undecidability results, showing that both Peano arithmetic and first-order logic are undecidable. Later work by Emil Post and Stephen Cole Kleene in the 1940s extended the scope of computability theory and introduced the concept of degrees of unsolvability.

The results of the first few decades of the twentieth century also had an impact upon analytic philosophy and philosophical logic, particularly from the 1950s onwards, in subjects such as modal logic, temporal logic, deontic logic, and relevance logic.

6.7.5   Logic after WWII

After World War II, mathematical logic branched into four inter-related but separate areas of research: model theory, proof theory, computability theory, and set theory.[124]

In set theory, the method of forcing revolutionized the field by providing a robust method for constructing models and obtaining independence results. Paul Cohen introduced this method in 1962 to prove the independence of the continuum hypothesis and the axiom of choice from Zermelo–Fraenkel set theory.[125] His technique, which was simplified and extended soon after its introduction, has since been applied to many other problems in all areas of mathematical logic.

Computability theory had its roots in the work of Turing, Church, Kleene, and Post in the 1930s and 40s. It developed into a study of abstract computability, which became known as recursion theory.[126] The priority method, discovered independently by Albert Muchnik and Richard Friedberg in the 1950s, led to major advances in the understanding of the degrees of unsolvability and related structures. Research into higher-order computability theory demonstrated its connections to set theory. The fields of constructive analysis and computable analysis were developed to study the effective content of classical mathematical theorems; these in turn inspired the program of reverse mathematics. A separate branch of computability theory, computational complexity theory, was also characterized in logical terms as a result of investigations into descriptive complexity.

Model theory applies the methods of mathematical logic to study models of particular mathematical theories. Alfred Tarski published much pioneering work in the field, which is named after a series of papers he published under the title *Contributions to the theory of models*. In the 1960s, Abraham Robinson used model-theoretic techniques to develop calculus and analysis based on infinitesimals, a problem that first had been proposed by Leibniz.

In proof theory, the relationship between classical mathematics and intuitionistic mathematics was clarified via tools such as the realizability method invented by Georg Kreisel and Gödel's *Dialectica* interpretation. This work inspired the contemporary area of proof mining. The Curry-Howard correspondence emerged as a deep analogy between logic and computation, including a correspondence between systems of natural deduction and typed lambda calculi used in computer science. As a result, research into this class of formal systems began to address both logical and computational aspects; this area of research came to be known as modern type theory. Advances were also made in ordinal analysis and the study of independence results in arithmetic such as the Paris–Harrington theorem.

This was also a period, particularly in the 1950s and afterwards, when the ideas of mathematical logic begin to influence philosophical thinking. For example, tense logic is a formalised system for representing, and reasoning about, propositions qualified in terms of time. The philosopher Arthur Prior played a significant role in its development in the 1960s. Modal logics extend the scope of formal logic to include the elements of modality (for example, possibility and necessity). The ideas of Saul Kripke, particularly about possible worlds, and the formal system now called Kripke semantics have had a profound impact on analytic philosophy.[127] His best known and most influential work is *Naming and Necessity* (1980).[128] Deontic logics are closely related to modal logics: they attempt to capture the logical features of obligation, permission and related concepts. Although some basic novelties syncretizing mathematical and philosophical logic were shown by Bolzano in the early 1800s, it was Ernst Mally, a pupil of Alexius Meinong, who was to propose the first formal deontic system in his *Grundgesetze des Sollens*, based on the syntax of Whitehead's and Russell's propositional calculus.

Another logical system founded after World War II was fuzzy logic by Azerbaijani mathematician Lotfi Asker Zadeh in 1965.

6.8 See also

- Timeline of mathematical logic

6.9 Notes

[1] Oxford Companion p. 498; Bochenski, Part I Introduction, *passim*

[2] Oxford Companion p. 500

[3] Kneale, p. 2

[4] Kneale p. 3

[5] H. F. J. Horstmanshoff, Marten Stol, Cornelis Tilburg (2004), *Magic and Rationality in Ancient Near Eastern and Graeco-Roman Medicine*, p. 99, Brill Publishers, ISBN 90-04-13666-5.

[6] D. Brown (2000), *Mesopotamian Planetary Astronomy-Astrology* , Styx Publications, ISBN 90-5693-036-2.

[7] Heath, *Mathematics in Aristotle*, cited in Kneale, p. 5

[8] Kneale p. 15

[9] Kneale, p. 16

[10] Encyclopedia Britannica

[11] Kneale p. 17

[12] "forming an opinion is talking, and opinion is speech that is held not with someone else or aloud but in silence with oneself" *Theaetetus* 189E–190A

[13] Kneale p. 20. For example, the proof given in the *Meno* that the square on the diagonal is double the area of the original square presumably involves the forms of the square and the triangle, and the necessary relation between them

[14] Kneale p. 21

[15] Zalta, Edward N. "Aristotle's Logic". Stanford University, 18 March 2000. Retrieved 13 March 2010.

[16] See e.g. Aristotle's logic, Stanford Encyclopedia of Philosophy

[17] Bochenski p. 63

[18] "Throughout later antiquity two great schools of logic were distinguished, the Peripatetic which was derived from Aristotle, and the Stoic which was developed by Chrysippus from the teachings of the Megarians" – Kneale p. 113

[19] *Oxford Companion*, article "Chrysippus", p. 134

[20] Stanford Encyclopedia of Philosophy: Susanne Bobzien, *Ancient Logic*

[21] K. Huelser, Die Fragmente zur Dialektik der Stoiker, 4 vols, Stuttgart 1986-7

[22] Kneale 117–158

[23] *Metaphysics* Eta 3, 1046b 29

[24] Boethius, *Commentary on the Perihermenias*, Meiser p. 234

[25] Epictetus, *Dissertationes* ed. Schenkel ii. 19. I.

[26] Alexander p. 177

[27] Sextus Empiricus, *Adv. Math.* viii, Section 113

[28] Sextus Empiricus, *Hypotyp.* ii. 110, comp.

[29] Cicero, *Academica*, ii. 47, *de Fato*, 6.

[30] See e.g. Lukasiewicz p. 21

[31] Sextus Bk viii., Sections 11, 12

[32] Bochenski p. 446

[33] S. C. Vidyabhusana (1971). *A History of Indian Logic: Ancient, Mediaeval, and Modern Schools.*

[34] R. P. Kangle (1986). *The Kautiliya Arthashastra* (1.2.11). Motilal Banarsidass.

[35] Bochenski p. 417 and *passim*

[36] Bochenski pp. 431–7

[37] Matilal, Bimal Krishna (1998). *The Character of Logic in India.* Albany, NY: State University of New York Press. pp. 12, 18. ISBN 9780791437407.

[38] Bochenksi p. 441

[39] Kisor Kumar Chakrabarti (June 1976). "Some Comparisons Between Frege's Logic and Navya-Nyaya Logic". *Philosophy and Phenomenological Research* (International Phenomenological Society) **36** (4): 554–563. doi:10.2307/2106873. JSTOR 2106873. This paper consists of three parts. The first part deals with Frege's distinction between sense and reference of proper names and a similar distinction in Navya-Nyaya logic. In the second part we have compared Frege's definition of number to the Navya-Nyaya definition of number. In the third part we have shown how the study of the so-called 'restrictive conditions for universals' in Navya-Nyaya logic anticipated some of the developments of modern set theory.

[40] Boole, Mary Everest "Collected Works" eds E M Cobham and E S Dummer London, Daniel 1931. Letter also published in the Ceylon National Review in 1909, and published as a separate pamphlet "The Psychologic Aspect of Imperialism" in 1911.

[41] Jonardon Ganeri (2001). *Indian logic: a reader.* Routledge. p. vii. ISBN 0-7007-1306-9

[42] Matilal, 17

[43] See e.g. Routledge Encyclopedia of Philosophy Online Version 2.0, article 'Islamic philosophy'

[44] History of logic: Arabic logic, *Encyclopædia Britannica.*

[45] Feldman, Seymour (1964-11-26). "Rescher on Arabic Logic". *The Journal of Philosophy* (Journal of Philosophy, Inc.) **61** (22): 724–734. doi:10.2307/2023632. ISSN 0022-362X. JSTOR 2023632. [726]. Long, A. A.; D. N. Sedley (1987). *The Hellenistic Philosophers. Vol 1: Translations of the principal sources with philosophical commentary.* Cambridge: Cambridge University Press. ISBN 0-521-27556-3.

[46] Dag Nikolaus Hasse (September 19, 2008). "Influence of Arabic and Islamic Philosophy on the Latin West". Stanford Encyclopedia of Philosophy. Retrieved 2009-10-13.

[47] Richard F. Washell (1973), "Logic, Language, and Albert the Great", *Journal of the History of Ideas* **34** (3), pp. 445–450 [445].

[48] Goodman, Lenn Evan (2003), *Islamic Humanism*, p. 155, Oxford University Press, ISBN 0-19-513580-6.

[49] Goodman, Lenn Evan (1992); *Avicenna*, p. 188, Routledge, ISBN 0-415-01929-X.

[50] Kneale p. 229

[51] Kneale: p. 266; Ockham: Summa Logicae i. 14; Avicenna: *Avicennae Opera* Venice 1508 f87rb

[52] Muhammad Iqbal, *The Reconstruction of Religious Thought in Islam*, "The Spirit of Muslim Culture" (cf. and)

[53] Tony Street (July 23, 2008). "Arabic and Islamic Philosophy of Language and Logic". Stanford Encyclopedia of Philosophy. Retrieved 2008-12-05.

[54] Dr. Lotfollah Nabavi, Sohrevardi's Theory of Decisive Necessity and kripke's QSS System, *Journal of Faculty of Literature and Human Sciences.*

[55] Dr. Abu Shadi Al-Roubi (1982), "Ibn Al-Nafis as a philosopher", *Symposium on Ibn al-Nafis*, Second International Conference on Islamic Medicine: Islamic Medical Organization, Kuwait (cf. Ibn al-Nafis As a Philosopher, *Encyclopedia of Islamic World*).

[56] See pp. 253–254 of Street, Tony (2005). "Logic". In Peter Adamson and Richard C. Taylor (edd.). *The Cambridge Companion to Arabic Philosophy*. Cambridge University Press. pp. 247–265. ISBN 978-0-521-52069-0.

[57] Ruth Mas (1998). "Qiyas: A Study in Islamic Logic" (PDF). *Folia Orientalia* **34**: 113–128. ISSN 0015-5675.

[58] John F. Sowa; Arun K. Majumdar (2003). "Analogical reasoning". *Conceptual Structures for Knowledge Creation and Communication, Proceedings of ICCS 2003*. Berlin: Springer-Verlag., pp. 16-36

[59] Nicholas Rescher and Arnold vander Nat, "The Arabic Theory of Temporal Modal Syllogistic", in George Fadlo Hourani (1975), *Essays on Islamic Philosophy and Science*, pp. 189–221, State University of New York Press, ISBN 0-87395-224-3.

[60] Boehner p. xiv

[61] Kneale p. 198

[62] Stephen Dumont, article "Peter Abelard" in Gracia and Noone p. 492

[63] Kneale, pp. 202–3

[64] See e.g. Kneale p. 225

[65] Boehner p. 1

[66] Boehner pp. 19–76

[67] Boehner p. 29

[68] Boehner p. 30

[69] Ebbesen 1981

[70] Boehner pp. 54–5

[71] *Oxford Companion* p. 504, article "Traditional logic"

[72] Buroker xxiii

[73] Farrington, 1964, 89

[74] N. Abbagnano, "Psychologism" in P. Edwards (ed) *The Encyclopaedia of Philosophy*, MacMillan, 1967

[75] Of the German literature in this period, Robert Adamson wrote "*Logics* swarm as bees in springtime..."; Robert Adamson, *A Short History of Logic*, Wm. Blackwood & Sons, 1911, page 242

[76] Carl von Prantl (1855-1867), *Geschichte von Logik in Abendland*, Leipsig: S. Hirzl, anastatically reprinted in 1997, Hildesheim: Georg Olds.

[77] See e.g. Psychologism, Stanford Encyclopedia of Philosophy

[78] Wilhelm Wundt, *Logik* (1880–1883); quoted in Edmund Husserl, *Logical Investigations,* translated J.N. Findlay, Routledge, 2008, Volume 1, pp. 115–116.

[79] Theodor Lipps, *Grundzüge der Logik* (1893); quoted in Edmund Husserl, *Logical Investigations,* translated J.N. Findlay, Routledge, 2008, Volume 1, p. 40

[80] Christoph von Sigwart, *Logik* (1873–78); quoted in Edmund Husserl, *Logical Investigations,* translated J.N. Findlay, Routledge, 2008, Volume 1, p. 51

[81] Benno Erdmann, *Logik* (1892); quoted in Edmund Husserl, *Logical Investigations,* translated J.N. Findlay, Routledge, 2008, Volume 1, p. 96

[82] Dermot Moran, "Introduction"; Edmund Husserl, *Logical Investigations,* translated J.N. Findlay, Routledge, 2008, Volume 1, p. xxi

[83] Michael Dummett, "Preface"; Edmund Husserl, *Logical Investigations,* translated J.N. Findlay, Routledge, 2008, Volume 1, p. xvii

[84] Josiah Royce, "Recent Logical Enquiries and their Psychological Bearings" (1902) in John J. McDermott (ed) *The Basic Writings of Josiah Royce* Volume 2, Fordham University Press, 2005, p. 661

[85] Bochenski, p. 266

[86] Peirce 1896

[87] See Bochenski p. 269

[88] *Oxford Companion* p. 499

[89] Edith Sylla (1999), "Oxford Calculators", in *The Cambridge Dictionary of Philosophy*, Cambridge, Cambridgeshire: Cambridge.

[90] El. philos. sect. I de corp 1.1.2.

[91] Bochenski p. 274

[92] Rutherford, Donald, 1995, "Philosophy and language" in Jolley, N., ed., *The Cambridge Companion to Leibniz*. Cambridge Univ. Press.

[93] Wiener, Philip, 1951. *Leibniz: Selections*. Scribner.

[94] *Essai de dialectique rationelle*, 211n, quoted in Bochenski p. 277.

[95] Bolzano, Bernard (1972). George, Rolf, ed. *The Theory of Science: Die Wissenschaftslehre oder Versuch einer Neuen Darstellung der Logik*. Translated by George Rolf. University of California Press. p. 209. ISBN 9780520017870.

[96] See e.g. Bochenski p. 296 and *passim*

[97] Before publishing, he wrote to De Morgan, who was just finishing his work *Formal Logic*. De Morgan suggested they should publish first, and thus the two books appeared at the same time, possibly even reaching the bookshops on the same day. cf. Kneale p. 404

[98] Kneale p. 404

[99] Kneale p. 407

[100] Boole (1847) p. 16

[101] Boole 1847 pp. 58–9

[102] Beaney p. 11

[103] Kneale p. 422

[104] Peirce, "A Boolean Algebra with One Constant", 1880 MS, *Collected Papers* v. 4, paragraphs 12–20, reprinted *Writings* v. 4, pp. 218-21. Google Preview.

[105] *Trans. Amer. Math. Soc., xiv (1913)*, pp. 481–8. This is now known as the Sheffer stroke

[106] Bochenski 296

[107] See CP III

[108] George Boole. 1854/2003. The Laws of Thought, facsimile of 1854 edition, with an introduction by J. Corcoran. Buffalo: Prometheus Books (2003). Reviewed by James van Evra in Philosophy in Review.24 (2004) 167–169.

[109] JOHN CORCORAN, Aristotle's Prior Analytics and Boole's Laws of Thought, History and Philosophy of Logic, vol. 24 (2003), pp. 261–288.

[110] Kneale p. 435

[111] Jevons, *The Principles of Science*, London 1879, p. 156, quoted in *Grundlagen* 15

[112] Beaney p. 10 – the completeness of Frege's system was eventually proved by Jan Łukasiewicz in 1934

[113] See for example the argument by the medieval logician William of Ockham that singular propositions are universal, in Summa Logicae III. 8 (??)

[114] "On concept and object" p. 198; Geach p. 48

[115] BLC p. 14, quoted in Beaney p. 12

[116] See e.g. The Internet Encyclopedia of Philosophy, article "Frege"

[117] See e.g. Potter 2004

[118] Zermelo 1908

[119] Feferman 1999 p. 1

[120] Girard, Jean-Yves; Paul Taylor; Yves Lafont (1990) [1989]. *Proofs and Types*. Cambridge University Press (Cambridge Tracts in Theoretical Computer Science, 7). ISBN 0-521-37181-3.

[121] Alex Sakharov, "Cut Elimination Theorem", *MathWorld*.

[122] Feferman and Feferman 2004, p. 122, discussing "The Impact of Tarski's Theory of Truth".

[123] Feferman 1999, p. 1

[124] See e.g. Barwise, *Handbook of Mathematical Logic*

[125] The Independence of the Continuum Hypothesis, II Paul J. Cohen Proceedings of the National Academy of Sciences of the United States of America, Vol. 51, No. 1. (Jan. 15, 1964), pp. 105-110.

[126] Many of the foundational papers are collected in *The Undecidable* (1965) edited by Martin Davis

[127] Jerry Fodor, "Water's water everywhere", *London Review of Books*, 21 October 2004

[128] See *Philosophical Analysis in the Twentieth Century: Volume 2: The Age of Meaning*, Scott Soames: "*Naming and Necessity* is among the most important works ever, ranking with the classical work of Frege in the late nineteenth century, and of Russell, Tarski and Wittgenstein in the first half of the twentieth century". Cited in Byrne, Alex and Hall, Ned. 2004. 'Necessary Truths'. *Boston Review* October/November 2004

6.10   References

Primary Sources

- Alexander of Aphrodisias, *In Aristotelis An. Pr. Lib. I Commentarium*, ed. Wallies, Berlin, C.I.A.G. vol. II/1, 1882.

- Avicenna, *Avicennae Opera* Venice 1508.

- Boethius *Commentary on the Perihermenias*, Secunda Editio, ed. Meiser, Leipzig, Teubner, 1880.

- Bolzano, Bernard *Wissenschaftslehre*, (1837) 4 Bde, Neudr., hrsg. W. Schultz, Leipzig I-II 1929, III 1930, IV 1931 (*Theory of Science*, four volumes, translated by Rolf George and Paul Rusnock, New York: Oxford University Press, 2014).

- Bolzano, Bernard *Theory of Science* (Edited, with an introduction, by Jan Berg. Translated from the German by Burnham Terrell – D. Reidel Publishing Company, Dordrecht and Boston 1973).

- Boole, George (1847) *The Mathematical Analysis of Logic* (Cambridge and London); repr. in *Studies in Logic and Probability*, ed. R. Rhees (London 1952).

- Boole, George (1854) *The Laws of Thought* (London and Cambridge); repr. as *Collected Logical Works*. Vol. 2, (Chicago and London: Open Court, 1940).

- Epictetus, *Epicteti Dissertationes ab Arriano digestae*, edited by Heinrich Schenkl, Leipzig, Teubner. 1894.

- Frege, G., *Boole's Logical Calculus and the Concept Script*, 1882, in *Posthumous Writings* transl. P. Long and R. White 1969, pp. 9–46.

- Gergonne, Joseph Diaz, (1816) *Essai de dialectique rationelle*, in *Annales de mathem, pures et appl*. 7, 1816/7, 189–228.

- Jevons, W.S. *The Principles of Science*, London 1879.

- *Ockham's Theory of Terms*: Part I of the Summa Logicae, translated and introduced by Michael J. Loux (Notre Dame, IN: University of Notre Dame Press 1974). Reprinted: South Bend, IN: St. Augustine's Press, 1998.

- *Ockham's Theory of Propositions*: Part II of the Summa Logicae, translated by Alfred J. Freddoso and Henry Schuurman and introduced by Alfred J. Freddoso (Notre Dame, IN: University of Notre Dame Press, 1980). Reprinted: South Bend, IN: St. Augustine's Press, 1998.

- Peirce, C.S., (1896), "The Regenerated Logic", *The Monist*, vol. VII, No. 1, p pp. 19–40, The Open Court Publishing Co., Chicago, IL, 1896, for the Hegeler Institute. Reprinted (CP 3.425–455). *Internet Archive The Monist* 7.

- Sextus Empiricus, *Against the Logicians*. (Adversus Mathematicos VII and VIII). Richard Bett (trans.) Cambridge: Cambridge University Press, 2005. ISBN 0-521-53195-0.

- Zermelo, Ernst (1908). "Untersuchungen über die Grundlagen der Mengenlehre I". *Mathematische Annalen* **65** (2): 261–281. doi:10.1007/BF01449999. English translation in Heijenoort, Jean van (1967). "Investigations in the foundations of set theory". *From Frege to Gödel: A Source Book in Mathematical Logic, 1879–1931*. Source Books in the History of the Sciences. Harvard Univ. Press. pp. 199–215. ISBN 978-0-674-32449-7..

Secondary Sources

- Barwise, Jon, (ed.), *Handbook of Mathematical Logic*, Studies in Logic and the Foundations of Mathematics, Amsterdam, North Holland, 1982 ISBN 978-0-444-86388-1 .

- Beaney, Michael, *The Frege Reader*, London: Blackwell 1997.

- Bochenski, I.M., *A History of Formal Logic*, Indiana, Notre Dame University Press, 1961.

- Philotheus Boehner, *Medieval Logic*, Manchester 1950.

- Jill Vance Buroker (transl. and introduction), A. Arnauld, P. Nicole *Logic or the Art of Thinking*, Cambridge University Press, 1996, ISBN 0-521-48249-6.

- Church, Alonzo, 1936-8. "A bibliography of symbolic logic". *Journal of Symbolic Logic 1*: 121–218; *3*:178–212.

- Ebbesen, S. "Early supposition theory (12th–13th Century)" *Histoire, Épistémologie, Langage* 3/1: 35–48 (1981).

- Farrington, B., *The Philosophy of Francis Bacon*, Liverpool 1964.

- Feferman, Anita B. (1999). "Alfred Tarski". *American National Biography*. 21. Oxford University Press. pp. 330–332. ISBN 978-0-19-512800-0.

- Feferman, Anita B.; Feferman, Solomon (2004). *Alfred Tarski: Life and Logic*. Cambridge University Press. ISBN 978-0-521-80240-6. OCLC 54691904.

- Gabbay, Dov and John Woods, eds, *Handbook of the History of Logic* 2004. 1. Greek, Indian and Arabic logic; 2. Mediaeval and Renaissance logic; 3. The rise of modern logic: from Leibniz to Frege; 4. British logic in the Nineteenth century; 5. Logic from Russell to Church; 6. Sets and extensions in the Twentieth century; 7. Logic and the modalities in the Twentieth century; 8. The many-valued and nonmonotonic turn in logic; 9. Computational Logic; 10. Inductive logic; 11. Logic: A history of its central concepts; Elsevier, ISBN 0-444-51611-5.

- Geach, P.T. *Logic Matters*, Blackwell 1972.

- Goodman, Lenn Evan (2003). *Islamic Humanism*. Oxford University Press, ISBN 0-19-513580-6.

- Goodman, Lenn Evan (1992). *Avicenna*. Routledge, ISBN 0-415-01929-X.

- Grattan-Guinness, Ivor, 2000. *The Search for Mathematical Roots 1870–1940*. Princeton University Press.

- Gracia, J.G. and Noone, T.B., *A Companion to Philosophy in the Middle Ages*, London 2003.

- Haaparanta, Leila (ed.) 2009. *The Development of Modern Logic* Oxford University Press.

- Heath, T.L., 1949. *Mathematics in Aristotle* Oxford University Press.

- Heath, T.L., 1931, *A Manual of Greek Mathematics*, Oxford (Clarendon Press).

- Honderich, Ted (ed.). The Oxford Companion to Philosophy (New York: Oxford University Press, 1995) ISBN 0-19-866132-0.

- Kneale, William and Martha, 1962. *The development of logic*. Oxford University Press, ISBN 0-19-824773-7.

- Lukasiewicz, *Aristotle's Syllogistic*, Oxford University Press 1951.

- Michael Potter (2004), *Set Theory and its Philosophy*, Oxford Univ. Press.

6.11 External links

- History of Logic in Relationship to Ontology with an annotated bibliography on the history of logic

- Ancient Logic entry by Susanne Bobzien in the *Stanford Encyclopedia of Philosophy*

- Peter of Spain entry by Joke Spruyt in the *Stanford Encyclopedia of Philosophy*

- Paul Spade's "Thoughts Words and Things" An Introduction to Late Mediaeval Logic and Semantic Theory

- Insights, Images, and Bios of 145 logicians by David Marans

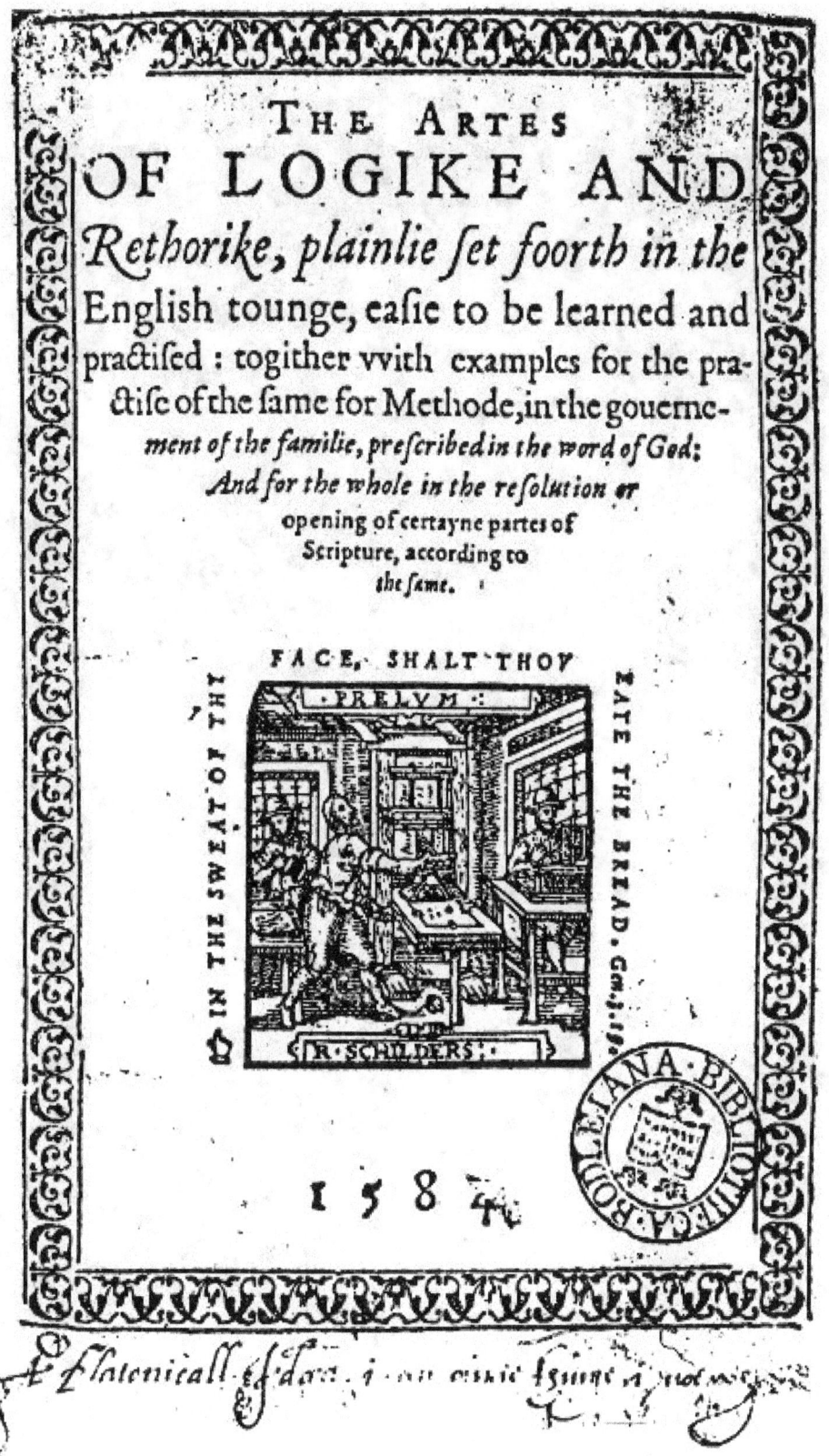

Dudley Fenner's Art of Logic *(1584)*

Georg Wilhelm Friedrich Hegel

Life of Raymond Lull. 14th-century manuscript.

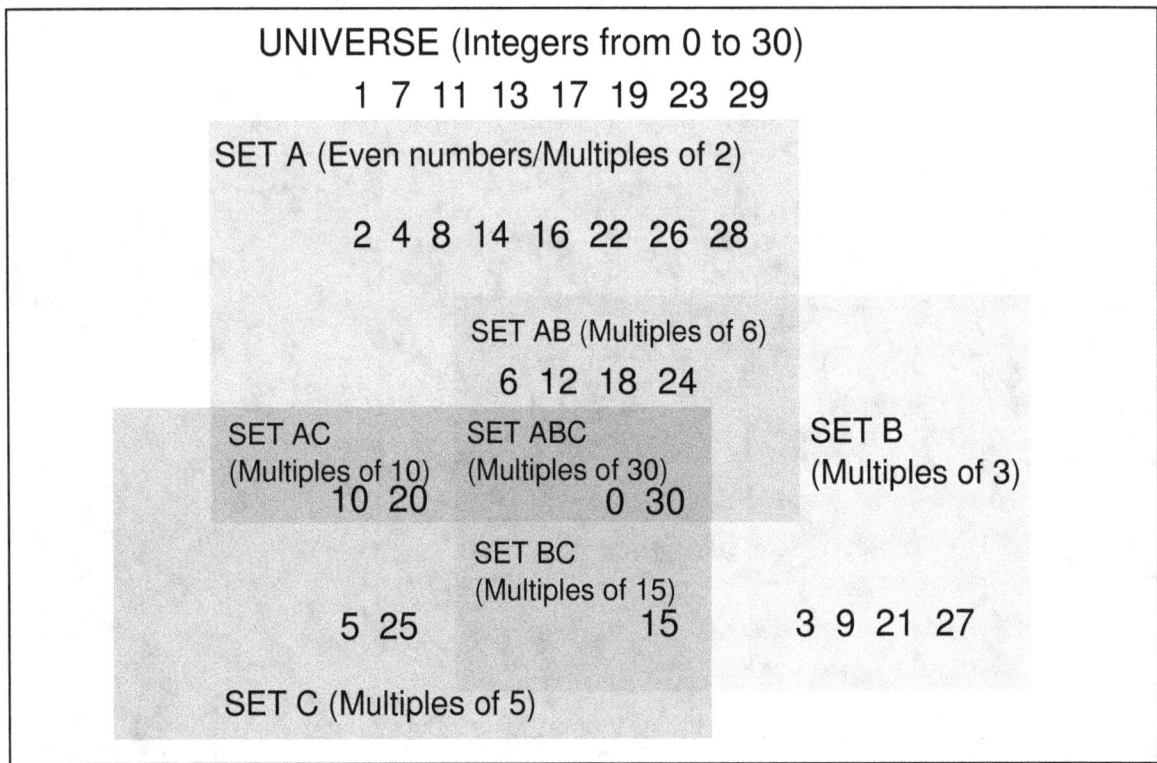

Boolean multiples

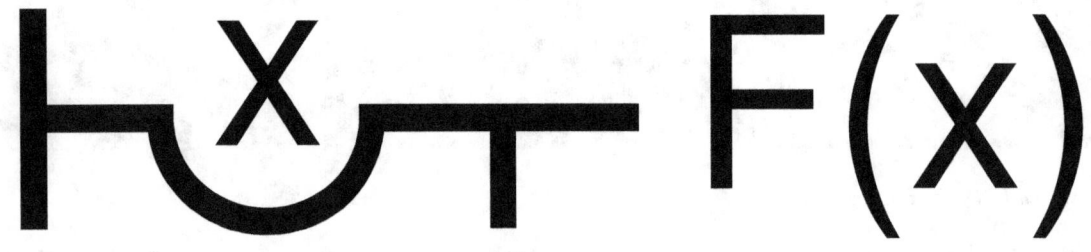

Frege's "Concept Script"

Alfred Tarski

Saul Kripke

Chapter 7

History of mathematical notation

The **history of mathematical notation**[1] includes the commencement, progress, and cultural diffusion of mathematical symbols and the conflict of the methods of notation confronted in a notation's move to popularity or inconspicuousness. Mathematical notation[2] comprises the symbols used to write mathematical equations and formulas. Notation generally implies a set of well-defined representations of quantities and symbols operators.[3] The history includes Hindu-Arabic numerals, letters from the Roman, Greek, Hebrew, and German alphabets, and a host of symbols invented by mathematicians over the past several centuries.

The development of mathematical notation can be divided in stages.[4][5] The "*rhetorical*" stage is where calculations are performed by words and no symbols are used.[6] The "*syncopated*" stage is where frequently used operations and quantities are represented by symbolic syntactical abbreviations. From ancient times through the post-classical age,[note 1] bursts of mathematical creativity were often followed by centuries of stagnation. As the early modern age opened and the worldwide spread of knowledge began, written examples of mathematical developments came to light. The "*symbolic*" stage is where comprehensive systems of notation supersede rhetoric. Beginning in Italy in the 16th century, new mathematical developments, interacting with new scientific discoveries, were made at an increasing pace that continues through the present day. This symbolic system was in use by medieval Indian mathematicians and in Europe since the middle of the 17th century,[7] and has continued to develop in the contemporary era.

The area of study known as the history of mathematics is primarily an investigation into the origin of discoveries in mathematics and, the focus here, the investigation into the mathematical methods and notation of the past.

7.1 History

See also: Timeline of mathematics and Foundations of mathematics

7.1.1 Rhetorical stage

See also: Measurement

Although the history commences with that of the Ionian schools, there is no doubt that those Ancient Greeks who paid attention to it were largely indebted to the previous investigations of the Ancient Egyptians and Ancient Phoenicians. Numerical notation distinctive feature, i.e. symbols having local as well as intrinsic values (arithmetic), implies a state of civilization at the period of its invention. Our knowledge of the mathematical attainments of these early peoples, to which this section is devoted, is imperfect and the following brief notes be regarded as a summary of the conclusions which seem most probable, and the history of mathematics begins with the symbolic sections.

Many areas of mathematics began with the study of real world problems, before the underlying rules and concepts were identified and defined as abstract structures. For example, geometry has its origins in the calculation of distances and areas in the real world; algebra started with methods of solving problems in arithmetic.

There can be no doubt that most early peoples which have left records knew something of numeration and mechanics, and that a few were also acquainted with the elements of land-surveying. In particular, the Egyptians paid attention to geometry and numbers, and the Phoenicians to practical arithmetic, book-keeping, navigation, and land-surveying. The results attained by these people seem to have been accessible, under certain conditions, to travelers. It is probable that the knowledge of the Egyptians and Phoenicians was largely the result of observation and measurement, and represented the accumulated experience of many ages.

Beginning of notation

See also: Ancient history, History of writing ancient numbers and History of science in early cultures

Written mathematics began with numbers expressed as tally marks, with each tally representing a single unit. The numerical symbols consisted probably of strokes or notches cut in wood or stone, and intelligible alike to all nations.[note 2] For example, one notch in a bone represented one animal, or person, or anything else. The peoples with whom the Greeks of Asia Minor (amongst whom notation in western history begins) were likely to have come into frequent contact were those inhabiting the eastern littoral of the Mediterranean: and Greek tradition uniformly assigned the special development of geometry to the Egyptians, and that of the science of numbers[note 3] either to the Egyptians or to the Phoenicians.

The Ancient Egyptians had a symbolic notation which was the numeration by Hieroglyphics.[8][9] The Egyptian mathematics had a symbol for one, ten, one-hundred, one-thousand, ten-thousand, one-hundred-thousand, and one-million. Smaller digits were placed on the left of the number, as they are in Hindu-Arabic numerals. Later, the Egyptians used hieratic instead of hieroglyphic script to show numbers. Hieratic was more like cursive and replaced several groups of symbols with individual ones. For example, the four vertical lines used to represent four were replaced by a single horizontal line. This is found in the Rhind Mathematical Papyrus (c. 2000–1800 BC) and the Moscow Mathematical Papyrus (c. 1890 BC). The system the Egyptians used was discovered and modified by many other civilizations in the Mediterranean. The Egyptians also had symbols for basic operations: legs going forward represented addition, and legs walking backward to represent subtraction.

The Mesopotamians had symbols for each power of ten.[10] Later, they wrote their numbers in almost exactly the same way done in modern times. Instead of having symbols for each power of ten, they would just put the coefficient of that number. Each digit was at separated by only a space, but by the time of Alexander the Great, they had created a symbol that represented zero and was a placeholder. The Mesopotamians also used a sexagesimal system, that is base sixty. It is this system that is used in modern times when measuring time and angles. Babylonian mathematics is derived from more than 400 clay tablets unearthed since the 1850s.[11] Written in Cuneiform script, tablets were inscribed whilst the clay was moist, and baked hard in an oven or by the heat of the sun. Some of these appear to be graded homework. The earliest evidence of written mathematics dates back to the ancient Sumerians and the system of metrology from 3000 BC. From around 2500 BC onwards, the Sumerians wrote multiplication tables on clay tablets and dealt with geometrical exercises and division problems. The earliest traces of the Babylonian numerals also date back to this period.[12]

The majority of Mesopotamian clay tablets date from 1800 to 1600 BC, and cover topics which include fractions, algebra, quadratic and cubic equations, and the calculation of regular reciprocal pairs.[13] The tablets also include multiplication tables and methods for solving linear and quadratic equations. The Babylonian tablet YBC 7289 gives an approximation of $\sqrt{2}$ accurate to five decimal places. Babylonian mathematics were written using a sexagesimal (base-60) numeral system. From this derives the modern day usage of 60 seconds in a minute, 60 minutes in an hour, and 360 (60 x 6) degrees in a circle, as well as the use of minutes and seconds of arc to denote fractions of a degree. Babylonian advances in mathematics were facilitated by the fact that 60 has many divisors: the reciprocal of any integer which is a multiple of divisors of 60 has a finite expansion in base 60. (In decimal arithmetic, only reciprocals of multiples of 2 and 5 have finite decimal expansions.) Also, unlike the Egyptians, Greeks, and Romans, the Babylonians had a true place-value system, where digits written in the left column represented larger values, much as in the decimal system. They lacked, however, an equivalent of the decimal point, and so the place value of a symbol often had to be inferred from the context.

7.1.2   Syncopated stage

See also: Fundamental theorem of arithmetic and Naive set theory

The history of mathematics cannot with certainty be traced back to any school or period before that of the Ionian Greeks, but the subsequent history may be divided into periods, the distinctions between which are tolerably well marked. Greek mathematics, which originated with the study of geometry, tended from its commencement to be deductive and scientific. Since the fourth century AD, Pythagoras has commonly been given credit for discovering the Pythagorean theorem, a theorem in geometry that states that in a right-angled triangle the area of the square on the hypotenuse (the side opposite the right angle) is equal to the sum of the areas of the squares of the other two sides.[note 5] The ancient mathematical texts are available with the prior mentioned Ancient Egyptians notation and with Plimpton 322 (Babylonian mathematics c. 1900 BC). The study of mathematics as a subject in its own right begins in the 6th century BC with the Pythagoreans, who coined the term "mathematics" from the ancient Greek μάθημα (*mathema*), meaning "subject of instruction".[14]

Plato's influence has been especially strong in mathematics and the sciences. He helped to distinguish between pure and applied mathematics by widening the gap between "arithmetic", now called number theory and "logistic", now called arithmetic. Greek mathematics greatly refined the methods (especially through the introduction of deductive reasoning and mathematical rigor in proofs) and expanded the subject matter of mathematics.[15] Aristotle is credited with what later would be called the law of excluded middle.

Abstract Mathematics[16] is what treats of magnitude[note 6] or quantity, absolutely and generally conferred, without regard to any species of particular magnitude, such as Arithmetic and Geometry, In this sense, abstract mathematics is opposed to mixed mathematics; wherein simple and abstract properties, and the relations of quantities primitively considered in mathematics, are applied to sensible objects, and by that means become intermixed with physical considerations; Such are Hydrostatics, Optics, Navigation, &c.[16]

Archimedes is generally considered to be the greatest mathematician of antiquity and one of the greatest of all time.[17][18] He used the method of exhaustion to calculate the area under the arc of a parabola with the summation of an infinite series, and gave a remarkably accurate approximation of pi.[19] He also defined the spiral bearing his name, formulae for the volumes of surfaces of revolution and an ingenious system for expressing very large numbers.

In the historical development of geometry, the steps in the abstraction of geometry were made by the ancient Greeks. Euclid's Elements being the earliest extant documentation of the axioms of plane geometry— though Proclus tells of an earlier axiomatisation by Hippocrates of Chios.[20] Euclid's *Elements* (c. 300 BC) is one of the oldest extant Greek mathematical treatises[note 7] and consisted of 13 books written in Alexandria; collecting theorems proven by other mathematicians, supplemented by some original work.[note 8] The document is a successful collection of definitions, postulates (axioms), propositions (theorems and constructions), and mathematical proofs of the propositions. Euclid's first theorem is a lemma that possesses properties of prime numbers. The influential thirteen books cover Euclidean geometry, geometric algebra, and the ancient Greek version of algebraic systems and elementary number theory. It was ubiquitous in the Quadrivium and is instrumental in the development of logic, mathematics, and science.

Diophantus of Alexandria was author of a series of books called *Arithmetica*, many of which are now lost. These texts deal with solving algebraic equations. Boethius provided a place for mathematics in the curriculum in the 6th century when he coined the term *quadrivium* to describe the study of arithmetic, geometry, astronomy, and music. He wrote *De institutione arithmetica*, a free translation from the Greek of Nicomachus's *Introduction to Arithmetic*; *De institutione musica*, also derived from Greek sources; and a series of excerpts from Euclid's *Elements*. His works were theoretical, rather than practical, and were the basis of mathematical study until the recovery of Greek and Arabic mathematical works.[21][22]

Acrophonic and Milesian numeration

The Greeks employed Attic numeration,[23] which was based on the system of the Egyptians and was later adapted and used by the Romans. Greek numerals one through four were vertical lines, as in the hieroglyphics. The symbol for five was the Greek letter Π (pi), which is the letter of the Greek word for five, *pente*. Numbers six through nine were *pente* with vertical lines next to it. Ten was represented by the letter (Δ) of the word for ten, *deka*, one hundred by the letter from the word for hundred, etc.

The Ionian numeration used their entire alphabet including three archaic letters. The numeral notation of the Greeks, though far less convenient than that now in use, was formed on a perfectly regular and scientific plan,[24] and could be used with tolerable effect as an instrument of calculation, to which purpose the Roman system was totally inapplicable. The Greeks divided the twenty-four letters of their alphabet into three classes, and, by adding another symbol to each class, they had characters to represent the units, tens, and hundreds. (Jean Baptiste Joseph Delambre's Astronomie Ancienne, t. ii.)

This system appeared in the third century BC, before the letters digamma (F), koppa (ϰ), and sampi (Λ) became obsolete. When lowercase letters became differentiated from upper case letters, the lower case letters were used as the symbols for notation. Multiples of one thousand were written as the nine numbers with a stroke in front of them: thus one thousand was ",α", two-thousand was ",β", etc. M (for μύριοι, as in "myriad") was used to multiply numbers by ten thousand. For example, the number 88,888,888 would be written as M,ηωπη*ηωπη[25]

Greek mathematical reasoning was almost entirely geometric (albeit often used to reason about non-geometric subjects such as number theory), and hence the Greeks had no interest in algebraic symbols. The great exception was Diophantus of Alexandria, the great algebraist.[26] His *Arithmetica* was one of the texts to use symbols in equations. It was not completely symbolic, but was much more so than previous books. An unknown number was called s.[27] The square of s was Δ^y ; the cube was K^y ; the fourth power was $\Delta^y\Delta$; and the fifth power was ΔK^y .[28][note 9]

Chinese mathematical notation

Main article: Suzhou numerals
See also: Chinese numerals
 The Chinese used numerals that look much like the tally system.[29] Numbers one through four were horizontal lines. Five was an X between two horizontal lines; it looked almost exactly the same as the Roman numeral for ten. Nowadays, the huāmǎ system is only used for displaying prices in Chinese markets or on traditional handwritten invoices.

In the history of the Chinese, there were those who were familiar with the sciences of arithmetic, geometry, mechanics, optics, navigation, and astronomy. Mathematics in China emerged independently by the 11th century BC.[30] It is indeed almost certain that the Chinese were acquainted with several geometrical or rather architectural implements;[note 10] with mechanical machines;[note 11] that they knew of the characteristic property of the magnetic needle; and were aware that astronomical events occurred in cycles. Chinese of that time had made attempts to classify or extend the rules of arithmetic or geometry which they knew, and to explain the causes of the phenomena with which they were acquainted beforehand. The Chinese independently developed very large and negative numbers, decimals, a place value decimal system, a binary system, algebra, geometry, and trigonometry.

Chinese mathematics made early contributions, including a place value system.[31][32] The geometrical theorem known to the ancient Chinese were acquainted was applicable in certain cases (namely the ratio of sides).[note 12] It is that geometrical theorems which can be demonstrated in the quasi-experimental way of superposition were also known to them. In arithmetic their knowledge seems to have been confined to the art of calculation by means of the swan-pan, and the power of expressing the results in writing. Our knowledge of the early attainments of the Chinese, slight though it is, is more complete than in the case of most of their contemporaries. It is thus instructive, and serves to illustrate the fact, that it can be known a nation may possess considerable skill in the applied arts with but our knowledge of the later mathematics on which those arts are founded can be scarce. Knowledge of Chinese mathematics before 254 BC is somewhat fragmentary, and even after this date the manuscript traditions are obscure. Dates centuries before the classical period are generally considered conjectural by Chinese scholars unless accompanied by verified archaeological evidence.

As in other early societies the focus was on astronomy in order to perfect the agricultural calendar, and other practical tasks, and not on establishing formal systems.The Chinese Board of Mathematics duties were confined to the annual preparation of an almanac, the dates and predictions in which it regulated. Ancient Chinese mathematicians did not develop an axiomatic approach, but made advances in algorithm development and algebra. The achievement of Chinese algebra reached its zenith in the 13th century, when Zhu Shijie invented method of four unknowns.

As a result of obvious linguistic and geographic barriers, as well as content, Chinese mathematics and that of the mathematics of the ancient Mediterranean world are presumed to have developed more or less independently up to the time when *The Nine Chapters on the Mathematical Art* reached its final form, while the *Writings on Reckoning* and *Huainanzi* are roughly contemporary with classical Greek mathematics. Some exchange of ideas across Asia through known cultural exchanges from at least Roman times is likely. Frequently, elements of the mathematics of early societies correspond to rudimentary results found later in branches of modern mathematics such as geometry or number theory. The Pythagorean theorem for example, has been attested to the time of the Duke of Zhou. Knowledge of Pascal's triangle has also been shown to have existed in China centuries before Pascal,[33] such as by Shen Kuo.

The state of trigonometry in China slowly began to change and advance during the Song Dynasty (960–1279), where Chinese mathematicians began to express greater emphasis for the need of spherical trigonometry in calendarical science and astronomical calculations.[34] The polymath Chinese scientist, mathematician and official Shen Kuo (1031–1095) used trigonometric functions to solve mathematical problems of chords and arcs.[34] Sal Restivo writes that Shen's work in the lengths of arcs of circles provided the basis for spherical trigonometry developed in the 13th century by the mathematician and astronomer Guo Shoujing (1231–1316).[35] As the historians L. Gauchet and Joseph Needham state, Guo Shoujing used spherical trigonometry in his calculations to improve the calendar system and Chinese astronomy.[34][36] The mathematical science of the Chinese would incorporate the work and teaching of Arab missionaries with knowledge of spherical trigonometry who had come to China in the course of the thirteenth century.

Indian mathematical notation

Although the origin of our present system of numerical notation is ancient, there is no doubt that it was in use among the Hindus over two thousand years ago. The algebraic notation of the Indian mathematician, Brahmagupta, was syncopated. Addition was indicated by placing the numbers side by side, subtraction by placing a dot over the subtrahend (the number to be subtracted), and division by placing the divisor below the dividend, similar to our notation but without the bar. Multiplication, evolution, and unknown quantities were represented by abbreviations of appropriate terms.[37] The Hindu-Arabic numeral system and the rules for the use of its operations, in use throughout the world today, likely evolved over the course of the first millennium AD in India and was transmitted to the west via Islamic mathematics.[38][39]

Hindu-Arabic numerals and notations

Main article: History of the Hindu-Arabic numeral system
 Despite their name, Arabic numerals actually started in India. The reason for this misnomer is Europeans saw the numerals used in an Arabic book, *Concerning the Hindu Art of Reckoning*, by Mohommed ibn-Musa al-Khwarizmi. Al-Khwārizmī wrote several important books on the Hindu-Arabic numerals and on methods for solving equations. His book *On the Calculation with Hindu Numerals*, written about 825, along with the work of Al-Kindi,[note 13] were instrumental in spreading Indian mathematics and Indian numerals to the West. Al-Khwarizmi did not claim the numerals as Arabic, but over several Latin translations, the fact that the numerals were Indian in origin was lost. The word *algorithm* is derived from the Latinization of Al-Khwārizmī's name, Algoritmi, and the word *algebra* from the title of one of his works, *Al-Kitāb al-mukhtaṣar fī ḥīsāb al-ǧabr wa'l-muqābala* (*The Compendious Book on Calculation by Completion and Balancing*).

Islamic mathematics developed and expanded the mathematics known to Central Asian civilizations.[40] Al-Khwārizmī gave an exhaustive explanation for the algebraic solution of quadratic equations with positive roots,[41] and Al-Khwārizmī was to teach algebra in an elementary form and for its own sake.[42] Al-Khwārizmī also discussed the fundamental method of "reduction" and "balancing", referring to the transposition of subtracted terms to the other side of an equation, that is, the cancellation of like terms on opposite sides of the equation. This is the operation which al-Khwārizmī originally described as *al-jabr*.[43] His algebra was also no longer concerned "with a series of problems to be resolved, but an exposition which starts with primitive terms in which the combinations must give all possible prototypes for equations, which henceforward explicitly constitute the true object of study." Al-Khwārizmī also studied an equation for its own sake and "in a generic manner, insofar as it does not simply emerge in the course of solving a problem, but is specifically called on to define an infinite class of problems."[44]

Al-Karaji, in his treatise *al-Fakhri*, extends the methodology to incorporate integer powers and integer roots of unknown quantities.[note 14][45] The historian of mathematics, F. Woepcke,[46] praised Al-Karaji for being "the first who introduced the theory of algebraic calculus." Also in the 10th century, Abul Wafa translated the works of Diophantus into Arabic. Ibn al-Haytham would develop analytic geometry. Al-Haytham derived the formula for the sum of the fourth powers, using a method that is readily generalizable for determining the general formula for the sum of any integral powers. Al-Haytham performed an integration in order to find the volume of a paraboloid, and was able to generalize his result for the integrals of polynomials up to the fourth degree.[note 15][47] In the late 11th century, Omar Khayyam would develop algebraic geometry, wrote *Discussions of the Difficulties in Euclid*,[note 16] and wrote on the general geometric solution to cubic equations. Nasir al-Din Tusi (Nasireddin) made advances in spherical trigonometry. Muslim mathematicians during this period include the addition of the decimal point notation to the Arabic numerals.

Many Greek and Arabic texts on mathematics were then translated into Latin, which led to further development of mathematics in medieval Europe. In the 12th century, scholars traveled to Spain and Sicily seeking scientific Arabic texts, including al-Khwārizmī's[note 17] and the complete text of Euclid's *Elements*.[note 18][48][49] One of the European books that advocated using the numerals was *Liber Abaci*, by Leonardo of Pisa, better known as Fibonacci. *Liber Abaci* is better known for the mathematical problem Fibonacci wrote in it about a population of rabbits. The growth of the population ended up being a Fibonacci sequence, where a term is the sum of the two preceding terms.

Abū al-Hasan ibn Alī al-Qalasādī (1412–1482) was the last major medieval Arab algebraist, who improved on the algebraic notation earlier used by Ibn al-Yāsamīn in the 12th century and, in the Maghreb, by Ibn al-Banna in the 13th century.[50] In contrast to the syncopated notations of their predecessors, Diophantus and Brahmagupta, which lacked symbols for mathematical operations,[51] al-Qalasadi's algebraic notation was the first to have symbols for these functions and was thus "the first steps toward the introduction of algebraic symbolism." He represented mathematical symbols using characters from the Arabic alphabet.[50]

7.1.3 Symbolic stage

Symbols by popular introduction date

Further information: Table of mathematical symbols by introduction date

Early arithmetic and multiplication

See also: Early modern age, Probability, Statistics, Notation in probability and statistics, History of probability, History of statistics and Scientific revolution

The 14th century saw the development of new mathematical concepts to investigate a wide range of problems.[52] The two widely used arithmetic symbols are addition and subtraction, + and −. The plus sign was used by 1360 by Nicole Oresme[53][note 19] in his work *Algorismus proportionum*.[54] It is thought an abbreviation for "et", meaning "and" in Latin, in much the same way the ampersand sign also began as "et". Oresme at the University of Paris and the Italian Giovanni di Casali independently provided graphical demonstrations of the distance covered by a body undergoing uniformly accelerated motion, asserting that the area under the line depicting the constant acceleration and represented the total distance traveled.[55] The minus sign was used in 1489 by Johannes Widmann in *Mercantile Arithmetic* or *Behende und hüpsche Rechenung auff allen Kauffmanschafft,*.[56] Widmann used the minus symbol with the plus symbol, to indicate deficit and surplus, respectively.[57] In *Summa de arithmetica, geometria, proportioni e proportionalità,*[note 20][58] Luca Pacioli used symbols for plus and minus symbols and contained algebra.[note 21]

In the 15th century, Ghiyath al-Kashi computed the value of π to the 16th decimal place. Kashi also had an algorithm for calculating nth roots.[note 22] In 1533, Regiomontanus's table of sines and cosines were published.[59] Scipione del Ferro and Niccolò Fontana Tartaglia discovered solutions for cubic equations. Gerolamo Cardano published them in his 1545 book *Ars Magna*, together with a solution for the quartic equations, discovered by his student Lodovico Ferrari. The radical symbol[note 23] for square root was introduced by Christoph Rudolff.[note 24] Michael Stifel's important work *Arithmetica integra*[60] contained important innovations in mathematical notation. In 1556, Nicolo Tartaglia used parentheses for precedence grouping. In 1557 Robert Recorde published The Whetstone of Witte which used the equal sign (=) as well as plus and minus signs for the English reader. In 1564, Gerolamo Cardano analyzed games of chance beginning the early stages of probability theory. In 1572 Rafael Bombelli published his *L'Algebra* in which he showed how to deal with the imaginary quantities that could appear in Cardano's formula for solving cubic equations. Simon Stevin's book *De Thiende* ('the art of tenths'), published in Dutch in 1585, contained a systematic treatment of decimal notation, which influenced all later work on the real number system. The New algebra (1591) of François Viète introduced the modern notational manipulation of algebraic expressions. For navigation and accurate maps of large areas, trigonometry grew to be a major branch of mathematics. Bartholomaeus Pitiscus coin the word "trigonometry", publishing his *Trigonometria* in 1595.

John Napier is best known as the inventor of logarithms[note 25][61] and made common the use of the decimal point in arithmetic and mathematics.[62][63] After Napier, Edmund Gunter created the logarithmic scales (lines, or rules) upon which slide rules are based, it was William Oughtred who used two such scales sliding by one another to perform direct multiplication and division; and he is credited as the inventor of the slide rule in 1622. In 1631 Oughtred introduced the multiplication sign (×) his proportionality sign,[note 26] and abbreviations *sin* and *cos* for the sine and cosine functions.[64] Albert Girard also used the abbreviations 'sin', 'cos' and 'tan' for the trigonometric functions in his treatise.

Johannes Kepler was one of the pioneers of the mathematical applications of infinitesimals.[note 27] René Descartes is credited as the father of analytical geometry, the bridge between algebra and geometry,[note 28] crucial to the discovery of infinitesimal calculus and analysis. In the 17th century, Descartes introduced Cartesian co-ordinates which allowed the development of analytic geometry.[note 29] Blaise Pascal influenced mathematics throughout his life. His *Traité du triangle arithmétique* ("Treatise on the Arithmetical Triangle") of 1653 described a convenient tabular presentation for binomial coefficients.[note 30] Pierre de Fermat and Blaise Pascal would investigate probability.[note 31] John Wallis introduced the infinity symbol.[note 32] He similarly used this notation for infinitesimals.[note 33] In 1657, Christiaan Huygens published the treatise on probability, *On Reasoning in Games of Chance*.[note 34][65]

Johann Rahn introduced the division symbol (obelus) and the therefore sign in 1659. William Jones used π in *Synopsis palmariorum mathesios*[66] in 1706 because it is the letter of the Greek word perimetron (περιμετρον), which means perimeter in Greek. This usage was popularized in 1737 by Euler. In 1734, Pierre Bouguer used double horizontal bar below the inequality sign.[67]

Derivatives notation: Leibniz and Newton   See also: Leibniz's notation and Leibniz–Newton calculus controversy

The study of linear algebra emerged from the study of determinants, which were used to solve systems of linear equations. Calculus had two main systems of notation, each created by one of the creators: that developed by Isaac Newton and the notation developed by Gottfried Leibniz. Leibniz's is the notation used most often today. Newton's was simply a dot or dash placed above the function.[note 35] In modern usage, this notation generally denotes derivatives of physical quantities with respect to time, and is used frequently in the science of mechanics. Leibniz, on the other hand, used the letter d as a prefix to indicate differentiation, and introduced the notation representing derivatives as if they were a special type of fraction.[note 36] This notation makes explicit the variable with respect to which the derivative of the function is taken. Leibniz also created the integral symbol.[note 37] The symbol is an elongated S, representing the Latin word *Summa*, meaning "sum". When finding areas under curves, integration is often illustrated by dividing the area into infinitely many tall, thin rectangles, whose areas are added. Thus, the integral symbol is an elongated s, for sum.

High division operators and functions

See also: modern age

Letters of the alphabet in this time were to be used as symbols of quantity; and although much diversity existed with respect to the choice of letters, there were to be several universally recognized rules in the following history.[24] Here thus in the history of equations the first letters of the alphabet were indicatively known as coefficients, the last letters the unknown terms (an *incerti ordinis*). In algebraic geometry, again, a similar rule was to be observed, the last letters of the alphabet there denoting the variable or current coordinates. Certain letters, such as π, e, etc., were by universal consent appropriated as symbols of the frequently occurring numbers 3.14159 ..., and 2.7182818 ....,[note 38] etc., and their use in any other acceptation was to be avoided as much as possible.[24] Letters, too, were to be employed as symbols of operation, and with them other previously mention arbitrary operation characters. The letters d, elongated S were to be appropriated as operative symbols in the differential calculus and integral calculus, Δ and $\sum$ in the calculus of differences.[24] In functional notation, a letter, as a symbol of operation, is combined with another which is regarded as a symbol of quantity.[24][note 39]

Beginning in 1718, Thomas Twinin used the division slash (solidus), deriving it from the earlier Arabic horizontal fraction bar. Pierre-Simon, marquis de Laplace developed the widely used Laplacian differential operator.[note 40] In 1750, Gabriel Cramer developed "*Cramer's Rule*" for solving linear systems.

Euler and prime notations   Leonhard Euler was one of the most prolific mathematicians in history, and also a prolific inventor of canonical notation. His contributions include his use of e to represent the base of natural logarithms. It is not known exactly why e was chosen, but it was probably because the four letters of the alphabet were already commonly used to represent variables and other constants. Euler used π to represent pi consistently. The use of π was suggested by William Jones, who used it as shorthand for perimeter. Euler used i to represent the square root of negative one,[note 41] although he earlier used it as an *infinite number.* [note 42][note 43] For summation, Euler used sigma, Σ.[note 44] For functions, Euler used the notation $f(x)$ to represent a function of x . In 1730, Euler wrote the gamma function.[note 45] In 1736, Euler produces his paper on the Seven Bridges of Königsberg[68] regarding topology.

The mathematician, William Emerson[69] would develop the proportionality sign.[note 46][note 47][70][71] Much later in the abstract expressions of the value of various proportional phenomena, the parts-per notation would became useful as a set of pseudo units to describe small values of miscellaneous dimensionless quantities. Marquis de Condorcet, in 1768, advanced the partial differential sign.[note 48] In 1771, Alexandre-Théophile Vandermonde deduced the importance of topological features when discussing the properties of knots related to the geometry of position. Between 1772 and 1788, Joseph-Louis Lagrange re-formulated the formulas and calculations of Classical "Newtonian" mechanics, called Lagrangian mechanics. The prime symbol for derivatives was also made by Lagrange.

Gauss, Hamilton, and Matrix notations   At the turn of the 19th century, Carl Friedrich Gauss developed the identity sign for congruence relation and, in Quadratic reciprocity, the integral part. Gauss contributed functions of complex variables, in geometry, and on the convergence of series. He gave the satisfactory proofs of the fundamental theorem of algebra and of the quadratic reciprocity law. Gauss developed the theory of solving linear systems by using Gaussian elimination, which was initially listed as an advancement in geodesy.[72] He would also develop the product sign. Also in this time, Niels Henrik Abel and Évariste Galois[note 50] conducted their work on the solvability of equations, linking group theory and field theory.

After the 1800s, Christian Kramp would promote factorial notation during his research in generalized factorial function which applied to non-integers.[73] Joseph Diaz Gergonne introduced the set inclusion signs.[note 51] Peter Gustav Lejeune Dirichlet developed Dirichlet L-functions to give the proof of Dirichlet's theorem on arithmetic progressions and began analytic number theory.[note 52] In 1828, Gauss proved his Theorema Egregium (*remarkable theorem* in Latin), establishing property of surfaces. In the 1830s, George Green developed Green's function. In 1829. Carl Gustav Jacob Jacobi publishes Fundamenta nova theoriae functionum ellipticarum with his elliptic theta functions. By 1841, Karl Weierstrass, the "father of modern analysis", elaborated on the concept of absolute value and the determinant of a matrix.

Matrix notation would be more fully developed by Arthur Cayley in his three papers, on subjects which had been suggested by reading the Mécanique analytique[74] of Lagrange and some of the works of Laplace. Cayley defined matrix multiplication and matrix inverses. Cayley used a single letter to denote a matrix,[75] thus treating a matrix as an aggregate object. He also realized the connection between matrices and determinants,[76] and wrote "*There would be many things to say about this theory of matrices which should, it seems to me, precede the theory of determinants*".[77]

William Rowan Hamilton would introduce the nabla symbol[note 54] for vector differentials.[78][79] This was previously used by Hamilton as a general-purpose operator sign.[80] Hamilton reformulated Newtonian mechanics, now called Hamiltonian mechanics. This work has proven central to the modern study of classical field theories such as electromagnetism. This was also important to the development of quantum mechanics.[note 55] In mathematics, he is perhaps best known as the inventor of quaternion notation[note 56] and biquaternions. Hamilton also introduced the word "tensor" in 1846.[81][note 57] James Cockle would develop the tessarines[note 58] and, in 1849, coquaternions. In 1848, James Joseph Sylvester introduced into matrix algebra the term matrix.[note 59]

Maxwell, Clifford, and Ricci notations   In 1864 James Clerk Maxwell reduced all of the then current knowledge of electromagnetism into a linked set of differential equations with 20 equations in 20 variables, contained in "*A Dynamical Theory of the Electromagnetic Field*".[83] The method of calculation which it is necessary to employ was given by Lagrange, and afterwards developed, with some modifications, by Hamilton's equations. It is usually referred to as Hamilton's principle; when the equations in the original form are used they are known as Lagrange's equations. In 1871, he presented the *Remarks on the mathematical classification of physical quantities.*[84] Also in 1871, Richard Dedekind called a set of real or complex numbers which is closed under the four arithmetic operations a "field".

In 1878, William Kingdon Clifford publishes his Elements of Dynamic.[85] Clifford would develop split-biquaternions,[note 60] which he called *algebraic motors*. Clifford eliminated quaternion study by separating the dot product and cross product of two vectors from the complete quaternion notation.[note 61] This approach made vector calculus available to engineers and others working in three dimensions and skeptical of the lead–lag effect[note 62] in the fourth dimension.[note 63] Between 1880 and 1887, Oliver Heaviside developed the operational calculus[86] (involving the D notation for the differential operator, which he is credited with creating), a method of solving differential equations by transforming them into ordinary algebraic equations which caused a great deal of controversy when introduced, owing to the lack of rigour in his derivation of it.[note 64] The common vector notation are used when working with vectors, which are spatial or more abstract members of vector spaces. The angle notation (or phasor notation) is a notation used in electronics.

In 1881, Leopold Kronecker defined what he called a "domain of rationality", which is a field extension of the field of rational numbers in modern terms.[87] In 1882, Hüseyin Tevfik Paşa wrote the book titled "Linear Algebra".[88][89] Lord Kelvin's aetheric atom theory (1860s) led Peter Guthrie Tait, in 1885, to publish a topological table of knots with up to ten crossings known as the Tait conjectures. In 1893, Heinrich M. Weber gave the clear definition of an abstract field.[note 65] Tensor calculus was developed by Gregorio Ricci-Curbastro between 1887–96, presented in 1892 under the title *absolute differential calculus*,[90] and the contemporary usage of "tensor" was stated by Woldemar Voigt in 1898.[91] In 1895, Henri Poincaré published *Analysis Situs*.[92] In 1897, Charles Proteus Steinmetz would publish *Theory and Calculation of Alternating Current Phenomena*, with the assistance of Ernst J. Berg.[93]

From formula mathematics to tensors   In 1895 Giuseppe Peano issued his *Formulario mathematico*,[94] an effort to digest mathematics into terse text based on special symbols. He would provide a definition of a vector space and linear map. He would also introduce the intersection sign, the union sign, the membership sign (is an element of), and existential quantifier[note 67] (there exists). Peano would pass to Bertrand Russell his work in 1900 at a Paris conference; it so impressed Russell that Russell too was taken with the drive to render mathematics more concisely. The result was Principia Mathematica written with Alfred North Whitehead. This treatise marks a watershed in modern literature where symbol became dominant.[note 68] Ricci-Curbastro and Tullio Levi-Civita popularized the tensor index notation around 1900.[95]

Mathematical logic and abstraction

At the beginning of this period, Felix Klein's "Erlangen program" identified the underlying theme of various geometries, defining each of them as the study of properties invariant under a given group of symmetries. This level of abstraction revealed connections between geometry and abstract algebra. Georg Cantor[note 69] would introduce the aleph symbol for cardinal numbers of transfinite sets.[note 70] His notation for the cardinal numbers was the Hebrew letter ℵ (aleph) with a natural number subscript; for the ordinals he employed the Greek letter ω (omega). This notation is still in use today in ordinal notation of a finite sequence of symbols from a finite alphabet which names an ordinal number according to some scheme which gives meaning to the language. His theory created a great deal of controversy. Cantor would, in his study of Fourier series, consider point sets in Euclidean space.

After the turn of the 20th century, Josiah Willard Gibbs would in physical chemistry introduce middle dot for dot product and the multiplication sign for cross products. He would also supply notation for the scalar and vector products, which was introduced in *Vector Analysis*. In 1904, Ernst Zermelo promotes axiom of choice and his proof of the well-ordering theorem.[96] Bertrand Russell would shortly afterward introduce logical disjunction (OR) in 1906. Also in 1906, Poincaré would publish *On the Dynamics of the Electron*[97] and Maurice Fréchet introduced metric space.[98] Later, Gerhard Kowalewski and Cuthbert Edmund Cullis[99][100][101] would successively introduce matrices notation, parenthetical matrix and box matrix notation respectively. After 1907, mathematicians[note 71] studied knots from the point of view of the knot group and invariants from homology theory.[note 72] In 1908, Joseph Wedderburn's structure theorems were formulated for finite-dimensional algebras over a field. Also in 1908, Ernst Zermelo proposed "definite" property and the first axiomatic set theory, Zermelo set theory. In 1910 Ernst Steinitz published the influential paper *Algebraic Theory of Fields*.[note 73][note 74] In 1911, Steinmetz would publish *Theory and Calculation of Transient Electric Phenomena and Oscillations*.

Albert Einstein, in 1916, introduced the Einstein notation[note 75] which summed over a set of indexed terms in a formula, thus exerting notational brevity. Arnold Sommerfeld would create the contour integral sign in 1917. Also in 1917, Dimitry Mirimanoff proposes axiom of regularity. In 1919, Theodor Kaluza would solve general relativity equations using five dimensions, the results would have electromagnetic equations emerge.[102] This would be published in 1921 in "Zum

Unitätsproblem der Physik".[103] In 1922, Abraham Fraenkel and Thoralf Skolem independently proposed replacing the axiom schema of specification with the axiom schema of replacement. Also in 1922, Zermelo–Fraenkel set theory was developed. In 1923, Steinmetz would publish *Four Lectures on Relativity and Space*. Around 1924, Jan Arnoldus Schouten would develop the modern notation and formalism for the Ricci calculus framework during the absolute differential calculus applications to general relativity and differential geometry in the early twentieth century.[note 76][104][105][106] In 1925, Enrico Fermi would describe a system comprising many identical particles that obey the Pauli exclusion principle, afterwards developing a diffusion equation (Fermi age equation). In 1926, Oskar Klein would develop the Kaluza–Klein theory. In 1928, Emil Artin abstracted ring theory with Artinian rings. In 1933, Andrey Kolmogorov introduces the *Kolmogorov axioms*. In 1937, Bruno de Finetti deduced the "operational subjective" concept.

Mathematical symbolism   See also: Category theory, Model theory, Table of logic symbols and Logic alphabet

Mathematical abstraction began as a process of extracting the underlying essence of a mathematical concept,[107][108] removing any dependence on real world objects with which it might originally have been connected,[109] and generalizing it so that it has wider applications or matching among other abstract descriptions of equivalent phenomena. Two abstract areas of modern mathematics are category theory and model theory. Bertrand Russell,[110] said, "*Ordinary language is totally unsuited for expressing what physics really asserts, since the words of everyday life are not sufficiently abstract. Only mathematics and mathematical logic can say as little as the physicist means to say*". Though, one can substituted mathematics for real world objects, and wander off through equation after equation, and can build a concept structure which has no relation to reality.[111]

Symbolic logic studies the purely formal properties of strings of symbols. The interest in this area springs from two sources. First, the notation used in symbolic logic can be seen as representing the words used in philosophical logic. Second, the rules for manipulating symbols found in symbolic logic can be implemented on a computing machine. Symbolic logic is usually divided into two subfields, propositional logic and predicate logic. Other logics of interest include temporal logic, modal logic and fuzzy logic. The area of symbolic logic called propositional logic, also called *propositional calculus*, studies the properties of sentences formed from constants[note 77] and logical operators. The corresponding logical operations are known, respectively, as conjunction, disjunction, material conditional, biconditional, and negation. These operators are denoted as keywords[note 78] and by symbolic notation.

Some of the introduced mathematical logic notation during this time included the set of symbols used in Boolean algebra. This was created by George Boole in 1854. Boole himself did not see logic as a branch of mathematics, but it has come to be encompassed anyway. Symbols found in Boolean algebra include $\land$ (AND), $\lor$ (OR), and $\neg$ (NOT). With these symbols, and letters to represent different truth values, one can make logical statements such as $a \lor \neg a = 1$, that is "(*a* is true OR *a* is NOT true) is true", meaning it is true that *a* is either true or not true (i.e. false). Boolean algebra has many practical uses as it is, but it also was the start of what would be a large set of symbols to be used in logic.[note 79] Predicate logic, originally called *predicate calculus*, expands on propositional logic by the introduction of variables[note 80] and by sentences containing variables, called predicates.[note 81] In addition, predicate logic allows quantifiers.[note 82] With these logic symbols and additional quantifiers from predicate logic,[note 83] valid proofs can be made that are irrationally artificial,[note 84] but syntactical.[note 85]

Gödel incompleteness notation   See also: Proof sketch for Gödel's first incompleteness theorem

While proving his incompleteness theorems,[note 86] Kurt Gödel created an alternative to the symbols normally used in logic. He used Gödel numbers, which were numbers that represented operations with set numbers, and variables with the prime numbers greater than 10. With Gödel numbers, logic statements can be broken down into a number sequence. Gödel then took this one step farther, taking the *n* prime numbers and putting them to the power of the numbers in the sequence. These numbers were then multiplied together to get the final product, giving every logic statement its own number.[113][note 87]

Contemporary notation and topics

See also: Contemporary era

Early 20th-century notation   Abstraction of notation is an ongoing process and the historical development of many mathematical topics exhibits a progression from the concrete to the abstract. Various set notations would be developed for fundamental object sets. Around 1924, David Hilbert and Richard Courant published "Methods of mathematical physics. Partial differential equations".[114] In 1926, Oskar Klein and Walter Gordon proposed the Klein–Gordon equation to describe relativistic particles.[note 88] The first formulation of a quantum theory describing radiation and matter interaction is due to Paul Adrien Maurice Dirac, who, during 1920, was first able to compute the coefficient of spontaneous emission of an atom.[115] In 1928, the relativistic Dirac equation was formulated by Dirac to explain the behavior of the relativistically moving electron.[note 89] Dirac described the quantification of the electromagnetic field as an ensemble of harmonic oscillators with the introduction of the concept of creation and annihilation operators of particles. In the following years, with contributions from Wolfgang Pauli, Eugene Wigner, Pascual Jordan, and Werner Heisenberg, and an elegant formulation of quantum electrodynamics due to Enrico Fermi,[116] physicists came to believe that, in principle, it would be possible to perform any computation for any physical process involving photons and charged particles.

In 1931, Alexandru Proca developed the Proca equation (Euler–Lagrange equation)[note 90] for the vector meson theory of nuclear forces and the relativistic quantum field equations. John Archibald Wheeler in 1937 develops S-matrix. Studies by Felix Bloch with Arnold Nordsieck,[117] and Victor Weisskopf,[118] in 1937 and 1939, revealed that such computations were reliable only at a first order of perturbation theory, a problem already pointed out by Robert Oppenheimer.[119] At higher orders in the series infinities emerged, making such computations meaningless and casting serious doubts on the internal consistency of the theory itself. With no solution for this problem known at the time, it appeared that a fundamental incompatibility existed between special relativity and quantum mechanics.

In the 1930s, the double-struck capital Z for integer number sets was created by Edmund Landau. Nicolas Bourbaki created the double-struck capital Q for rational number sets. In 1935, Gerhard Gentzen made universal quantifiers. In 1936, Tarski's undefinability theorem is stated by Alfred Tarski and proved.[note 91] In 1938, Gödel proposes the constructible universe in the paper "*The Consistency of the Axiom of Choice and of the Generalized Continuum-Hypothesis*". André Weil and Nicolas Bourbaki would develop the empty set sign in 1939. That same year, Nathan Jacobson would coin the double-struck capital C for complex number sets.

Around the 1930s, Voigt notation[note 92] would be developed for multilinear algebra as a way to represent a symmetric tensor by reducing its order. Schönflies notation[note 93] became one of two conventions used to describe point groups (the other being Hermann–Mauguin notation). Also in this time, van der Waerden notation[120][121] became popular for the usage of two-component spinors (Weyl spinors) in four spacetime dimensions. Arend Heyting would introduce Heyting algebra and Heyting arithmetic.

The arrow, e.g., →, was developed for function notation in 1936 by Øystein Ore to denote images of specific elements.[note 94][note 95] Later, in 1940, it took its present form, e.g., $f: X \to Y$, through the work of Witold Hurewicz. Werner Heisenberg, in 1941, proposed the S-matrix theory of particle interactions.

Bra–ket notation (Dirac notation) is a standard notation for describing quantum states, composed of angle brackets and vertical bars. It can also be used to denote abstract vectors and linear functionals. It is so called because the inner product (or dot product on a complex vector space) of two states is denoted by a ⟨bra|ket⟩[note 96] consisting of a left part, ⟨φ|, and a right part, |ψ⟩. The notation was introduced in 1939 by Paul Dirac,[122] though the notation has precursors in Grassmann's use of the notation $[\varphi|\psi]$ for his inner products nearly 100 years previously.[123]

Bra–ket notation is widespread in quantum mechanics: almost every phenomenon that is explained using quantum mechanics—including a large portion of modern physics—is usually explained with the help of bra–ket notation. The notation establishes an encoded abstract representation-independence, producing a versatile specific representation (e.g., x, or p, or eigenfunction base) without much ado, or excessive reliance on, the nature of the linear spaces involved. The overlap expression ⟨φ|ψ⟩ is typically interpreted as the probability amplitude for the state ψ to collapse into the state ϕ. The Feynman slash notation (Dirac slash notation[124]) was developed by Richard Feynman for the study of Dirac fields in quantum field theory.

In 1948, Valentine Bargmann and Eugene Wigner proposed the relativistic Bargmann–Wigner equations to describe free particles and the equations are in the form of multi-component spinor field wavefunctions. In 1950, William Vallance Douglas Hodge presented "The topological invariants of algebraic varieties" at the Proceedings of the International Congress of Mathematicians. Between 1954 and 1957, Eugenio Calabi worked on the Calabi conjecture for Kähler metrics and the development of Calabi–Yau manifolds. In 1957, Tullio Regge formulated the mathematical property of potential scattering in the Schrödinger equation.[note 97] Stanley Mandelstam, along with Regge, did the initial development of the Regge theory of strong interaction phenomenology. In 1958, Murray Gell-Mann and Richard Feynman, along with George Sudarshan and Robert Marshak, deduced the chiral structures of the weak interaction in physics. Geoffrey Chew, along with others, would promote matrix notation for the strong interaction, and the associated bootstrap principle, in 1960. In the 1960s, set-builder notation was developed for describing a set by stating the properties that its members must satisfy. Also in the 1960s, tensors are abstracted within category theory by means of the concept of monoidal category. Later, multi-index notation eliminates conventional notions used in multivariable calculus, partial differential equations, and the theory of distributions, by abstracting the concept of an integer index to an ordered tuple of indices.

Modern mathematical notation   See also: Approximation theory, Universal property, Tensor algebra, Free algebra and Abstract algebra

In the modern mathematics of special relativity, electromagnetism and wave theory, the d'Alembert operator[note 98][note 99] is the Laplace operator of Minkowski space. The Levi-Civita symbol[note 100] is used in tensor calculus.

After the full Lorentz covariance formulations that were finite at any order in a perturbation series of quantum electrodynamics, Sin-Itiro Tomonaga, Julian Schwinger and Richard Feynman were jointly awarded with a Nobel prize in physics in 1965.[125] Their contributions, and those of Freeman Dyson, were about covariant and gauge invariant formulations of quantum electrodynamics that allow computations of observables at any order of perturbation theory. Feynman's mathematical technique, based on his diagrams, initially seemed very different from the field-theoretic, operator-based approach of Schwinger and Tomonaga, but Freeman Dyson later showed that the two approaches were equivalent. Renormalization, the need to attach a physical meaning at certain divergences appearing in the theory through integrals, has subsequently become one of the fundamental aspects of quantum field theory and has come to be seen as a criterion for a theory's general acceptability. Quantum electrodynamics has served as the model and template for subsequent quantum field theories. Peter Higgs, Jeffrey Goldstone, and others, Sheldon Glashow, Steven Weinberg and Abdus Salam independently showed how the weak nuclear force and quantum electrodynamics could be merged into a single electroweak force. In the late 1960s, the particle zoo was composed of the then known elementary particles before the discovery of quarks.

A step towards the Standard Model was Sheldon Glashow's discovery, in 1960, of a way to combine the electromagnetic and weak interactions.[126] In 1967, Steven Weinberg[127] and Abdus Salam[128] incorporated the Higgs mechanism[129][130][131] into Glashow's electroweak theory, giving it its modern form. The Higgs mechanism is believed to give rise to the masses of all the elementary particles in the Standard Model. This includes the masses of the W and Z bosons, and the masses of the fermions - i.e. the quarks and leptons. Also in 1967, Bryce DeWitt published his equation under the name "*Einstein–Schrödinger equation*" (later renamed the "*Wheeler–DeWitt equation*").[132] In 1969, Yoichiro Nambu, Holger Bech Nielsen, and Leonard Susskind descried space and time in terms of strings. In 1970, Pierre Ramond develop two-dimensional supersymmetries. Michio Kaku and Keiji Kikkawa would afterwards formulate string variations. In 1972, Michael Artin, Alexandre Grothendieck, Jean-Louis Verdier propose the Grothendieck universe.[133]

After the neutral weak currents caused by Z boson exchange were discovered at CERN in 1973,[134][135][136][137] the electroweak theory became widely accepted and Glashow, Salam, and Weinberg shared the 1979 Nobel Prize in Physics for discovering it. The theory of the strong interaction, to which many contributed, acquired its modern form around 1973–74. With the establishment of quantum chromodynamics, a finalized a set of fundamental and exchange particles, which allowed for the establishment of a "standard model" based on the mathematics of gauge invariance, which successfully described all forces except for gravity, and which remains generally accepted within the domain to which it is designed to be applied. In the late 1970s, William Thurston introduced hyperbolic geometry into the study of knots with the hyperbolization theorem. The orbifold notation system, invented by Thurston, has been developed for representing types of symmetry groups in two-dimensional spaces of constant curvature. In 1978, Shing-Tung Yau deduced that the Calabi conjecture have Ricci flat metrics. In 1979, Daniel Friedan showed that the equations of motions of string theory are abstractions of Einstein equations of General Relativity.

The first superstring revolution is composed of mathematical equations developed between 1984 and 1986. In 1984, Vaughan Jones deduced the Jones polynomial and subsequent contributions from Edward Witten, Maxim Kontsevich, and others, revealed deep connections between knot theory and mathematical methods in statistical mechanics and quantum field theory. According to string theory, all particles in the "particle zoo" have a common ancestor, namely a vibrating string. In 1985, Philip Candelas, Gary Horowitz,[138] Andrew Strominger, and Edward Witten would publish "Vacuum configurations for superstrings"[139] Later, the tetrad formalism (tetrad index notation) would be introduced as an approach to general relativity that replaces the choice of a coordinate basis by the less restrictive choice of a local basis for the tangent bundle.[note 103][140]

In the 1990s, Roger Penrose would propose Penrose graphical notation (tensor diagram notation) as a, usually handwritten, visual depiction of multilinear functions or tensors.[141] Penrose would also introduce abstract index notation.[note 104] In 1995, Edward Witten suggested M-theory and subsequently used it to explain some observed dualities, initiating the second superstring revolution.[note 105]

John Conway would further various notations, including the Conway chained arrow notation, the Conway notation of knot theory, and the Conway polyhedron notation. The Coxeter notation system classifies symmetry groups, describing the angles between with fundamental reflections of a Coxeter group. It uses a bracketed notation, with modifiers to indicate certain subgroups. The notation is named after H. S. M. Coxeter and Norman Johnson more comprehensively defined it.

Combinatorial LCF notation[note 106] has been developed for the representation of cubic graphs that are Hamiltonian.[142][143] The cycle notation is the convention for writing down a permutation in terms of its constituent cycles.[144] This is also called circular notation and the permutation called a *cyclic* or *circular* permutation.[145]

Computers and markup notation   Main articles: History of computing and Timeline of computing
See also: Symbolic computation, Symbolic dynamics, Computational complexity theory, Internet shorthand notation, ASCII math notation, MathML, Basic Linear Algebra Subprograms, Numerical linear algebra, List of numerical libraries, List of numerical analysis software, DOT language, Lisp (programming language), Object-oriented programming and Earley algorithm

In 1931, IBM produces the IBM 601 Multiplying Punch; it is an electromechanical machine that could read two numbers, up to 8 digits long, from a card and punch their product onto the same card.[146] In 1934, Wallace Eckert used a rigged IBM 601 Multiplying Punch to automate the integration of differential equations.[147] In 1936, Alan Turing publishes "On Computable Numbers, With an Application to the Entscheidungsproblem".[148][note 107] John von Neumann, pioneer of the digital computer and of computer science,[note 108] in 1945, writes the incomplete *First Draft of a Report on the EDVAC*. In 1962, Kenneth E. Iverson developed an integral part notation that became known as Iverson Notation for manipulating arrays that he taught to his students, and described in his book *A Programming Language*. In 1970, E.F. Codd proposed relational algebra as a relational model of data for database query languages. In 1971, Stephen Cook publishes "The complexity of theorem proving procedures"[149] In the 1970s within computer architecture, Quote notation was developed for a representing number system of rational numbers. Also in this decade, the Z notation (just like the APL language, long before it) uses many non-ASCII symbols, the specification includes suggestions for rendering the Z notation symbols in ASCII and in LaTeX. There are presently various C mathematical functions (Math.h) and numerical libraries. They are libraries used in software development for performing numerical calculations. These calculations can be handled by symbolic executions; analyzing a program to determine what inputs cause each part of a program to execute. Mathematica and SymPy are examples of computational software programs based on symbolic mathematics.

7.1.4   Future of mathematical notation

Main article: Future of mathematics
 In the history of mathematical notation, ideographic symbol notation has come full circle with the rise of computer visualization systems. The notations can be applied to abstract visualizations, such as for rendering some projections of a Calabi-Yau manifold. Examples of abstract visualization which properly belong to the mathematical imagination can be found in computer graphics. The need for such models abounds, for example, when the measures for the subject of study are actually random variables and not really ordinary mathematical functions.

7.2   See also

Main relevance Abuse of notation, Well-formed formula, Big O notation (L-notation), Dowker notation, Hungarian notation, Infix notation, Positional notation, Polish notation (Reverse Polish notation), Sign-value notation, Subtractive notation, infix notation, History of writing numbers

Numbers and quantities List of numbers, Irrational and suspected irrational numbers, γ, $\zeta(3)$, $\sqrt{2}$, $\sqrt{3}$, $\sqrt{5}$, φ, ρ, δS, α, e, π, δ, Physical constants, c, ε_0, h, G, Greek letters used in mathematics, science, and engineering

General relevance Order of operations, Scientific notation (Engineering notation), Actuarial notation

Dot notation Chemical notation (Lewis dot notation (Electron dot notation)), Dot-decimal notation

Arrow notation Knuth's up-arrow notation, infinitary combinatorics (Arrow notation (Ramsey theory))

Geometries Projective geometry, Affine geometry, Finite geometry

Lists and outlines Outline of mathematics (Mathematics history topics and Mathematics topics (Mathematics categories)), Mathematical theories (First-order theories, Theorems and Disproved mathematical ideas), Mathematical proofs (Incomplete proofs), Mathematical identities, Mathematical series, Mathematics reference tables, Mathematical logic topics, Mathematics-based methods, Mathematical functions, Transforms and Operators, Points in mathematics, Mathematical shapes, Knots (Prime knots and Mathematical knots and links), Inequalities, Mathematical concepts named after places, Mathematical topics in classical mechanics, Mathematical topics in quantum theory, Mathematical topics in relativity, String theory topics, Unsolved problems in mathematics, Mathematical jargon, Mathematical examples, Mathematical abbreviations, List of mathematical symbols

Misc. Hilbert's problems, Mathematical coincidence, Chess notation, Line notation, Musical notation (Dotted note), Whyte notation, Dice notation, recursive categorical syntax

People Mathematicians (Amateur mathematicians and Female mathematicians), Thomas Bradwardine, Thomas Harriot, Felix Hausdorff, Gaston Julia, Helge von Koch, Paul Lévy, Aleksandr Lyapunov, Benoit Mandelbrot, Lewis Fry Richardson, Wacław Sierpiński, Saunders Mac Lane, Paul Cohen, Gottlob Frege, G. S. Carr, Robert Recorde, Bartel Leendert van der Waerden, G. H. Hardy, E. M. Wright, James R. Newman, Carl Gustav Jacob Jacobi, Roger Joseph Boscovich, Eric W. Weisstein, Mathematical probabilists, Statisticians

7.3   Further reading

General

- A Short Account of the History of Mathematics. By Walter William Rouse Ball.

- A Primer of the History of Mathematics. By Walter William Rouse Ball.

- A History of Elementary Mathematics: With Hints on Methods of Teaching. By Florian Cajori.

- A History of Elementary Mathematics. By Florian Cajori.

- A History of Mathematics. By Florian Cajori.

- A Short History of Greek Mathematics. By James Gow.

- On the Development of Mathematical Thought During the Nineteenth Century. By John Theodore Merz.

- A New Mathematical and Philosophical Dictionary. By Peter Barlow.

- Historical Introduction to Mathematical Literature. By George Abram Miller

- A Brief History of Mathematics. By Karl Fink, Wooster Woodruff Beman, David Eugene Smith

- History of Modern Mathematics. By David Eugene Smith.

- History of modern mathematics. By David Eugene Smith, Mansfield Merriman.

Other

- Principia Mathematica, Volume 1 & Volume 2. By Alfred North Whitehead, Bertrand Russell.

- The Mathematical Principles of Natural Philosophy, Volume 1, Issue 1. By Sir Isaac Newton, Andrew Motte, William Davis, John Machin, William Emerson.

- General investigations of curved surfaces of 1827 and 1825. By Carl Friedrich Gaus.

7.4   Notes

[1] Or the Middle Ages.

[2] Such characters, in fact, are preserved with little alteration in the Roman notation, an account of which may be found in John Leslie's Philosophy of Arithmetic.

[3] Number theory is branch of pure mathematics devoted primarily to the study of the integers. Number theorists study prime numbers as well as the properties of objects made out of integers (e.g., rational numbers) or defined as generalizations of the integers (e.g., algebraic integers).

[4] Greek: μή μου τοὺς κύκλους τάραττε

[5] That is, $a^2 + b^2 = c^2$.

[6] Magnitude (mathematics), the relative size of an object ; Magnitude (vector), a term for the size or length of a vector; Scalar (mathematics), a quantity defined only by its magnitude; Euclidean vector, a quantity defined by both its magnitude and its direction; Order of magnitude, the class of scale having a fixed value ratio to the preceding class.

[7] Autolycus' On the Moving Sphere is another ancient mathematical manuscript of the time.

[8] Proclus, a Greek mathematician who lived several centuries after Euclid, wrote in his commentary of the Elements: "Euclid, who put together the Elements, collecting many of Eudoxus' theorems, perfecting many of Theaetetus', and also bringing to irrefragable demonstration the things which were only somewhat loosely proved by his predecessors".

[9] The expression:
$2x^4 + 3x^3 - 4x^2 + 5x - 6$
would be written as:
SS2 C3 x5 M S4 u6

[10] such as the rule, square, compasses, water level (reed level), and plumb-bob.

[11] such as the wheel and axle

[12] The area of the square described on the hypotenuse of a right-angled triangle is equal to the sum of the areas of the squares described on the sides

[13] Al-Kindi also introduced cryptanalysis and frequency analysis.

[14] Something close to a proof by mathematical induction appears in a book written by Al-Karaji around 1000 AD, who used it to prove the binomial theorem, Pascal's triangle, and the sum of integral cubes.

[15] He thus came close to finding a general formula for the integrals of polynomials, but he was not concerned with any polynomials higher than the fourth degree.

[16] a book about what he perceived as flaws in Euclid's *Elements*, especially the parallel postulate

[17] translated into Latin by Robert of Chester

[18] translated in various versions by Adelard of Bath, Herman of Carinthia, and Gerard of Cremona

[19] His own personal use started around 1351.

[20] Summa de Arithmetica: Geometria Proportioni et Proportionalita. *Tr.* Sum of Arithmetic: Geometry in proportions and proportionality.

[21] Much of the work originated from Piero Della Francesca whom he appropriated and purloined.

[22] This was a special case of the methods given many centuries later by Ruffini and Horner.

[23] That is, $\sqrt{}$.

[24] Because, it is thought, it resembled a lowercase "r" (for "radix").

[25] Published in Description of the Marvelous Canon of Logarithms

[26] That is,
::

[27] see Law of Continuity.

[28] Using Cartesian coordinates on the plane, the distance between two points (x_1, y_1) and (x_2, y_2) is defined by the formula:
$d = \sqrt{(x_2 - x_1)^2 + (y_2 - y_1)^2}$,
which can be viewed as a version of the Pythagorean theorem.

[29] Further steps in abstraction were taken by Lobachevsky, Bolyai, Riemann, and Gauss who generalised the concepts of geometry to develop non-Euclidean geometries.

[30] Now called Pascal's triangle.

[31] For example, the "problem of points".

[32] That is, ∞ .

[33] For example, $\frac{1}{\infty}$.

[34] Original title, "*De ratiociniis in ludo aleae*"

[35] For example, the derivative of the function x would be written as $\dot{x}$. The second derivative of x would be written as $\ddot{x}$, etc.

[36] For example, the derivative of the function x with respect to the variable t in Leibniz's notation would be written as $\frac{dx}{dt}$.

[37] That is, $\int_{-N}^{N} f(x)\, dx$.

[38] See also: List of representations of e

[39] Thus $f(x)$ denotes the mathematical result of the performance of the operation f upon the subject x . If upon this result the same operation were repeated, the new result would be expressed by $f[f(x)]$, or more concisely by $f^2(x)$, and so on. The quantity x itself regarded as the result of the same operation f upon some other function; the proper symbol for which is, by analogy, $f^{-1}(x)$. Thus f and f^{-1} are symbols of inverse operations, the former cancelling the effect of the latter on the subject x . $f(x)$ and $f^{-1}(x)$ in a similar manner are termed inverse functions.

[40] That is, $\Delta f(p)$

[41] That is, $\sqrt{-1}$

[42] Today, the symbol created by John Wallis, ∞ , is used for infinity.

[43] As in, $\sum_{n=1}^{\infty} \frac{1}{n^2}$

[44] Capital-sigma notation uses a symbol that compactly represents summation of many similar terms: the *summation symbol*, Σ, an enlarged form of the upright capital Greek letter Sigma. This is defined as:

Where, i represents the *index of summation*; ai is an indexed variable representing each successive term in the series; m is the *lower bound of summation*, and n is the *upper bound of summation*. The "$i = m$" under the summation symbol means that the index i starts out equal to m. The index, i, is incremented by 1 for each successive term, stopping when $i = n$.

[45] That is, $n! = \int_0^1 (-\ln s)^n \, ds$.
valid for n > 0.

[46] That is,
$\propto$

[47] Proportionality is the ratio of one quantity to another, especially the ratio of a part compared to a whole. In a mathematical context, a proportion is the statement of equality between two ratios; See Proportionality (mathematics), the relationship of two variables whose ratio is constant. See also aspect ratio, geometric proportions.

[48] The *curly d* or *Jacobi's delta*.

[49] About the proof of Wilson's theorem. *Disquisitiones Arithmeticae* (1801) Article 76

[50] Galois theory and Galois geometry is named after him.

[51] That is, "subset of" and "superset of"; This would later be redeveloped by Ernst Schröder.

[52] A science of numbers that uses methods from mathematical analysis to solve problems about the integers.

[53] quoted in Robert Percival Graves' "*Life of Sir William Rowan Hamilton*" (3 volumes, 1882, 1885, 1889)

[54] That is, ∇ (or, later called *del*, ∇)

[55] See Hamiltonian (quantum mechanics).

[56] That is, $i^2 = j^2 = k^2 = ijk = -1$

[57] Though his use describes something different from what is now meant by a tensor. Namely, the norm operation in a certain type of algebraic system (now known as a Clifford algebra).

[58] That is,
$t = w + xi + yj + zk, \quad w, x, y, z \in \mathbb{R}$
where
$ij = ji = k, \quad i^2 = -1, \quad j^2 = +1.$

[59] This is Latin for "womb".

[60] That is, $q = w + xi + yj + zk$

[61] Clifford intersected algebra with Hamilton's quaternions by replacing Hermann Grassmann's rule *epep* = 0 by the rule *epep* = 1. For more details, see exterior algebra.

[62] See: Phasor, Group (mathematics), Signal velocity, Polyphase system, Harmonic oscillator, and RLC series circuit

[63] Or the concept of a fourth spatial dimension. See also: Spacetime, the unification of time and space as a four-dimensional continuum; and, Minkowski space, the mathematical setting for special relativity.

[64] He famously said, "Mathematics is an experimental science, and definitions do not come first, but later on." He was replying to criticism over his use of operators that were not clearly defined. On another occasion he stated somewhat more defensively, "I do not refuse my dinner simply because I do not understand the process of digestion."

[65] See also: Mathematic fields and Field extension

[66] Comment after the proof that 1+1=2, completed in Principia mathematica, by Alfred North Whitehead ... and Bertrand Russell. Volume II, 1st edition (1912)

[67] This raises questions of the pure existence theorems.

[68] Peano's *Formulario Mathematico*, though less popular than Russell's work, continued through five editions. The fifth appeared in 1908 and included 4200 formulas and theorems.

[69] Inventor of set theory

[70] *Transfinite arithmetic* is the generalization of elementary arithmetic to infinite quantities like infinite sets; See Transfinite numbers, Transfinite induction, and Transfinite interpolation. See also Ordinal arithmetic.

[71] Such as Max Dehn, J. W. Alexander, and others.

[72] Such as the Alexander polynomial.

[73] (German: Algebraische Theorie der Körper)

[74] In this paper Steinitz axiomatically studied the properties of fields and defined many important field theoretic concepts like prime field, perfect field and the transcendence degree of a field extension.

[75] The indices range over set {1, 2, 3},
$$y = \sum_{i=1}^{3} c_i x^i = c_1 x^1 + c_2 x^2 + c_3 x^3$$
is reduced by the convention to:
$$y = c_i x^i.$$
Upper indices are not exponents but are indices of coordinates, coefficients or basis vectors.
See also: Ricci calculus

[76] Ricci calculus constitutes the rules of index notation and manipulation for tensors and tensor fields. See also: Synge J.L., Schild A. (1949). *Tensor Calculus*. first Dover Publications 1978 edition. pp. 6–108.

[77] Here a logical constant is a symbol in symbolic logic that has the same meaning in all models, such as the symbol "=" for "equals".
A *constant*, in a mathematical context, is a number that arises naturally in mathematics, such as π or e; Such mathematics constant value do not change. It can mean polynomial constant term (the term of degree 0) or the constant of integration, a free parameter arising in integration.
Related, the physical constant are a physical quantity generally believed to be universal and unchanging. Programming constants are a values that, unlike a variable, cannot be reassociated with a different value.

[78] Though not an index term, keywords are terms that represent information. A keyword is a word with special meaning (this is a semantic definition), while syntactically these are terminal symbols in the phrase grammar. See reserved word for the related concept.

[79] Most of these symbols can be found in propositional calculus, a formal system described as $\mathcal{L} = \mathcal{L}\,(A,\,\Omega,\,Z,\,I)$. A is the set of elements, such as the a in the example with Boolean algebra above. Ω is the set that contains the subsets that contain operations, such as $\vee$ or $\wedge$. Z contains the inference rules, which are the rules dictating how inferences may be logically made, and I contains the axioms. See also: Basic and Derived Argument Forms.

[80] Usually denoted by x, y, z, or other lowercase letters
Here a symbols that represents a quantity in a mathematical expression, a mathematical variable as used in many sciences.
Variables can be symbolic name associated with a value and whose associated value may be changed, known in computer science as a variable reference. A *variable* can also be the operationalized way in which the attribute is represented for further data processing (e.g., a logical set of attributes). See also: Dependent and independent variables in statistics.

[81] Usually denoted by an uppercase letter followed by a list of variables, such as $P(x)$ or $Q(y,z)$
Here a mathematical logic predicate, a fundamental concept in first-order logic. Grammatical predicates are grammatical components of a sentence.
Related is the syntactic predicate in parser technology which are guidelines for the parser process. In computer programming, a branch predication allows a choice to execute or not to execute a given instruction based on the content of a machine register.

[82] Representing ALL and EXISTS

[83] e.g. $\exists$ for "there exists" and $\forall$ for "for all"

[84] See also: Dialetheism, Contradiction, and Paradox

[85] Related, facetious abstract nonsense describes certain kinds of arguments and methods related to category theory which resembles comical literary non sequitur devices (not illogical non sequiturs).

[86] Gödel's incompleteness theorems shows that Hilbert's program to find a complete and consistent set of axioms for all mathematics is impossible, giving a contested negative answer to Hilbert's second problem

[87] For example, take the statement "There exists a number x such that it is not y". Using the symbols of propositional calculus, this would become: $(\exists x)(x = \neg y)$.
      If the Gödel numbers replace the symbols, it becomes: $\{8, 4, 11, 9, 8, 11, 5, 1, 13, 9\}$.
      There are ten numbers, so the ten prime numbers are found and these are: $\{2, 3, 5, 7, 11, 13, 17, 19, 23, 29\}$.
      Then, the Gödel numbers are made the powers of the respective primes and multiplied, giving: $2^8 \times 3^4 \times 5^{11} \times 7^9 \times 11^8 \times 13^{11} \times 17^5 \times 19^1 \times 23^{13} \times 29^9$.
      The resulting number is approximately $3.096262735 \times 10^{78}$.

[88] The Klein–Gordon equation is:

[89] The Dirac equation in the form originally proposed by Dirac is:

      where, $\psi = \psi(\mathbf{x}, t)$ is the wave function for the electron, $\mathbf{x}$ and t are the space and time coordinates, m is the rest mass of the electron, p is the momentum, understood to be the momentum operator in the Schrödinger theory, c is the speed of light, and $\hbar = h/2\pi$ is the reduced Planck constant.

[90] That is,

[91] The theorem applies more generally to any sufficiently strong formal system, showing that truth in the standard model of the system cannot be defined within the system.

[92] Named to honor Voigt's 1898 work.

[93] Named after Arthur Moritz Schoenflies

[94] See Galois connections.

[95] Oystein Ore would also write "Number Theory and Its History".

[96] $\langle \phi | \psi \rangle$

[97] That the scattering amplitude can be thought of as an analytic function of the angular momentum, and that the position of the poles determine power-law growth rates of the amplitude in the purely mathematical region of large values of the cosine of the scattering angle.

[98] That is, $\square$

[99] Also known as the d'Alembertian or wave operator.

[100] Also known as, "permutation symbol" (see: permutation), "antisymmetric symbol" (see: antisymmetric), or "alternating symbol"

[101] Note that "masses" (e.g., the coherent non-definite body shape) of particles are periodically *reevaluated* by the scientific community. The values may have been adjusted; adjustment by operations carried out on instruments in order that it provides given indications corresponding to given values of the measurand. In engineering, mathematics, and geodesy, the optimal parameter such estimation of a mathematical model so as to best fit a data set.

[102] For the consensus, see Particle Data Group.

[103] A locally defined set of four linearly independent vector fields called a tetrad

[104] His usage of the Einstein summation was in order to offset the inconvenience in describing contractions and covariant differentiation in modern abstract tensor notation, while maintaining explicit covariance of the expressions involved.

[105] See also: String theory landscape and Swampland

[106] Devised by Joshua Lederberg and extended by Coxeter and Frucht

[107] And, in 1938, "On Computable Numbers, with an Application to the Entscheidungsproblem: A correction" (Proceedings of the London Mathematical Society, 2 (1937) 43 (6): 544–6, doi:10.1112/plms/s2-43.6.544).

[108] Among von Neumann's other contributions include the application of operator theory to quantum mechanics, in the development of functional analysis, and on various forms of operator theory.

7.5 References and citations

General

- Florian Cajori (1929) *A History of Mathematical Notations*, 2 vols. Dover reprint in 1 vol., 1993. ISBN 0-486-67766-4.

Citations

[1] Florian Cajori. A History of Mathematical Notations: Two Volumes in One. Cosimo, Inc., Dec 1, 2011

[2] A Dictionary of Science, Literature, & Art, Volume 2. Edited by William Thomas Brande, George William Cox. Pg 683

[3] "Notation - from Wolfram MathWorld". Mathworld.wolfram.com. Retrieved 2014-06-24.

[4] Diophantos of Alexandria: A Study in the History of Greek Algebra. By Sir Thomas Little Heath. Pg 77.

[5] Mathematics: Its Power and Utility. By Karl J. Smith. Pg 86.

[6] The Commercial Revolution and the Beginnings of Western Mathematics in Renaissance Florence, 1300-1500. Warren Van Egmond. 1976. Page 233.

[7] Solomon Gandz. "The Sources of al-Khowarizmi's Algebra"

[8] Encyclopædia Americana. By Thomas Gamaliel Bradford. Pg 314

[9] Mathematical Excursion, Enhanced Edition: Enhanced Webassign Edition By Richard N. Aufmann, Joanne Lockwood, Richard D. Nation, Daniel K. Cleg. Pg 186

[10] Mathematics in Egypt and Mesopotamia

[11] Boyer, C. B. *A History of Mathematics*, 2nd ed. rev. by Uta C. Merzbach. New York: Wiley, 1989 ISBN 0-471-09763-2 (1991 pbk ed. ISBN 0-471-54397-7). "Mesopotamia" p. 25.

[12] Duncan J. Melville (2003). Third Millennium Chronology, *Third Millennium Mathematics*. St. Lawrence University.

[13] Aaboe, Asger (1998). *Episodes from the Early History of Mathematics*. New York: Random House. pp. 30–31.

[14] Heath. *A Manual of Greek Mathematics*. p. 5.

[15] Sir Thomas L. Heath, *A Manual of Greek Mathematics*, Dover, 1963, p. 1: "In the case of mathematics, it is the Greek contribution which it is most essential to know, for it was the Greeks who made mathematics a science."

[16] The new encyclopædia; or, Universal dictionary of arts and sciences. By Encyclopaedia Perthensi. Pg 49

[17] Calinger, Ronald (1999). *A Contextual History of Mathematics*. Prentice-Hall. p. 150. ISBN 0-02-318285-7. Shortly after Euclid, compiler of the definitive textbook, came Archimedes of Syracuse (ca. 287 212 BC), the most original and profound mathematician of antiquity.

[18] "Archimedes of Syracuse". The MacTutor History of Mathematics archive. January 1999. Retrieved 2008-06-09.

[19] O'Connor, J.J. and Robertson, E.F. (February 1996). "A history of calculus". University of St Andrews. Archived from the original on 15 July 2007. Retrieved 2007-08-07.

[20] "Proclus' Summary". Gap.dcs.st-and.ac.uk. Retrieved 2014-06-24.

[21] Caldwell, John (1981) "The *De Institutione Arithmetica* and the *De Institutione Musica*", pp. 135–54 in Margaret Gibson, ed., *Boethius: His Life, Thought, and Influence*, (Oxford: Basil Blackwell).

[22] Folkerts, Menso, *"Boethius" Geometrie II*, (Wiesbaden: Franz Steiner Verlag, 1970).

[23] Mathematics and Measurement By Oswald Ashton Wentworth Dilk. Pg 14

[24] A dictionary of science, literature and art, ed. by W.T. Brande. Pg 683

[25] Boyer, Carl B. *A History of Mathematics*, 2nd edition, John Wiley & Sons, Inc., 1991.

[26] Diophantine Equations. Submitted by: Aaron Zerhusen, Chris Rakes, & Shasta Meece. MA 330-002. Dr. Carl Eberhart. February 16, 1999.

[27] A History of Greek Mathematics: From Aristarchus to Diophantus. By Sir Thomas Little Heath. Pg 456

[28] A History of Greek Mathematics: From Aristarchus to Diophantus. By Sir Thomas Little Heath. Pg 458

[29] The American Mathematical Monthly, Volume 16. Pg 131

[30] "Overview of Chinese mathematics". Groups.dcs.st-and.ac.uk. Retrieved 2014-06-24.

[31] George Gheverghese Joseph, *The Crest of the Peacock: Non-European Roots of Mathematics*,Penguin Books, London, 1991, pp.140—148

[32] Georges Ifrah, *Universalgeschichte der Zahlen*, Campus, Frankfurt/New York, 1986, pp.428—437

[33] "Frank J. Swetz and T. I. Kao: Was Pythagoras Chinese?". Psupress.psu.edu. Retrieved 2014-06-24.

[34] Needham, Joseph (1986). Science and Civilization in China: Volume 3, Mathematics and the Sciences of the Heavens and the Earth. Taipei: Caves Books, Ltd..

[35] Sal Restivo

[36] Marcel Gauchet, 151.

[37] Boyer, C. B. A History of Mathematics, 2nd ed. rev. by Uta C. Merzbach. New York: Wiley, 1989 ISBN 0-471-09763-2 (1991 pbk ed. ISBN 0-471-54397-7). "China and India" p. 221. (cf., "he was the first one to give a *general* solution of the linear Diophantine equation $ax + by = c$, where a, b, and c are integers. [...] It is greatly to the credit of Brahmagupta that he gave *all* integral solutions of the linear Diophantine equation, whereas Diophantus himself had been satisfied to give one particular solution of an indeterminate equation. Inasmuch as Brahmagupta used some of the same examples as Diophantus, we see again the likelihood of Greek influence in India – or the possibility that they both made use of a common source, possibly from Babylonia. It is interesting to note also that the algebra of Brahmagupta, like that of Diophantus, was syncopated. Addition was indicated by juxtaposition, subtraction by placing a dot over the subtrahend, and division by placing the divisor below the dividend, as in our fractional notation but without the bar. The operations of multiplication and evolution (the taking of roots), as well as unknown quantities, were represented by abbreviations of appropriate words.")

[38] Robert Kaplan, "The Nothing That Is: A Natural History of Zero", Allen Lane/The Penguin Press, London, 1999

[39] ""The ingenious method of expressing every possible number using a set of ten symbols (each symbol having a place value and an absolute value) emerged in India. The idea seems so simple nowadays that its significance and profound importance is no longer appreciated. Its simplicity lies in the way it facilitated calculation and placed arithmetic foremost amongst useful inventions. the importance of this invention is more readily appreciated when one considers that it was beyond the two greatest men of Antiquity, Archimedes and Apollonius." - Pierre-Simon Laplace". History.mcs.st-and.ac.uk. Retrieved 2014-06-24.

[40] A.P. Juschkewitsch, "Geschichte der Mathematik im Mittelalter", Teubner, Leipzig, 1964

[41] Boyer, C. B. A History of Mathematics, 2nd ed. rev. by Uta C. Merzbach. New York: Wiley, 1989 ISBN 0-471-09763-2 (1991 pbk ed. ISBN 0-471-54397-7). "The Arabic Hegemony" p. 230. (cf., "The six cases of equations given above exhaust all possibilities for linear and quadratic equations having positive root. So systematic and exhaustive was al-Khwārizmī's exposition that his readers must have had little difficulty in mastering the solutions.")

[42] Gandz and Saloman (1936), *The sources of Khwarizmi's algebra*, Osiris i, pp. 263–77: "In a sense, Khwarizmi is more entitled to be called "the father of algebra" than Diophantus because Khwarizmi is the first to teach algebra in an elementary form and for its own sake, Diophantus is primarily concerned with the theory of numbers".

[43] Boyer, C. B. A History of Mathematics, 2nd ed. rev. by Uta C. Merzbach. New York: Wiley, 1989 ISBN 0-471-09763-2 (1991 pbk ed. ISBN 0-471-54397-7). "The Arabic Hegemony" p. 229. (cf., "It is not certain just what the terms *al-jabr* and *muqabalah* mean, but the usual interpretation is similar to that implied in the translation above. The word *al-jabr* presumably meant something like "restoration" or "completion" and seems to refer to the transposition of subtracted terms to the other side of an equation; the word *muqabalah* is said to refer to "reduction" or "balancing" - that is, the cancellation of like terms on opposite sides of the equation.")

[44] Rashed, R.; Armstrong, Angela (1994). *The Development of Arabic Mathematics*. Springer. pp. 11–12. ISBN 0-7923-2565-6. OCLC 29181926.

[45] Victor J. Katz (1998). *History of Mathematics: An Introduction*, pp. 255–59. Addison-Wesley. ISBN 0-321-01618-1.

[46] F. Woepcke (1853). *Extrait du Fakhri, traité d'Algèbre par Abou Bekr Mohammed Ben Alhacan Alkarkhi*. Paris.

[47] Victor J. Katz (1995), "Ideas of Calculus in Islam and India", *Mathematics Magazine* **68** (3): 163–74.

[48] Marie-Thérèse d'Alverny, "Translations and Translators", pp. 421–62 in Robert L. Benson and Giles Constable, *Renaissance and Renewal in the Twelfth Century*, (Cambridge: Harvard University Press, 1982).

[49] Guy Beaujouan, "The Transformation of the Quadrivium", pp. 463–87 in Robert L. Benson and Giles Constable, *Renaissance and Renewal in the Twelfth Century*, (Cambridge: Harvard University Press, 1982).

[50] O'Connor, John J.; Robertson, Edmund F., "Abu'l Hasan ibn Ali al Qalasadi", *MacTutor History of Mathematics archive*, University of St Andrews.

[51] Boyer, C. B. A History of Mathematics, 2nd ed. rev. by Uta C. Merzbach. New York: Wiley, 1989 ISBN 0-471-09763-2 (1991 pbk ed. ISBN 0-471-54397-7). "Revival and Decline of Greek Mathematics" p. 178 (cf., "The chief difference between Diophantine syncopation and the modern algebraic notation is the lack of special symbols for operations and relations, as well as of the exponential notation.")

[52] Grant, Edward and John E. Murdoch (1987), eds., *Mathematics and Its Applications to Science and Natural Philosophy in the Middle Ages*, (Cambridge: Cambridge University Press) ISBN 0-521-32260-X.

[53] Mathematical Magazine, Volume 1. Artemas Martin, 1887. Pg 124

[54] Der Algorismus proportionum des Nicolaus Oresme: Zum ersten Male nach der Lesart der Handschrift R.40.2. der Königlichen Gymnasial-bibliothek zu Thorn. Nicole Oresme. S. Calvary & Company, 1868.

[55] Clagett, Marshall (1961) *The Science of Mechanics in the Middle Ages*, (Madison: University of Wisconsin Press), pp. 332–45, 382–91.

[56] *Later early modern version*: A New System of Mercantile Arithmetic: Adapted to the Commerce of the United States, in Its Domestic and Foreign Relations with Forms of Accounts and Other Writings Usually Occurring in Trade. By Michael Walsh. Edmund M. Blunt (proprietor.), 1801.

[57] Miller, Jeff (4 June 2006). "Earliest Uses of Symbols of Operation". *Gulf High School*. Retrieved 24 September 2006.

[58] Arithmetical Books from the Invention of Printing to the Present Time. By Augustus De Morgan. p 2.

[59] Grattan-Guinness, Ivor (1997). *The Rainbow of Mathematics: A History of the Mathematical Sciences*. W.W. Norton. ISBN 0-393-32030-8.

[60] Arithmetica integra. By Michael Stifel, Philipp Melanchton. Norimbergæ: Apud Iohan Petreium, 1544.

[61] The History of Mathematics By Anne Roone. Pg 40

[62] Memoirs of John Napier of Merchiston. By Mark Napier

[63] An Account of the Life, Writings, and Inventions of John Napier, of Merchiston. By David Stewart Erskine Earl of Buchan, Walter Minto

[64] Florian Cajori (1919). *A History of Mathematics*. Macmillan.

[65] Jan Gullberg, Mathematics from the birth of numbers, W. W. Norton & Company; ISBN 978-0-393-04002-9 . pg 963-965,

[66] Synopsis Palmariorum Matheseos. By William Jones. 1706. (Alt: Synopsis Palmariorum Matheseos: or, a New Introduction to the Mathematics. archive.org.)

[67] When Less is More: Visualizing Basic Inequalities.By Claudi Alsina, Roger B. Nelse. Pg 18.

[68] Euler, Leonhard, Solutio problematis ad geometriam situs pertinentis

[69] The elements of geometry. By William Emerson

[70] The Doctrine of Proportion, Arithmetical and Geometrical. Together with a General Method of Arening by Proportional Quantities. By William Emerson.

[71] The Mathematical Correspondent. By George Baron. 83

[72] Vitulli, Marie. "A Brief History of Linear Algebra and Matrix Theory". *Department of Mathematics*. University of Oregon. Retrieved 2012-01-24.

[73] "Kramp biography". History.mcs.st-and.ac.uk. Retrieved 2014-06-24.

[74] Mécanique analytique: Volume 1, Volume 2. By Joseph Louis Lagrange. Ms. Ve Courcier, 1811.

[75] The collected mathematical papers of Arthur Cayley. Volume 11. Page 243.

[76] Historical Encyclopedia of Natural and Mathematical Sciences, Volume 1. By Ari Ben-Menahem. Pg 2070.

[77] Vitulli, Marie. "A Brief History of Linear Algebra and Matrix Theory". Department of Mathematics. University of Oregon. Originally at: darkwing.uoregon.edu/~{}vitulli/441.sp04/LinAlgHistory.html

[78] The Words of Mathematics. By Steven Schwartzman. 6.

[79] Electro-Magnetism: Theory and Applications. By A. Pramanik. 38

[80] History of Nabla and Other Math Symbols. homepages.math.uic.edu/~{}hanson.

[81] Hamilton, William Rowan (1854–1855). Wilkins, David R., ed. "On some Extensions of Quaternions" (PDF). *Philosophical Magazine* (7–9): 492–499, 125–137, 261–269, 46–51, 280–290. ISSN 0302-7597.

[82] "James Clerk Maxwell". IEEE Global History Network. Retrieved 25 March 2013.

[83] Maxwell, James Clerk (1865). "A dynamical theory of the electromagnetic field" (PDF). *Philosophical Transactions of the Royal Society of London* **155**: 459–512. doi:10.1098/rstl.1865.0008. (This article accompanied a December 8, 1864 presentation by Maxwell to the Royal Society.)

[84] Proceedings of the London Mathematical Society, Volume 3. London Mathematical Society, 1871. Pg. 224

[85] *Books I, II, III (1878)* on Internet Archive; *Book IV (1887)* on Internet Archive

[86] The Heaviside Operational Calculus www.quadritek.com/bstj/vol01-1922/articles/bstj1-2-43.pdf

[87] Cox, David A. (2012). *Galois Theory*. Pure and Applied Mathematics **106** (2nd ed.). John Wiley & Sons. p. 348. ISBN 1118218426.

[88] "TÜBİTAK ULAKBİM DergiPark". Journals.istanbul.edu.tr. Retrieved 2014-06-24.

[89] "Linear Algebra : Hussein Tevfik : Free Download & Streaming : Internet Archive". Archive.org. Retrieved 2014-06-24.

[90] Ricci Curbastro, G. (1892). "Résumé de quelques travaux sur les systèmes variables de fonctions associés à une forme différentielle quadratique". *Bulletin des Sciences Mathématiques* **2** (16): 167–189.

[91] Voigt, Woldemar (1898). *Die fundamentalen physikalischen Eigenschaften der Krystalle in elementarer Darstellung*. Leipzig: Von Veit.

[92] Poincaré, Henri, "Analysis situs", Journal de l'École Polytechnique ser 2, 1 (1895) pp. 1–123

[93] Whitehead, John B., Jr. (1901). "Review: *Alternating Current Phenomena*, by C. P. Steinmetz" (PDF). *Bull. Amer. Math. Soc.* (3rd ed.) **7** (9): 399–408. doi:10.1090/s0002-9904-1901-00825-7.

[94] There are many editions. Here are two:

- (French) Published 1901 by Gauthier-Villars, Paris. 230p. OpenLibrary OL15255022W, PDF.
- (Italian) Published 1960 by Edizione cremonese, Roma. 463p. OpenLibrary OL16587658M.

[95] Ricci, Gregorio; Levi-Civita, Tullio (March 1900), "Méthodes de calcul différentiel absolu et leurs applications" (PDF), *Mathematische Annalen* (Springer) **54** (1–2): 125–201, doi:10.1007/BF01454201

[96] Zermelo, Ernst (1904). "Beweis, dass jede Menge wohlgeordnet werden kann" (REPRINT). *Mathematische Annalen* **59** (4): 514–16. doi:10.1007/BF01445300.

[97] 🔟 *On the Dynamics of the Electron (July)*. Wikisource.

[98] Fréchet, Maurice, "Sur quelques points du calcul fonctionnel", PhD dissertation, 1906

[99] Cuthbert Edmund Cullis (Author) (2011-06-05). "Matrices and determinoids Volume 2: Cuthbert Edmund Cullis: Amazon.com: Books". Amazon.com. Retrieved 2014-06-24.

[100] Can be assigned a given matrix: About a class of matrices. (Gr. Ueber eine Klasse von Matrizen: die sich einer gegebenen Matrix zuordnen lassen.) by Isay Schur

[101] An Introduction To The Modern Theory Of Equations. By Florian Cajori.

[102] Proceedings of the Prussian Academy of Sciences (1918). Pg 966.

[103] Sitzungsberichte der Preussischen Akademie der Wissenschaften (1918) (Tr. Proceedings of the Prussian Academy of Sciences (1918)). archive.org; See also: Kaluza–Klein theory .

[104] J.A. Wheeler, C. Misner, K.S. Thorne (1973). *Gravitation*. W.H. Freeman & Co. pp. 85–86, §3.5. ISBN 0-7167-0344-0.

[105] R. Penrose (2007). *The Road to Reality*. Vintage books. ISBN 0-679-77631-1.

[106] Schouten, Jan A. (1924). R. Courant, ed. *Der Ricci-Kalkül – Eine Einführung in die neueren Methoden und Probleme der mehrdimensionalen Differentialgeometrie (Ricci Calculus – An introduction in the latest methods and problems in multi-dimmensional differential geometry)*. Grundlehren der mathematischen Wissenschaften (in German) **10**. Berlin: Springer Verlag.

[107] Robert B. Ash. A Primer of Abstract Mathematics. Cambridge University Press, Jan 1, 1998

[108] The New American Encyclopedic Dictionary. Edited by Edward Thomas Roe, Le Roy Hooker, Thomas W. Handford. Pg 34

[109] The Mathematical Principles of Natural Philosophy, Volume 1. By Sir Isaac Newton, John Machin. Pg 12.

[110] In The Scientific Outlook (1931)

[111] Mathematics simplified and made attractive: or, The laws of motion explained. By Thomas Fisher. Pg 15. (cf. *But an abstraction not founded upon, and not consonant with Nature and (Logical) Truth, would be a falsity, an insanity.*)

[112] Proposition VI, *On Formally Undecidable Propositions in* Principia Mathematica *and Related Systems I* (1931)

[113] Casti, John L. *5 Golden Rules*. New York: MJF Books, 1996.

[114] Gr. *Methoden Der Mathematischen Physik*

[115] P.A.M. Dirac (1927). "The Quantum Theory of the Emission and Absorption of Radiation". *Proceedings of the Royal Society of London A* **114**: 243–265. Bibcode:1927RSPSA.114..243D. doi:10.1098/rspa.1927.0039.

[116] E. Fermi (1932). "Quantum Theory of Radiation". *Reviews of Modern Physics* **4**: 87–132. Bibcode:1932RvMP....4...87F. doi:10.1103/RevModPhys.4.87.

[117] F. Bloch; A. Nordsieck (1937). "Note on the Radiation Field of the Electron". *Physical Review* **52**: 54–59. Bibcode:1937PhRv...52...54B. doi:10.1103/PhysRev.52.54.

[118] V. F. Weisskopf (1939). "On the Self-Energy and the Electromagnetic Field of the Electron". *Physical Review* **56**: 72–85. Bibcode:1939PhRv...56...72W. doi:10.1103/PhysRev.56.72.

[119] R. Oppenheimer (1930). "Note on the Theory of the Interaction of Field and Matter". *Physical Review* **35**: 461–477. Bibcode:1930PhRv...35..461O. doi:10.1103/PhysRev.35.461.

[120] Van der Waerden B.L. (1929). "Spinoranalyse". *Nachr. Ges. Wiss. Göttingen Math.-Phys.* **1929**: 100–109.

[121] Veblen O. (1933). "Geometry of two-component Spinors". *Proc. Natl. Acad. Sci. USA* **19**: 462–474. doi:10.1073/pnas.19.4.462.

[122] PAM Dirac (1939). "A new notation for quantum mechanics". *Mathematical Proceedings of the Cambridge Philosophical Society* **35** (3). pp. 416–418. doi:10.1017/S0305004100021162.

[123] H. Grassmann (1862). *Extension Theory*. History of Mathematics Sources. American Mathematical Society, London Mathematical Society, 2000 translation by Lloyd C. Kannenberg.

[124] Steven Weinberg (1964), *The quantum theory of fields, Volume 2, Cambridge University Press, 1995*, p. 358, ISBN 0-521-55001-7

[125] "The Nobel Prize in Physics 1965". Nobel Foundation. Retrieved 2008-10-09.

[126] S.L. Glashow (1961). "Partial-symmetries of weak interactions". *Nuclear Physics* **22**: 579–588. Bibcode:1961NucPh..22..579G. doi:10.1016/0029-5582(61)90469-2.

[127] S. Weinberg (1967). "A Model of Leptons". *Physical Review Letters* **19**: 1264–1266. Bibcode:1967PhRvL..19.1264W. doi:10.1103/PhysRevLett.19.1264.

[128] A. Salam (1968). N. Svartholm, ed. *Elementary Particle Physics: Relativistic Groups and Analyticity*. Eighth Nobel Symposium. Stockholm: Almquvist and Wiksell. p. 367.

[129] F. Englert, R. Brout (1964). "Broken Symmetry and the Mass of Gauge Vector Mesons". *Physical Review Letters* **13**: 321–323. Bibcode:1964PhRvL..13..321E. doi:10.1103/PhysRevLett.13.321.

[130] P.W. Higgs (1964). "Broken Symmetries and the Masses of Gauge Bosons". *Physical Review Letters* **13**: 508–509. Bibcode:1964PhRvL..13..508H. doi:10.1103/PhysRevLett.13.508.

[131] G.S. Guralnik, C.R. Hagen, T.W.B. Kibble (1964). "Global Conservation Laws and Massless Particles". *Physical Review Letters* **13**: 585–587. Bibcode:1964PhRvL..13..585G. doi:10.1103/PhysRevLett.13.585.

[132] http://www.physics.drexel.edu/~{}vkasli/phys676/Notes%20for%20a%20brief%20history%20of%20quantum%20gravity%20-%20Carlo%20Rovelli.pdf

[133] Bourbaki, Nicolas (1972). "Univers". In Michael Artin, Alexandre Grothendieck, Jean-Louis Verdier, eds. *Séminaire de Géométrie Algébrique du Bois Marie - 1963-64 - Théorie des topos et cohomologie étale des schémas - (SGA 4) - vol. 1 (Lecture notes in mathematics **269**)* (in French). Berlin; New York: Springer-Verlag. pp. 185–217.

[134] F.J. Hasert et al. (1973). "Search for elastic muon-neutrino electron scattering". *Physics Letters B* **46**: 121. Bibcode:1973PhLB...46..121H. doi:10.1016/0370-2693(73)90494-2.

[135] F.J. Hasert et al. (1973). "Observation of neutrino-like interactions without muon or electron in the gargamelle neutrino experiment". *Physics Letters B* **46**: 138. Bibcode:1973PhLB...46..138H. doi:10.1016/0370-2693(73)90499-1.

[136] F.J. Hasert et al. (1974). "Observation of neutrino-like interactions without muon or electron in the Gargamelle neutrino experiment". *Nuclear Physics B* **73**: 1. Bibcode:1974NuPhB..73....1H. doi:10.1016/0550-3213(74)90038-8.

[137] D. Haidt (4 October 2004). "The discovery of the weak neutral currents". *CERN Courier*. Retrieved 2008-05-08.

[138] http://web.physics.ucsb.edu/~{}gary/

[139] Nuclear Physics B 258: 46–74, Bibcode:1985NuPhB.258...46C, doi:10.1016/0550-3213(85)90602-9

[140] De Felice, F.; Clarke, C.J.S. (1990), *Relativity on Curved Manifolds*, p. 133

[141] "Quantum invariants of knots and 3-manifolds" by V. G. Turaev (1994), page 71

[142] Pisanski, Tomaž; Servatius, Brigitte (2013), "2.3.2 Cubic graphs and LCF notation", *Configurations from a Graphical Viewpoint*, Springer, p. 32, ISBN 9780817683641

[143] Frucht, R. (1976), "A canonical representation of trivalent Hamiltonian graphs", *Journal of Graph Theory* **1** (1): 45–60, doi:10.1002/jgt.3190010111

[144] Fraleigh 2002:89; Hungerford 1997:230

[145] Dehn, Edgar. Algebraic Equations, Dover. 1930:19

[146] "The IBM 601 Multiplying Punch". Columbia.edu. Retrieved 2014-06-24.

[147] "Interconnected Punched Card Equipment". Columbia.edu. 1935-10-24. Retrieved 2014-06-24.

[148] Proceedings of the London Mathematical Society 42 (2)

[149] Cook, Stephen (1971). "The complexity of theorem proving procedures". *Proceedings of the Third Annual ACM Symposium on Theory of Computing*. pp. 151–158.

7.6 External links

- Mathematical Notation: Past and Future

- History of Mathematical Notation

- Earliest Uses of Mathematical Notation

- Finger counting. files.chem.vt.edu.

- Some Common Mathematical Symbols and Abbreviations (with History). Isaiah Lankham, Bruno Nachtergaele, Anne Schilling.

Archimedes Thoughtful
by Fetti (1620) ---- The last words attributed to Archimedes are "Do not disturb my circles", [note 4] a reference to the circles in the mathematical drawing that he was studying when disturbed by the Roman soldier.

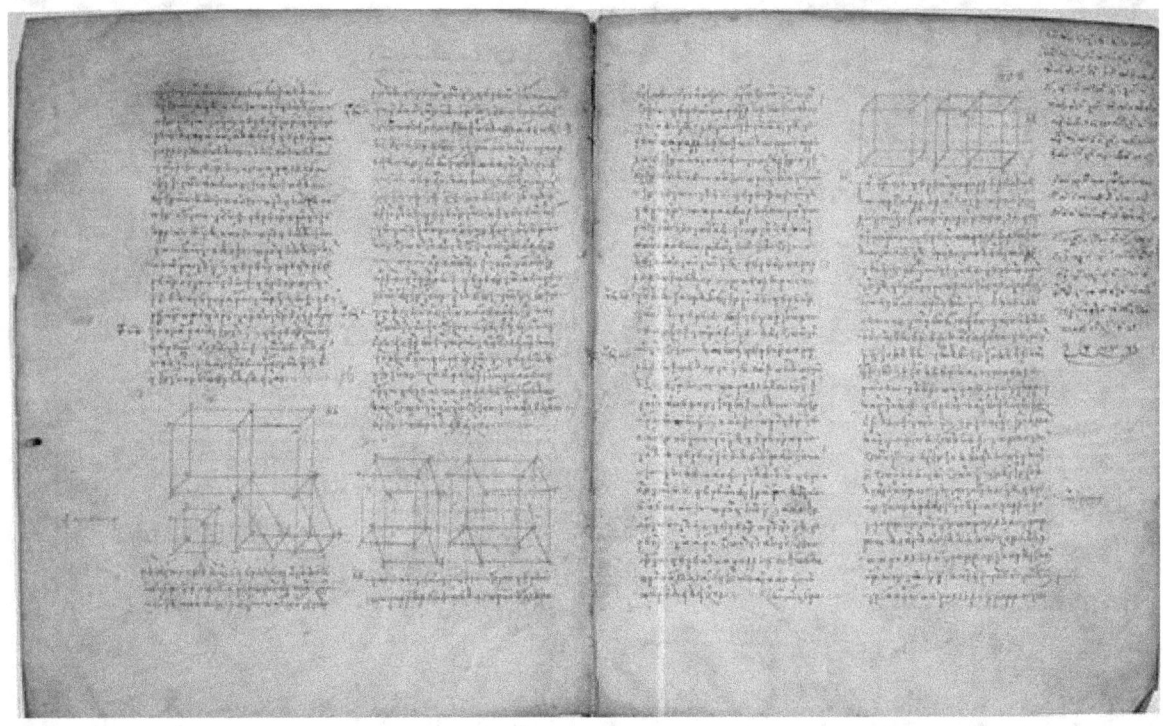

Euclid's Elements ---- The prop. 31, 32 and 33 of the book of Euclid XI, which is located in vol. 2 of the manuscript, the sheets 207 to - 208 recto.

The numbers 0–9 in Chinese huāmǎ (花码) numerals

Counting rod numerals

Modern artist's impression of Shen Kuo.

72

```
4  +  5        Wilt du das wyſ-
4 ——— 17       ſen oder deßgley-
3  +  30       chen/ So ſumier
4 ——— 19       die zenetner vnd
3  +  44       lb vnnd was auß
3  +  22       —iſt/das iſt mi-
```
zentner 3 ——— 11 lb nus dz ſetz beſon-
```
3  +  50        der vnnd werden
4 ——— 16        4 5 3 9 lb ( So
3  +  44         du die zendtner
3  +  29         zů lb gemachett
3 —+— 12         haſt vnnd das /
3  +  9       +  das iſt meer
```

darzů Addiereſt) vnd 75 minus. Nun
ſolc du für Holtz abſchlahen allweeg für
ain legel 24 lb. Vnd das iſt 13 mal 24.
vnd macht 312 lb darzů addier das ——
das iſt 75 lb vnd werden 387. Dye ſub-
trahier von 4539. Vnd Bleyben 4152
lb. Nun ſprich 100 lb das iſt ein zentner
pro 4 fl ⅛ wie kumien 4152 lb vnd kumē
171 fl 5 ß 4 heller₃ Vñ iſt recht gmacht

Pfeffer

Leonhard Euler's signature

James Clerk Maxwell ---- Maxwell's most prominent achievement was to formulate a set of equations that united previously unrelated observations, experiments, and equations of electricity, magnetism, and optics into a consistent theory.[82]

Albert Einstein in 1921

Paul Dirac, pictured here, made fundamental contributions to the early development of both quantum mechanics and quantum electro-dynamics.

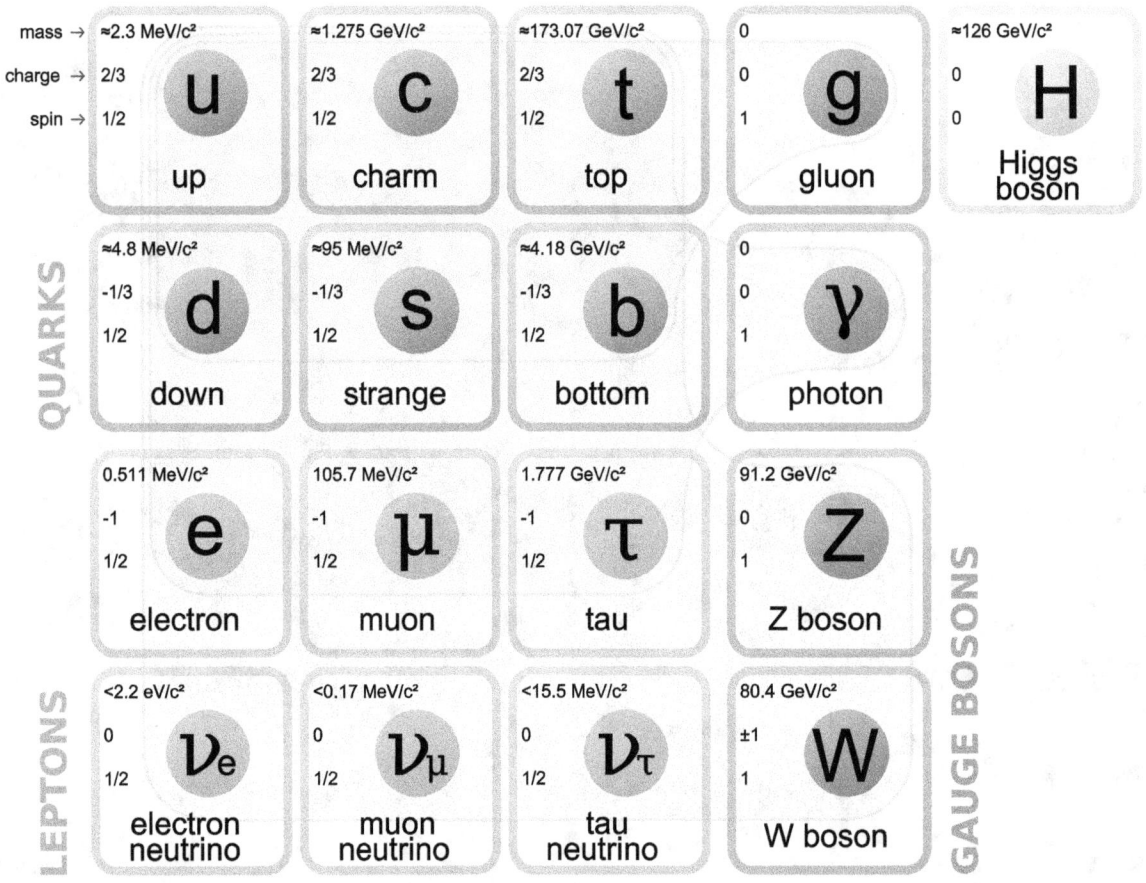

Standard model of elementary particles. ---- The fundamental fermions and the fundamental bosons. (c.2008)[note 101] Based on the proprietary publication, Review of Particle Physics.[note 102]

John H Conway, prolific mathematician of notation.

A section of a quintic Calabi–Yau three-fold (3D projection); recalling atomic vortex theory.

Chapter 8

Number theory

Not to be confused with Numerology.

Number theory (or **arithmetic**[note 1]) is a branch of pure mathematics devoted primarily to the study of the integers. It is sometimes called "The Queen of Mathematics" because of its foundational place in the discipline.[1] Number theorists study prime numbers as well as the properties of objects made out of integers (e.g., rational numbers) or defined as generalizations of the integers (e.g., algebraic integers).

Integers can be considered either in themselves or as solutions to equations (Diophantine geometry). Questions in number theory are often best understood through the study of analytical objects (e.g., the Riemann zeta function) that encode properties of the integers, primes or other number-theoretic objects in some fashion (analytic number theory). One may also study real numbers in relation to rational numbers, e.g., as approximated by the latter (Diophantine approximation).

The older term for number theory is *arithmetic*. By the early twentieth century, it had been superseded by "number theory".[note 2] (The word "arithmetic" is used by the general public to mean "elementary calculations"; it has also acquired other meanings in mathematical logic, as in *Peano arithmetic*, and computer science, as in *floating point arithmetic*.) The use of the term *arithmetic* for *number theory* regained some ground in the second half of the 20th century, arguably in part due to French influence.[note 3] In particular, *arithmetical* is preferred as an adjective to *number-theoretic*.

8.1 History

8.1.1 Origins

Dawn of arithmetic

The first historical find of an arithmetical nature is a fragment of a table: the broken clay tablet Plimpton 322 (Larsa, Mesopotamia, ca. 1800 BCE) contains a list of "Pythagorean triples", i.e., integers (a,b,c) such that $a^2+b^2=c^2$. The triples are too many and too large to have been obtained by brute force. The heading over the first column reads: "The *takiltum* of the diagonal which has been subtracted such that the width..."[2]

The table's layout suggests[3] that it was constructed by means of what amounts, in modern language, to the identity

$$\left(\tfrac{1}{2}\left(x-\tfrac{1}{x}\right)\right)^2 + 1 = \left(\tfrac{1}{2}\left(x+\tfrac{1}{x}\right)\right)^2,$$

which is implicit in routine Old Babylonian exercises.[4] If some other method was used,[5] the triples were first constructed and then reordered by c/a, presumably for actual use as a "table", i.e., with a view to applications.

It is not known what these applications may have been, or whether there could have been any; Babylonian astronomy, for example, truly flowered only later. It has been suggested instead that the table was a source of numerical examples for school problems.[6][note 4]

While Babylonian number theory—or what survives of Babylonian mathematics that can be called thus—consists of this single, striking fragment, Babylonian algebra (in the secondary-school sense of "algebra") was exceptionally well developed.[7] Late Neoplatonic sources[8] state that Pythagoras learned mathematics from the Babylonians. Much earlier sources[9] state that Thales and Pythagoras traveled and studied in Egypt.

Euclid IX 21—34 is very probably Pythagorean;[10] it is very simple material ("odd times even is even", "if an odd number measures [= divides] an even number, then it also measures [= divides] half of it"), but it is all that is needed to prove that $\sqrt{2}$ is irrational.[11] Pythagorean mystics gave great importance to the odd and the even.[12] The discovery that $\sqrt{2}$ is irrational is credited to the early Pythagoreans (pre-Theodorus).[13] By revealing (in modern terms) that numbers could be irrational, this discovery seems to have provoked the first foundational crisis in mathematical history; its proof or its divulgation are sometimes credited to Hippasus, who was expelled or split from the Pythagorean sect.[14] This forced a distinction between *numbers* (integers and the rationals—the subjects of arithmetic), on the one hand, and *lengths* and *proportions* (which we would identify with real numbers, whether rational or not), on the other hand.

The Pythagorean tradition spoke also of so-called polygonal or figurate numbers.[15] While square numbers, cubic numbers, etc., are seen now as more natural than triangular numbers, pentagonal numbers, etc., the study of the sums of triangular and pentagonal numbers would prove fruitful in the early modern period (17th to early 19th century).

We know of no clearly arithmetical material in ancient Egyptian or Vedic sources, though there is some algebra in both. The Chinese remainder theorem appears as an exercise [16] in Sun Zi's *Suan Ching*, also known as *The Mathematical Classic of Sun Zi* (3rd, 4th or 5th century CE.)[17] (There is one important step glossed over in Sun Zi's solution:[note 5] it is the problem that was later solved by Āryabhaṭa's kuṭṭaka – see below.)

There is also some numerical mysticism in Chinese mathematics,[note 6] but, unlike that of the Pythagoreans, it seems to have led nowhere. Like the Pythagoreans' perfect numbers, magic squares have passed from superstition into recreation.

Classical Greece and the early Hellenistic period

Aside from a few fragments, the mathematics of Classical Greece is known to us either through the reports of contemporary non-mathematicians or through mathematical works from the early Hellenistic period.[18] In the case of number theory, this means, by and large, *Plato* and *Euclid*, respectively.

Plato had a keen interest in mathematics, and distinguished clearly between arithmetic and calculation. (By *arithmetic* he meant, in part, theorising on number, rather than what *arithmetic* or *number theory* have come to mean.) It is through one of Plato's dialogues—namely, *Theaetetus*—that we know that Theodorus had proven that $\sqrt{3}, \sqrt{5}, ..., \sqrt{17}$ are irrational. Theaetetus was, like Plato, a disciple of Theodorus's; he worked on distinguishing different kinds of incommensurables, and was thus arguably a pioneer in the study of number systems. (Book X of Euclid's Elements is described by Pappus as being largely based on Theaetetus's work.)

Euclid devoted part of his *Elements* to prime numbers and divisibility, topics that belong unambiguously to number theory and are basic to it (Books VII to IX of Euclid's Elements). In particular, he gave an algorithm for computing the greatest common divisor of two numbers (the Euclidean algorithm; *Elements*, Prop. VII.2) and the first known proof of the infinitude of primes (*Elements*, Prop. IX.20).

In 1773, Lessing published an epigram he had found in a manuscript during his work as a librarian; it claimed to be a letter sent by Archimedes to Eratosthenes.[19][20] The epigram proposed what has become known as Archimedes' cattle problem; its solution (absent from the manuscript) requires solving an indeterminate quadratic equation (which reduces to what would later be misnamed Pell's equation). As far as we know, such equations were first successfully treated by the Indian school. It is not known whether Archimedes himself had a method of solution.

Diophantus

Very little is known about Diophantus of Alexandria; he probably lived in the third century CE, that is, about five hundred years after Euclid. Six out of the thirteen books of Diophantus's *Arithmetica* survive in the original Greek; four more books survive in an Arabic translation. The *Arithmetica* is a collection of worked-out problems where the task is invariably to find rational solutions to a system of polynomial equations, usually of the form $f(x,y)=z^2$ or $f(x,y,z)=w^2$. Thus, nowadays, we speak of *Diophantine equations* when we speak of polynomial equations to which rational or integer solutions must be found.

One may say that Diophantus was studying rational points — i.e., points whose coordinates are rational — on curves and algebraic varieties; however, unlike the Greeks of the Classical period, who did what we would now call basic algebra in geometrical terms, Diophantus did what we would now call basic algebraic geometry in purely algebraic terms. In modern language, what Diophantus did was to find rational parametrizations of varieties; that is, given an equation of the form (say) $f(x_1,x_2,x_3)=0$, his aim was to find (in essence) three rational functions g_1,g_2,g_3 such that, for all values of r and s , setting $x_i=g_i(r,s)$ for $i=1,2,3$ gives a solution to $f(x_1,x_2,x_3)=0$.

Diophantus also studied the equations of some non-rational curves, for which no rational parametrisation is possible. He managed to find some rational points on these curves (elliptic curves, as it happens, in what seems to be their first known occurrence) by means of what amounts to a tangent construction: translated into coordinate geometry (which did not exist in Diophantus's time), his method would be visualised as drawing a tangent to a curve at a known rational point, and then finding the other point of intersection of the tangent with the curve; that other point is a new rational point. (Diophantus also resorted to what could be called a special case of a secant construction.)

While Diophantus was concerned largely with rational solutions, he assumed some results on integer numbers, in particular that every integer is the sum of four squares (though he never stated as much explicitly).

Āryabhaṭa, Brahmagupta, Bhāskara

While Greek astronomy probably influenced Indian learning, to the point of introducing trigonometry,[21] it seems to be the case that Indian mathematics is otherwise an indigenous tradition;[22] in particular, there is no evidence that Euclid's Elements reached India before the 18th century.[23]

Āryabhaṭa (476–550 CE) showed that pairs of simultaneous congruences $n\equiv a_1 \pmod{m}_1$, $n\equiv a_2 \pmod{m}_2$ could be solved by a method he called *kuṭṭaka*, or *pulveriser*;[24] this is a procedure close to (a generalisation of) the Euclidean algorithm, which was probably discovered independently in India.[25] Āryabhaṭa seems to have had in mind applications to astronomical calculations.[21]

Brahmagupta (628 CE) started the systematic study of indefinite quadratic equations—in particular, the misnamed Pell equation, in which Archimedes may have first been interested, and which did not start to be solved in the West until the time of Fermat and Euler. Later Sanskrit authors would follow, using Brahmagupta's technical terminology. A general procedure (the chakravala, or "cyclic method") for solving Pell's equation was finally found by Jayadeva (cited in the eleventh century; his work is otherwise lost); the earliest surviving exposition appears in Bhāskara II's Bīja-gaṇita (twelfth century).[26]

Unfortunately, Indian mathematics remained largely unknown in the West until the late eighteenth century;[27] Brahmagupta and Bhāskara's work was translated into English in 1817 by Henry Colebrooke.[28]

Arithmetic in the Islamic golden age

In the early ninth century, the caliph Al-Ma'mun ordered translations of many Greek mathematical works and at least one Sanskrit work (the *Sindhind*, which may [29] or may not[30] be Brahmagupta's Brāhmasphuṭasiddhānta). Diophantus's main work, the *Arithmetica*, was translated into Arabic by Qusta ibn Luqa (820–912). Part of the treatise *al-Fakhri* (by al-Karajī, 953 – ca. 1029) builds on it to some extent. According to Rashed Roshdi, Al-Karajī's contemporary Ibn al-Haytham knew[31] what would later be called Wilson's theorem.

Western Europe in the Middle Ages

Other than a treatise on squares in arithmetic progression by Fibonacci — who lived and studied in north Africa and Constantinople during his formative years, ca. 1175–1200 — no number theory to speak of was done in western Europe during the Middle Ages. Matters started to change in Europe in the late Renaissance, thanks to a renewed study of the works of Greek antiquity. A catalyst was the textual emendation and translation into Latin of Diophantus's *Arithmetica* (Bachet, 1621, following a first attempt by Xylander, 1575).

8.1.2 Early modern number theory

Fermat

Pierre de Fermat (1601–1665) never published his writings; in particular, his work on number theory is contained almost entirely in letters to mathematicians and in private marginal notes.[32] He wrote down nearly no proofs in number theory; he had no models in the area.[33] He did make repeated use of mathematical induction, introducing the method of infinite descent.

One of Fermat's first interests was perfect numbers (which appear in Euclid, *Elements* IX) and amicable numbers;[note 7] this led him to work on integer divisors, which were from the beginning among the subjects of the correspondence (1636 onwards) that put him in touch with the mathematical community of the day.[34] He had already studied Bachet's edition of Diophantus carefully;[35] by 1643, his interests had shifted largely to Diophantine problems and sums of squares[36] (also treated by Diophantus).

Fermat's achievements in arithmetic include:

- Fermat's little theorem (1640),[37] stating that, if a is not divisible by a prime p, then $a^{p-1} \equiv 1 \pmod{p}$. [note 8]

- If a and b are coprime, then $a^2 + b^2$ is not divisible by any prime congruent to -1 modulo 4;[38] *and* Every prime congruent to 1 modulo 4 can be written in the form $a^2 + b^2$.[39] These two statements also date from 1640; in 1659, Fermat stated to Huygens that he had proven the latter statement by the method of infinite descent.[40] Fermat and Frenicle also did some work (some of it erroneous)[41] on other quadratic forms.

- Fermat posed the problem of solving $x^2 - Ny^2 = 1$ as a challenge to English mathematicians (1657). The problem was solved in a few months by Wallis and Brouncker.[42] Fermat considered their solution valid, but pointed out they had provided an algorithm without a proof (as had Jayadeva and Bhaskara, though Fermat would never know this.) He states that a proof can be found by descent.

- Fermat developed methods for (doing what in our terms amounts to) finding points on curves of genus 0 and 1. As in Diophantus, there are many special procedures and what amounts to a tangent construction, but no use of a secant construction.[43]

- Fermat states and proves (by descent) in the appendix to *Observations on Diophantus* (Obs. XLV)[44] that $x^4 + y^4 = z^4$ has no non-trivial solutions in the integers. Fermat also mentioned to his correspondents that $x^3 + y^3 = z^3$ has no non-trivial solutions, and that this could be proven by descent.[45] The first known proof is due to Euler (1753; indeed by descent).[46]

Fermat's claim ("Fermat's last theorem") to have shown there are no solutions to $x^n + y^n = z^n$ for all $n \geq 3$ (a fact the only known proof of which is beyond his methods) appears only in his annotations on the margin of his copy of Diophantus; he never claimed this to others[47] and thus would have had no need to retract it if he found any mistake in his supposed proof.

Euler

The interest of Leonhard Euler (1707–1783) in number theory was first spurred in 1729, when a friend of his, the amateur[note 9] Goldbach, pointed him towards some of Fermat's work on the subject.[48][49] This has been called the "rebirth" of modern number theory,[35] after Fermat's relative lack of success in getting his contemporaries' attention for the subject.[50] Euler's work on number theory includes the following:[51]

- *Proofs for Fermat's statements.* This includes Fermat's little theorem (generalised by Euler to non-prime moduli); the fact that $p = x^2 + y^2$ if and only if $p \equiv 1 \bmod 4$; initial work towards a proof that every integer is the sum of four squares (the first complete proof is by Joseph-Louis Lagrange (1770), soon improved by Euler himself[52]); the lack of non-zero integer solutions to $x^4 + y^4 = z^2$ (implying the case $n=4$ of Fermat's last theorem, the case $n=3$ of which Euler also proved by a related method).

- *Pell's equation*, first misnamed by Euler.[53] He wrote on the link between continued fractions and Pell's equation.[54]

- *First steps towards analytic number theory.* In his work of sums of four squares, partitions, pentagonal numbers, and the distribution of prime numbers, Euler pioneered the use of what can be seen as analysis (in particular, infinite series) in number theory. Since he lived before the development of complex analysis, most of his work is restricted to the formal manipulation of power series. He did, however, do some very notable (though not fully rigorous) early work on what would later be called the Riemann zeta function.[55]

- *Quadratic forms.* Following Fermat's lead, Euler did further research on the question of which primes can be expressed in the form x^2+Ny^2 , some of it prefiguring quadratic reciprocity.[56] [57][58]

- *Diophantine equations.* Euler worked on some Diophantine equations of genus 0 and 1.[59][60] In particular, he studied Diophantus's work; he tried to systematise it, but the time was not yet ripe for such an endeavour – algebraic geometry was still in its infancy.[61] He did notice there was a connection between Diophantine problems and elliptic integrals,[61] whose study he had himself initiated.

Lagrange, Legendre and Gauss

Joseph-Louis Lagrange (1736–1813) was the first to give full proofs of some of Fermat's and Euler's work and observations - for instance, the four-square theorem and the basic theory of the misnamed "Pell's equation" (for which an algorithmic solution was found by Fermat and his contemporaries, and also by Jayadeva and Bhaskara II before them.) He also studied quadratic forms in full generality (as opposed to mX^2+nY^2) — defining their equivalence relation, showing how to put them in reduced form, etc.

Adrien-Marie Legendre (1752–1833) was the first to state the law of quadratic reciprocity. He also conjectured what amounts to the prime number theorem and Dirichlet's theorem on arithmetic progressions. He gave a full treatment of the equation $ax^2+by^2+cz^2=0$ [62] and worked on quadratic forms along the lines later developed fully by Gauss.[63] In his old age, he was the first to prove "Fermat's last theorem" for $n = 5$ (completing work by Peter Gustav Lejeune Dirichlet, and crediting both him and Sophie Germain).[64]

In his *Disquisitiones Arithmeticae* (1798), Carl Friedrich Gauss (1777–1855) proved the law of quadratic reciprocity and developed the theory of quadratic forms (in particular, defining their composition). He also introduced some basic notation (congruences) and devoted a section to computational matters, including primality tests.[65] The last section of the *Disquisitiones* established a link between roots of unity and number theory:

> The theory of the division of the circle...which is treated in sec. 7 does not belong by itself to arithmetic, but its principles can only be drawn from higher arithmetic.[66]

In this way, Gauss arguably made a first foray towards both Évariste Galois's work and algebraic number theory.

8.1.3 Maturity and division into subfields

Starting early in the nineteenth century, the following developments gradually took place:

- The rise to self-consciousness of number theory (or *higher arithmetic*) as a field of study.[67]

- The development of much of modern mathematics necessary for basic modern number theory: complex analysis, group theory, Galois theory—accompanied by greater rigor in analysis and abstraction in algebra.

- The rough subdivision of number theory into its modern subfields—in particular, analytic and algebraic number theory.

Algebraic number theory may be said to start with the study of reciprocity and cyclotomy, but truly came into its own with the development of abstract algebra and early ideal theory and valuation theory; see below. A conventional starting point for analytic number theory is Dirichlet's theorem on arithmetic progressions (1837),[68] [69] whose proof introduced

L-functions and involved some asymptotic analysis and a limiting process on a real variable.[70] The first use of analytic ideas in number theory actually goes back to Euler (1730s),[71] [72] who used formal power series and non-rigorous (or implicit) limiting arguments. The use of *complex* analysis in number theory comes later: the work of Bernhard Riemann (1859) on the zeta function is the canonical starting point;[73] Jacobi's four-square theorem (1839), which predates it, belongs to an initially different strand that has by now taken a leading role in analytic number theory (modular forms).[74]

The history of each subfield is briefly addressed in its own section below; see the main article of each subfield for fuller treatments. Many of the most interesting questions in each area remain open and are being actively worked on.

8.2 Main subdivisions

8.2.1 Elementary tools

The term *elementary* generally denotes a method that does not use complex analysis. For example, the prime number theorem was first proven using complex analysis in 1896, but an elementary proof was found only in 1949 by Erdős and Selberg.[75] The term is somewhat ambiguous: for example, proofs based on complex Tauberian theorems (e.g. Wiener–Ikehara) are often seen as quite enlightening but not elementary, in spite of using Fourier analysis, rather than complex analysis as such. Here as elsewhere, an *elementary* proof may be longer and more difficult for most readers than a non-elementary one.

Number theory has the reputation of being a field many of whose results can be stated to the layperson. At the same time, the proofs of these results are not particularly accessible, in part because the range of tools they use is, if anything, unusually broad within mathematics.[76]

8.2.2 Analytic number theory

Main article: Analytic number theory
Analytic number theory may be defined

- in terms of its tools, as the study of the integers by means of tools from real and complex analysis;[68] or

- in terms of its concerns, as the study within number theory of estimates on size and density, as opposed to identities.[77]

Some subjects generally considered to be part of analytic number theory, e.g., sieve theory,[note 10] are better covered by the second rather than the first definition: some of sieve theory, for instance, uses little analysis,[note 11] yet it does belong to analytic number theory.

The following are examples of problems in analytic number theory: the prime number theorem, the Goldbach conjecture (or the twin prime conjecture, or the Hardy–Littlewood conjectures), the Waring problem and the Riemann Hypothesis. Some of the most important tools of analytic number theory are the circle method, sieve methods and L-functions (or, rather, the study of their properties). The theory of modular forms (and, more generally, automorphic forms) also occupies an increasingly central place in the toolbox of analytic number theory.[78]

One may ask analytic questions about algebraic numbers, and use analytic means to answer such questions; it is thus that algebraic and analytic number theory intersect. For example, one may define prime ideals (generalizations of prime numbers in the field of algebraic numbers) and ask how many prime ideals there are up to a certain size. This question can be answered by means of an examination of Dedekind zeta functions, which are generalizations of the Riemann zeta function, a key analytic object at the roots of the subject.[79] This is an example of a general procedure in analytic number theory: deriving information about the distribution of a sequence (here, prime ideals or prime numbers) from the analytic behavior of an appropriately constructed complex-valued function.[80]

8.2.3 Algebraic number theory

Main article: Algebraic number theory

An *algebraic number* is any complex number that is a solution to some polynomial equation $f(x) = 0$ with rational coefficients; for example, every solution x of $x^5 + (11/2)x^3 - 7x^2 + 9 = 0$ (say) is an algebraic number. Fields of algebraic numbers are also called *algebraic number fields*, or shortly *number fields*. Algebraic number theory studies algebraic number fields.[81] Thus, analytic and algebraic number theory can and do overlap: the former is defined by its methods, the latter by its objects of study.

It could be argued that the simplest kind of number fields (viz., quadratic fields) were already studied by Gauss, as the discussion of quadratic forms in *Disquisitiones arithmeticae* can be restated in terms of ideals and norms in quadratic fields. (A *quadratic field* consists of all numbers of the form $a + b\sqrt{d}$, where a and b are rational numbers and d is a fixed rational number whose square root is not rational.) For that matter, the 11th-century chakravala method amounts—in modern terms—to an algorithm for finding the units of a real quadratic number field. However, neither Bhāskara nor Gauss knew of number fields as such.

The grounds of the subject as we know it were set in the late nineteenth century, when *ideal numbers*, the *theory of ideals* and *valuation theory* were developed; these are three complementary ways of dealing with the lack of unique factorisation in algebraic number fields. (For example, in the field generated by the rationals and $\sqrt{-5}$, the number 6 can be factorised both as $6 = 2 \cdot 3$ and $6 = (1 + \sqrt{-5})(1 - \sqrt{-5})$; all of 2, 3, $1 + \sqrt{-5}$ and $1 - \sqrt{-5}$ are irreducible, and thus, in a naïve sense, analogous to primes among the integers.) The initial impetus for the development of ideal numbers (by Kummer) seems to have come from the study of higher reciprocity laws,[82] i.e., generalisations of quadratic reciprocity.

Number fields are often studied as extensions of smaller number fields: a field L is said to be an *extension* of a field K if L contains K. (For example, the complex numbers C are an extension of the reals R, and the reals R are an extension of the rationals Q.) Classifying the possible extensions of a given number field is a difficult and partially open problem. Abelian extensions—that is, extensions L of K such that the Galois group[note 12] Gal(L/K) of L over K is an abelian group—are relatively well understood. Their classification was the object of the programme of class field theory, which was initiated in the late 19th century (partly by Kronecker and Eisenstein) and carried out largely in 1900—1950.

An example of an active area of research in algebraic number theory is Iwasawa theory. The Langlands program, one of the main current large-scale research plans in mathematics, is sometimes described as an attempt to generalise class field theory to non-abelian extensions of number fields.

8.2.4 Diophantine geometry

Main articles: Diophantine geometry and Glossary of arithmetic and Diophantine geometry

The central problem of *Diophantine geometry* is to determine when a Diophantine equation has solutions, and if it does, how many. The approach taken is to think of the solutions of an equation as a geometric object.

For example, an equation in two variables defines a curve in the plane. More generally, an equation, or system of equations, in two or more variables defines a curve, a surface or some other such object in n-dimensional space. In Diophantine geometry, one asks whether there are any *rational points* (points all of whose coordinates are rationals) or *integral points* (points all of whose coordinates are integers) on the curve or surface. If there are any such points, the next step is to ask how many there are and how they are distributed. A basic question in this direction is: are there finitely or infinitely many rational points on a given curve (or surface)? What about integer points?

An example here may be helpful. Consider the Pythagorean equation $x^2 + y^2 = 1$; we would like to study its rational solutions, i.e., its solutions (x, y) such that x and y are both rational. This is the same as asking for all integer solutions to $a^2 + b^2 = c^2$; any solution to the latter equation gives us a solution $x = a/c$, $y = b/c$ to the former. It is also the same as asking for all points with rational coordinates on the curve described by $x^2 + y^2 = 1$. (This curve happens to be a circle of radius 1 around the origin.)

The rephrasing of questions on equations in terms of points on curves turns out to be felicitous. The finiteness or not of the number of rational or integer points on an algebraic curve—that is, rational or integer solutions to an equation $f(x,y) = 0$, where f is a polynomial in two variables—turns out to depend crucially on the *genus* of the curve. The *genus* can be defined as follows:[note 13] allow the variables in $f(x,y) = 0$ to be complex numbers; then $f(x,y) = 0$ defines a 2-dimensional surface in (projective) 4-dimensional space (since two complex variables can be decomposed into four real variables, i.e., four dimensions). Count the number of (doughnut) holes in the surface; call this number the *genus* of $f(x,y) = 0$. Other geometrical notions turn out to be just as crucial.

There is also the closely linked area of Diophantine approximations: given a number x, how well can it be approximated by rationals? (We are looking for approximations that are good relative to the amount of space that it takes to write the rational: call a/q (with $\gcd(a,q) = 1$) a good approximation to x if $|x-a/q| < \frac{1}{q^c}$, where c is large.) This question is of special interest if x is an algebraic number. If x cannot be well approximated, then some equations do not have integer or rational solutions. Moreover, several concepts (especially that of height) turn out to be crucial both in Diophantine geometry and in the study of Diophantine approximations. This question is also of special interest in transcendence theory: if a number can be better approximated than any algebraic number, then it is a transcendental number. It is by this argument that π and e have been shown to be transcendental.

Diophantine geometry should not be confused with the geometry of numbers, which is a collection of graphical methods for answering certain questions in algebraic number theory. *Arithmetic geometry*, on the other hand, is a contemporary term for much the same domain as that covered by the term *Diophantine geometry*. The term *arithmetic geometry* is arguably used most often when one wishes to emphasise the connections to modern algebraic geometry (as in, for instance, Faltings' theorem) rather than to techniques in Diophantine approximations.

8.3 Recent approaches and subfields

The areas below date as such from no earlier than the mid-twentieth century, even if they are based on older material. For example, as is explained below, the matter of algorithms in number theory is very old, in some sense older than the concept of proof; at the same time, the modern study of computability dates only from the 1930s and 1940s, and computational complexity theory from the 1970s.

8.3.1 Probabilistic number theory

Main article: Probabilistic number theory

Take a number at random between one and a million. How likely is it to be prime? This is just another way of asking how many primes there are between one and a million. Further: how many prime divisors will it have, on average? How many divisors will it have altogether, and with what likelihood? What is the probability that it have many more or many fewer divisors or prime divisors than the average?

Much of probabilistic number theory can be seen as an important special case of the study of variables that are almost, but not quite, mutually independent. For example, the event that a random integer between one and a million be divisible by two and the event that it be divisible by three are almost independent, but not quite.

It is sometimes said that probabilistic combinatorics uses the fact that whatever happens with probability greater than 0 must happen sometimes; one may say with equal justice that many applications of probabilistic number theory hinge on the fact that whatever is unusual must be rare. If certain algebraic objects (say, rational or integer solutions to certain equations) can be shown to be in the tail of certain sensibly defined distributions, it follows that there must be few of them; this is a very concrete non-probabilistic statement following from a probabilistic one.

At times, a non-rigorous, probabilistic approach leads to a number of heuristic algorithms and open problems, notably Cramér's conjecture.

8.3.2 Arithmetic combinatorics

Main articles: Arithmetic combinatorics and Additive number theory

Let A be a set of N integers. Consider the set $A + A = \{\, m + n \mid m, n \in A \,\}$ consisting of all sums of two elements of A. Is $A + A$ much larger than A? Barely larger? If $A + A$ is barely larger than A, must A have plenty of arithmetic structure, for example, does A resemble an arithmetic progression?

If we begin from a fairly "thick" infinite set A, does it contain many elements in arithmetic progression: a, $a + b$, $a + 2b$, $a + 3b$, ..., $a + 10b$, say? Should it be possible to write large integers as sums of elements of A?

These questions are characteristic of *arithmetic combinatorics*. This is a presently coalescing field; it subsumes *additive number theory* (which concerns itself with certain very specific sets A of arithmetic significance, such as the primes or the squares) and, arguably, some of the *geometry of numbers*, together with some rapidly developing new material. Its focus on issues of growth and distribution accounts in part for its developing links with ergodic theory, finite group theory, model theory, and other fields. The term *additive combinatorics* is also used; however, the sets A being studied need not be sets of integers, but rather subsets of non-commutative groups, for which the multiplication symbol, not the addition symbol, is traditionally used; they can also be subsets of rings, in which case the growth of $A + A$ and $A \cdot A$ may be compared.

8.3.3 Computations in number theory

Main article: Computational number theory

While the word *algorithm* goes back only to certain readers of al-Khwārizmī, careful descriptions of methods of solution are older than proofs: such methods (that is, algorithms) are as old as any recognisable mathematics—ancient Egyptian, Babylonian, Vedic, Chinese—whereas proofs appeared only with the Greeks of the classical period. An interesting early case is that of what we now call the Euclidean algorithm. In its basic form (namely, as an algorithm for computing the greatest common divisor) it appears as Proposition 2 of Book VII in *Elements*, together with a proof of correctness. However, in the form that is often used in number theory (namely, as an algorithm for finding integer solutions to an equation $ax+by=c$, or, what is the same, for finding the quantities whose existence is assured by the Chinese remainder theorem) it first appears in the works of Āryabhaṭa (5th–6th century CE) as an algorithm called *kuṭṭaka* ("pulveriser"), without a proof of correctness.

There are two main questions: "can we compute this?" and "can we compute it rapidly?". Anybody can test whether a number is prime or, if it is not, split it into prime factors; doing so rapidly is another matter. We now know fast algorithms for testing primality, but, in spite of much work (both theoretical and practical), no truly fast algorithm for factoring.

The difficulty of a computation can be useful: modern protocols for encrypting messages (e.g., RSA) depend on functions that are known to all, but whose inverses (a) are known only to a chosen few, and (b) would take one too long a time to figure out on one's own. For example, these functions can be such that their inverses can be computed only if certain large integers are factorized. While many difficult computational problems outside number theory are known, most working encryption protocols nowadays are based on the difficulty of a few number-theoretical problems.

On a different note — some things may not be computable at all; in fact, this can be proven in some instances. For instance, in 1970, it was proven, as a solution to Hilbert's 10th problem, that there is no Turing machine which can solve all Diophantine equations.[83] In particular, this means that, given a computably enumerable set of axioms, there are Diophantine equations for which there is no proof, starting from the axioms, of whether the set of equations has or does not have integer solutions. (We would necessarily be speaking of Diophantine equations for which there are no integer solutions, since, given a Diophantine equation with at least one solution, the solution itself provides a proof of the fact that a solution exists. We cannot prove, of course, that a particular Diophantine equation is of this kind, since this would imply that it has no solutions.)

8.4 Applications

The number-theorist Leonard Dickson (1874-1954) said "Thank God that number theory is unsullied by any application". Such a view is no longer applicable to number theory.[84] In 1974, Donald Knuth said "...virtually every theorem in elementary number theory arises in a natural, motivated way in connection with the problem of making computers do high-speed numerical calculations".[85] Elementary number theory is taught in discrete mathematics courses for computer scientists; and, on the other hand, number theory also has applications to the continuous in numerical analysis.[86] As well as the well-known applications to cryptography, there are also applications to many other areas of mathematics.[87][88]

8.5 Literature

Two of the most popular introductions to the subject are:

- G. H. Hardy; E. M. Wright (2008) [1938]. *An introduction to the theory of numbers* (rev. by D. R. Heath-Brown and J. H. Silverman, 6th ed.). Oxford University Press. ISBN 978-0-19-921986-5.

- Vinogradov, I. M. (2003) [1954]. *Elements of Number Theory* (reprint of the 1954 ed.). Mineola, NY: Dover Publications.

Hardy and Wright's book is a comprehensive classic, though its clarity sometimes suffers due to the authors' insistence on elementary methods.[89] Vinogradov's main attraction consists in its set of problems, which quickly lead to Vinogradov's own research interests; the text itself is very basic and close to minimal. Other popular first introductions are:

- Ivan M. Niven; Herbert S. Zuckerman; Hugh L. Montgomery (2008) [1960]. *An introduction to the theory of numbers* (reprint of the 5th edition 1991 ed.). John Wiley & Sons. ISBN 978-8-12-651811-1.

- Kenneth H. Rosen (2010). *Elementary Number Theory* (6th ed.). Pearson Education. ISBN 978-0-32-171775-7.

Popular choices for a second textbook include:

- Borevich, A. I.; Shafarevich, Igor R. (1966). *Number theory*. Pure and Applied Mathematics **20**. Boston, MA: Academic Press. ISBN 978-0-12-117850-5. MR 0195803.

- Serre, Jean-Pierre (1996) [1973]. *A course in arithmetic*. Graduate texts in mathematics **7**. Springer. ISBN 978-0-387-90040-7.

8.6 Prizes

The American Mathematical Society awards the *Cole Prize in Number Theory*. Moreover number theory is one of the three mathematical subdisciplines rewarded by the *Fermat Prize*.

8.7 See also

- Algebraic function field
- Finite field
- p-adic number

8.8   Notes

[1] Especially in older sources; see two following notes.

[2] Already in 1921, T. L. Heath had to explain: "By arithmetic, Plato meant, not arithmetic in our sense, but the science which considers numbers in themselves, in other words, what we mean by the Theory of Numbers." (Heath 1921, p. 13)

[3] Take, e.g. Serre 1973. In 1952, Davenport still had to specify that he meant *The Higher Arithmetic*. Hardy and Wright wrote in the introduction to *An Introduction to the Theory of Numbers* (1938): "We proposed at one time to change [the title] to *An introduction to arithmetic*, a more novel and in some ways a more appropriate title; but it was pointed out that this might lead to misunderstandings about the content of the book." (Hardy & Wright 2008)

[4] Robson 2001, p. 201. This is controversial. See Plimpton 322. Robson's article is written polemically (Robson 2001, p. 202) with a view to "perhaps [...] knocking [Plimpton 322] off its pedestal" (Robson 2001, p. 167); at the same time, it settles to the conclusion that

> [...] the question "how was the tablet calculated?" does not have to have the same answer as the question "what problems does the tablet set?" The first can be answered most satisfactorily by reciprocal pairs, as first suggested half a century ago, and the second by some sort of right-triangle problems (Robson 2001, p. 202).

Robson takes issue with the notion that the scribe who produced Plimpton 322 (who had to "work for a living", and would not have belonged to a "leisured middle class") could have been motivated by his own "idle curiosity" in the absence of a "market for new mathematics".(Robson 2001, pp. 199–200)

[5] Sun Zi, *Suan Ching*, Ch. 3, Problem 26, in Lam & Ang 2004, pp. 219–220:

> [26] Now there are an unknown number of things. If we count by threes, there is a remainder 2; if we count by fives, there is a remainder 3; if we count by sevens, there is a remainder 2. Find the number of things. *Answer*: 23.
> *Method*: If we count by threes and there is a remainder 2, put down 140. If we count by fives and there is a remainder 3, put down 63. If we count by sevens and there is a remainder 2, put down 30. Add them to obtain 233 and subtract 210 to get the answer. If we count by threes and there is a remainder 1, put down 70. If we count by fives and there is a remainder 1, put down 21. If we count by sevens and there is a remainder 1, put down 15. When [a number] exceeds 106, the result is obtained by subtracting 105.

[6] See, e.g., Sun Zi, *Suan Ching*, Ch. 3, Problem 36, in Lam & Ang 2004, pp. 223–224:

> [36] Now there is a pregnant woman whose age is 29. If the gestation period is 9 months, determine the sex of the unborn child. *Answer*: Male.
> *Method*: Put down 49, add the gestation period and subtract the age. From the remainder take away 1 representing the heaven, 2 the earth, 3 the man, 4 the four seasons, 5 the five phases, 6 the six pitch-pipes, 7 the seven stars [of the Dipper], 8 the eight winds, and 9 the nine divisions [of China under Yu the Great]. If the remainder is odd, [the sex] is male and if the remainder is even, [the sex] is female.

This is the last problem in Sun Zi's otherwise matter-of-fact treatise.

[7] Perfect and especially amicable numbers are of little or no interest nowadays. The same was not true in medieval times – whether in the West or the Arab-speaking world – due in part to the importance given to them by the Neopythagorean (and hence mystical) Nicomachus (ca. 100 CE), who wrote a primitive but influential "Introduction to Arithmetic". See van der Waerden 1961, Ch. IV.

[8] Here, as usual, given two integers a and b and a non-zero integer m, we write $a \equiv b \pmod{m}$ (read "a is congruent to b modulo m") to mean that m divides $a - b$, or, what is the same, a and b leave the same residue when divided by m. This notation is actually much later than Fermat's; it first appears in section 1 of Gauss's Disquisitiones Arithmeticae. Fermat's little theorem is a consequence of the fact that the order of an element of a group divides the order of the group. The modern proof would have been within Fermat's means (and was indeed given later by Euler), even though the modern concept of a group came long after Fermat or Euler. (It helps to know that inverses exist modulo p (i.e., given a not divisible by a prime p, there is an integer x such that $xa \equiv 1 \pmod{p}$); this fact (which, in modern language, makes the residues mod p into a group, and which was already known to Āryabhaṭa; see above) was familiar to Fermat thanks to its rediscovery by Bachet (Weil 1984, p. 7). Weil goes on to say that Fermat would have recognised that Bachet's argument is essentially Euclid's algorithm.

[9] Up to the second half of the seventeenth century, academic positions were very rare, and most mathematicians and scientists earned their living in some other way (Weil 1984, pp. 159, 161). (There were already some recognisable features of professional *practice*, viz., seeking correspondents, visiting foreign colleagues, building private libraries (Weil 1984, pp. 160–161). Matters started to shift in the late 17th century (Weil 1984, p. 161); scientific academies were founded in England (the Royal Society, 1662) and France (the Académie des sciences, 1666) and Russia (1724). Euler was offered a position at this last one in 1726; he accepted, arriving in St. Petersburg in 1727 (Weil 1984, p. 163 and Varadarajan 2006, p. 7). In this context, the term *amateur* usually applied to Goldbach is well-defined and makes some sense: he has been described as a man of letters who earned a living as a spy (Truesdell 1984, p. xv); cited in Varadarajan 2006, p. 9). Notice, however, that Goldbach published some works on mathematics and sometimes held academic positions.

[10] Sieve theory figures as one of the main subareas of analytic number theory in many standard treatments; see, for instance, Iwaniec & Kowalski 2004 or Montgomery & Vaughan 2007

[11] This is the case for small sieves (in particular, some combinatorial sieves such as the Brun sieve) rather than for large sieves; the study of the latter now includes ideas from harmonic and functional analysis.

[12] The Galois group of an extension *K/L* consists of the operations (isomorphisms) that send elements of L to other elements of L while leaving all elements of K fixed. Thus, for instance, *Gal(C/R)* consists of two elements: the identity element (taking every element $x + iy$ of C to itself) and complex conjugation (the map taking each element $x + iy$ to $x - iy$). The Galois group of an extension tells us many of its crucial properties. The study of Galois groups started with Évariste Galois; in modern language, the main outcome of his work is that an equation $f(x) = 0$ can be solved by radicals (that is, x can be expressed in terms of the four basic operations together with square roots, cubic roots, etc.) if and only if the extension of the rationals by the roots of the equation $f(x) = 0$ has a Galois group that is solvable in the sense of group theory. ("Solvable", in the sense of group theory, is a simple property that can be checked easily for finite groups.)

[13] It may be useful to look at an example here. Say we want to study the curve $y^2 = x^3 + 7$. We allow x and y to be complex numbers: $(a + bi)^2 = (c + di)^3 + 7$. This is, in effect, a set of two equations on four variables, since both the real and the imaginary part on each side must match. As a result, we get a surface (two-dimensional) in four-dimensional space. After we choose a convenient hyperplane on which to project the surface (meaning that, say, we choose to ignore the coordinate a), we can plot the resulting projection, which is a surface in ordinary three-dimensional space. It then becomes clear that the result is a torus, i.e., the surface of a doughnut (somewhat stretched). A doughnut has one hole; hence the genus is 1.

8.9 References

[1] Long 1972, p. 1.

[2] Neugebauer & Sachs 1945, p. 40. The term *takiltum* is problematic. Robson prefers the rendering "The holding-square of the diagonal from which 1 is torn out, so that the short side comes up...".Robson 2001, p. 192

[3] Robson 2001, p. 189. Other sources give the modern formula $(p^2-q^2, 2pq, p^2+q^2)$. Van der Waerden gives both the modern formula and what amounts to the form preferred by Robson.(van der Waerden 1961, p. 79)

[4] van der Waerden 1961, p. 184.

[5] Neugebauer (Neugebauer 1969, pp. 36–40) discusses the table in detail and mentions in passing Euclid's method in modern notation (Neugebauer 1969, p. 39).

[6] Friberg 1981, p. 302.

[7] van der Waerden 1961, p. 43.

[8] Iamblichus, *Life of Pythagoras*,(trans. e.g. Guthrie 1987) cited in van der Waerden 1961, p. 108. See also Porphyry, *Life of Pythagoras*, paragraph 6, in Guthrie 1987 Van der Waerden (van der Waerden 1961, pp. 87–90) sustains the view that Thales knew Babylonian mathematics.

[9] Herodotus (II. 81) and Isocrates (*Busiris* 28), cited in: Huffman 2011. On Thales, see Eudemus ap. Proclus, 65.7, (e.g. Morrow 1992, p. 52) cited in: O'Grady 2004, p. 1. Proclus was using a work by Eudemus of Rhodes (now lost), the *Catalogue of Geometers*. See also introduction, Morrow 1992, p. xxx on Proclus' reliability.

[10] Becker 1936, p. 533, cited in: van der Waerden 1961, p. 108.

[11] Becker 1936.

[12] van der Waerden 1961, p. 109.

[13] Plato, *Theaetetus*, p. 147 B, (e.g. Jowett 1871), cited in von Fritz 2004, p. 212: "Theodorus was writing out for us something about roots, such as the roots of three or five, showing that they are incommensurable by the unit;..." *See also* Spiral of Theodorus.

[14] von Fritz 2004.

[15] Heath 1921, p. 76.

[16] Sun Zi, *Suan Ching*, Chapter 3, Problem 26. This can be found in Lam & Ang 2004, pp. 219–220, which contains a full translation of the *Suan Ching* (based on Qian 1963). See also the discussion in Lam & Ang 2004, pp. 138–140.

[17] The date of the text has been narrowed down to 220–420 AD (Yan Dunjie) or 280–473 AD (Wang Ling) through internal evidence (= taxation systems assumed in the text). See Lam & Ang 2004, pp. 27–28.

[18] Boyer & Merzbach 1991, p. 82.

[19] Vardi 1998, p. 305-319.

[20] Weil 1984, pp. 17–24.

[21] Plofker 2008, p. 119.

[22] Any early contact between Babylonian and Indian mathematics remains conjectural (Plofker 2008, p. 42).

[23] Mumford 2010, p. 387.

[24] Āryabhaṭa, Āryabhatīya, Chapter 2, verses 32–33, cited in: Plofker 2008, pp. 134–140. See also Clark 1930, pp. 42–50. A slightly more explicit description of the kuṭṭaka was later given in Brahmagupta, *Brāhmasphuṭasiddhānta*, XVIII, 3–5 (in Colebrooke 1817, p. 325, cited in Clark 1930, p. 42).

[25] Mumford 2010, p. 388.

[26] Plofker 2008, p. 194.

[27] Plofker 2008, p. 283.

[28] Colebrooke 1817.

[29] Colebrooke 1817, p. lxv, cited in Hopkins 1990, p. 302. See also the preface in Sachau 1888 cited in Smith 1958, pp. 168

[30] Pingree 1968, pp. 97–125, and Pingree 1970, pp. 103–123, cited in Plofker 2008, p. 256.

[31] Rashed 1980, p. 305–321.

[32] Weil 1984, pp. 45–46.

[33] Weil 1984, p. 118. This was more so in number theory than in other areas (remark in Mahoney 1994, p. 284). Bachet's own proofs were "ludicrously clumsy" (Weil 1984, p. 33).

[34] Mahoney 1994, pp. 48, 53–54. The initial subjects of Fermat's correspondence included divisors ("aliquot parts") and many subjects outside number theory; see the list in the letter from Fermat to Roberval, 22.IX.1636, Tannery & Henry 1891, Vol. II, pp. 72, 74, cited in Mahoney 1994, p. 54.

[35] Weil 1984, pp. 1–2.

[36] Weil 1984, p. 53.

[37] Tannery & Henry 1891, Vol. II, p. 209, Letter XLVI from Fermat to Frenicle, 1640, cited in Weil 1984, p. 56

[38] Tannery & Henry 1891, Vol. II, p. 204, cited in Weil 1984, p. 63. All of the following citations from Fermat's *Varia Opera* are taken from Weil 1984, Chap. II. The standard Tannery & Henry work includes a revision of Fermat's posthumous *Varia Opera Mathematica* originally prepared by his son (Fermat 1679).

[39] Tannery & Henry 1891, Vol. II, p. 213.

[40] Tannery & Henry 1891, Vol. II, p. 423.

[41] Weil 1984, pp. 80, 91–92.

[42] Weil 1984, p. 92.

[43] Weil 1984, Ch. II, sect. XV and XVI.

[44] Tannery & Henry 1891, Vol. I, pp. 340–341.

[45] Weil 1984, p. 115.

[46] Weil 1984, pp. 115–116.

[47] Weil 1984, p. 104.

[48] Weil 1984, pp. 2, 172.

[49] Varadarajan 2006, p. 9.

[50] Weil 1984, p. 2 and Varadarajan 2006, p. 37

[51] Varadarajan 2006, p. 39 and Weil 1984, pp. 176–189

[52] Weil 1984, pp. 178–179.

[53] Weil 1984, p. 174. Euler was generous in giving credit to others (Varadarajan 2006, p. 14), not always correctly.

[54] Weil 1984, p. 183.

[55] Varadarajan 2006, pp. 45–55; see also chapter III.

[56] Varadarajan 2006, pp. 44–47.

[57] Weil 1984, pp. 177–179.

[58] Edwards 1983, pp. 285–291.

[59] Varadarajan 2006, pp. 55–56.

[60] Weil 1984, pp. 179–181.

[61] Weil 1984, p. 181.

[62] Weil 1984, pp. 327–328.

[63] Weil 1984, pp. 332–334.

[64] Weil 1984, pp. 337–338.

[65] Goldstein & Schappacher 2007, p. 14.

[66] From the preface of *Disquisitiones Arithmeticae*; the translation is taken from Goldstein & Schappacher 2007, p. 16

[67] See the discussion in section 5 of Goldstein & Schappacher 2007. Early signs of self-consciousness are present already in letters by Fermat: thus his remarks on what number theory is, and how "Diophantus's work [...] does not really belong to [it]" (quoted in Weil 1984, p. 25).

[68] Apostol 1976, p. 7.

[69] Davenport & Montgomery 2000, p. 1.

[70] See the proof in Davenport & Montgomery 2000, section 1

[71] Iwaniec & Kowalski 2004, p. 1.

[72] Varadarajan 2006, sections 2.5, 3.1 and 6.1.

[73] Granville 2008, pp. 322–348.

[74] See the comment on the importance of modularity in Iwaniec & Kowalski 2004, p. 1

[75] Goldfeld 2003.

[76] See, e.g., the initial comment in Iwaniec & Kowalski 2004, p. 1.

[77] Granville 2008, section 1: "The main difference is that in algebraic number theory [...] one typically considers questions with answers that are given by exact formulas, whereas in analytic number theory [...] one looks for *good approximations*."

[78] See the remarks in the introduction to Iwaniec & Kowalski 2004, p. 1: "However much stronger...".

[79] Granville 2008, section 3: "[Riemann] defined what we now call the Riemann zeta function [...] Riemann's deep work gave birth to our subject [...]"

[80] See, e.g., Montgomery & Vaughan 2007, p. 1.

[81] CITEREFMilne2014, p. 2.

[82] Edwards 2000, p. 79.

[83] Davis, Martin; Matiyasevich, Yuri; Robinson, Julia (1976). "Hilbert's Tenth Problem: Diophantine Equations: Positive Aspects of a Negative Solution". In Felix E. Browder. *Mathematical Developments Arising from Hilbert Problems.* Proceedings of Symposia in Pure Mathematics. XXVIII.2. American Mathematical Society. pp. 323–378. ISBN 0-8218-1428-1. Zbl 0346.02026. Reprinted in *The Collected Works of Julia Robinson*, Solomon Feferman, editor, pp.269–378, American Mathematical Society 1996.

[84] "The Unreasonable Effectiveness of Number Theory", Stefan Andrus Burr, George E. Andrews, American Mathematical Soc., 1992, ISBN 9780821855010

[85] Computer science and its relation to mathematics" DE Knuth - The American Mathematical Monthly, 1974

[86] "Applications of number theory to numerical analysis", Lo-keng Hua, Luogeng Hua, Yuan Wang, Springer-Verlag, 1981, ISBN 978-3-540-10382-0

[87] "Practical applications of algebraic number theory". Mathoverflow.net. Retrieved 2012-05-18.

[88] "Where is number theory used in the rest of mathematics?". Mathoverflow.net. 2008-09-23. Retrieved 2012-05-18.

[89] Apostol n.d..

8.10   Sources

- Apostol, Tom M. (1976). *Introduction to analytic number theory*. Undergraduate Texts in Mathematics. Springer. ISBN 978-0-387-90163-3.

- Apostol, Tom M. (n.d.). "An Introduction to the Theory of Numbers". (Review of Hardy & Wright.) Mathematical Reviews (MathSciNet) MR0568909. American Mathematical Society. (Subscription needed)

- Becker, Oskar (1936). "Die Lehre von Geraden und Ungeraden im neunten Buch der euklidischen Elemente". *Quellen und Studien zur Geschichte der Mathematik, Astronomie und Physik.* Abteilung B:Studien (in German) (Berlin: J. Springer Verlag) **3**: 533–53.

- Boyer, Carl Benjamin; Merzbach, Uta C. (1991) [1968]. *A History of Mathematics* (2nd ed.). New York: Wiley. ISBN 978-0-471-54397-8. 1968 edition at archive.org

- Clark, Walter Eugene (trans.) (1930). *The Āryabhaṭīya of Āryabhaṭa: An ancient Indian work on Mathematics and Astronomy.* University of Chicago Press.

- Colebrooke, Henry Thomas (1817). *Algebra, with Arithmetic and Mensuration, from the Sanscrit of Brahmegupta and Bháscara.* London: J. Murray.

- Davenport, Harold; Montgomery, Hugh L. (2000). *Multiplicative Number Theory*. Graduate texts in mathematics **74** (revised 3rd ed.). Springer. ISBN 978-0-387-95097-6.

- Edwards, Harold M. (November 1983). "Euler and Quadratic Reciprocity". *Mathematics Magazine* (Mathematical Association of America) **56** (5): 285–291. doi:10.2307/2690368. JSTOR 2690368.

- Edwards, Harold M. (2000) [1977]. *Fermat's Last Theorem: a Genetic Introduction to Algebraic Number Theory*. Graduate Texts in Mathematics **50** (reprint of 1977 ed.). Springer Verlag. ISBN 978-0-387-95002-0.

- Fermat, Pierre de (1679). *Varia Opera Mathematica* (in French and Latin). Toulouse: Joannis Pech.

- Friberg, Jöran (August 1981). "Methods and Traditions of Babylonian Mathematics: Plimpton 322, Pythagorean Triples and the Babylonian Triangle Parameter Equations". *Historia Mathematica* (Elsevier) **8** (3): 277–318. doi:10.1016/0315-0860(81)90069-0.

- von Fritz, Kurt (2004). "The Discovery of Incommensurability by Hippasus of Metapontum". In Christianidis, J. *Classics in the History of Greek Mathematics*. Berlin: Kluwer (Springer). ISBN 978-1-4020-0081-2.

- Gauss, Carl Friedrich; Waterhouse, William C. (trans.) (1966) [1801]. *Disquisitiones Arithmeticae*. Springer. ISBN 978-0-387-96254-2.

- Goldfeld, Dorian M. (2003). "Elementary Proof of the Prime Number Theorem: a Historical Perspective" (PDF).

- Goldstein, Catherine; Schappacher, Norbert (2007). "A book in search of a discipline". In Goldstein, C.; Schappacher, N.; Schwermer, Joachim. *The Shaping of Arithmetic after Gauss' "Disquisitiones Arithmeticae"*. Berlin & Heidelberg: Springer. pp. 3–66. ISBN 978-3-540-20441-1.

- Granville, Andrew (2008). "Analytic number theory". In Gowers, Timothy; Barrow-Green, June; Leader, Imre. *The Princeton Companion to Mathematics*. Princeton University Press. ISBN 978-0-691-11880-2.

- Porphyry; Guthrie, K. S. (trans.) (1920). *Life of Pythagoras*. Alpine, New Jersey: Platonist Press.

- Guthrie, Kenneth Sylvan (1987). *The Pythagorean Sourcebook and Library*. Grand Rapids, Michigan: Phanes Press. ISBN 978-0-933999-51-0.

- Hardy, Godfrey Harold; Wright, E. M. (2008) [1938]. *An Introduction to the Theory of Numbers* (Sixth ed.). Oxford University Press. ISBN 978-0-19-921986-5. MR 2445243.

- Heath, Thomas L. (1921). *A History of Greek Mathematics, Volume 1: From Thales to Euclid*. Oxford: Clarendon Press.

- Hopkins, J. F. P. (1990). "Geographical and Navigational Literature". In Young, M. J. L.; Latham, J. D.; Serjeant, R. B. *Religion, Learning and Science in the `Abbasid Period*. The Cambridge history of Arabic literature. Cambridge University Press. ISBN 978-0-521-32763-3.

- Huffman, Carl A. (8 August 2011). Zalta, Edward N., ed. "Pythagoras". *Stanford Encyclopaedia of Philosophy* (Fall 2011 ed.). Retrieved 7 February 2012.

- Iwaniec, Henryk; Kowalski, Emmanuel (2004). *Analytic Number Theory*. American Mathematical Society Colloquium Publications **53**. Providence, RI,: American Mathematical Society. ISBN 0-8218-3633-1.

- Plato; Jowett, Benjamin (trans.) (1871). *Theaetetus*.

- Lam, Lay Yong; Ang, Tian Se (2004). *Fleeting Footsteps: Tracing the Conception of Arithmetic and Algebra in Ancient China* (revised ed.). Singapore: World Scientific. ISBN 978-981-238-696-0.

- Long, Calvin T. (1972). *Elementary Introduction to Number Theory* (2nd ed.). Lexington, VA: D. C. Heath and Company. LCCN 77171950.

- Mahoney, M. S. (1994). *The Mathematical Career of Pierre de Fermat, 1601–1665* (Reprint, 2nd ed.). Princeton University Press. ISBN 978-0-691-03666-3.

- Milne, J. S. (2014). "Algebraic Number Theory". Available at www.jmilne.org/math.

- Montgomery, Hugh L.; Vaughan, Robert C. (2007). *Multiplicative Number Theory: I, Classical Theory,*. Cambridge University Press. ISBN 978-0-521-84903-6.

- Morrow, Glenn Raymond (trans., ed.); Proclus (1992). *A Commentary on Book 1 of Euclid's Elements.* Princeton University Press. ISBN 978-0-691-02090-7.

- Mumford, David (March 2010). "Mathematics in India: reviewed by David Mumford" (PDF). *Notices of the American Mathematical Society* **57** (3): 387. ISSN 1088-9477.

- Neugebauer, Otto E. (1969). *The Exact Sciences in Antiquity* (corrected reprint of the 1957 ed.). New York: Dover Publications. ISBN 978-0-486-22332-2.

- Neugebauer, Otto E.; Sachs, Abraham Joseph; Götze, Albrecht (1945). *Mathematical Cuneiform Texts.* American Oriental Series **29**. American Oriental Society etc.

- O'Grady, Patricia (September 2004). "Thales of Miletus". The Internet Encyclopaedia of Philosophy. Retrieved 7 February 2012.

- Pingree, David; Ya'qub, ibn Tariq (1968). "The Fragments of the Works of Ya'qub ibn Tariq". *Journal of Near Eastern Studies* (University of Chicago Press) **26**.

- Pingree, D.; al-Fazari (1970). "The Fragments of the Works of al-Fazari". *Journal of Near Eastern Studies* (University of Chicago Press) **28**.

- Plofker, Kim (2008). *Mathematics in India.* Princeton University Press. ISBN 978-0-691-12067-6.

- Qian, Baocong, ed. (1963). *Suanjing shi shu (Ten Mathematical Classics)* (in Chinese). Beijing: Zhonghua shuju.

- Rashed, Roshdi (1980). "Ibn al-Haytham et le théorème de Wilson". *Archive for History of Exact Sciences* **22** (4): 305–321. doi:10.1007/BF00717654.

- Robson, Eleanor (2001). "Neither Sherlock Holmes nor Babylon: a Reassessment of Plimpton 322" (PDF). *Historia Mathematica* (Elsevier) **28** (28): 167–206. doi:10.1006/hmat.2001.2317.

- Sachau, Eduard; Bīrūni, Muḥammad ibn Aḥmad (1888). *Alberuni's India: An Account of the Religion, Philosophy, Literature, Geography, Chronology, Astronomy and Astrology of India, Vol. 1.* London: Kegan, Paul, Trench, Trübner & Co.

- Serre, Jean-Pierre (1996) [1973]. *A Course in Arithmetic.* Graduate texts in mathematics **7**. Springer. ISBN 978-0-387-90040-7.

- Smith, D. E. (1958). *History of Mathematics, Vol I.* New York: Dover Publications.

- Tannery, Paul; Henry, Charles (eds.); Fermat, Pierre de (1891). *Oeuvres de Fermat.* (4 Vols.) (in French and Latin). Paris: Imprimerie Gauthier-Villars et Fils. Volume 1 Volume 2 Volume 3 Volume 4 (1912)

- Iamblichus; Taylor, Thomas (trans.) (1818). *Life of Pythagoras or, Pythagoric Life.* London: J. M. Watkins. For other editions, see Iamblichus#List of editions and translations

- Truesdell, C. A. (1984). "Leonard Euler, Supreme Geometer". In Hewlett, John (trans.). *Leonard Euler, Elements of Algebra* (reprint of 1840 5th ed.). New York: Springer-Verlag. ISBN 978-0-387-96014-2. This Google books preview of *Elements of algebra* lacks Truesdell's intro, which is reprinted (slightly abridged) in the following book:

- Truesdell, C. A. (2007). "Leonard Euler, Supreme Geometer". In Dunham, William. *The Genius of Euler: reflections on his life and work.* Volume 2 of MAA tercentenary Euler celebration. New York: Mathematical Association of America. ISBN 978-0-88385-558-4.

- Varadarajan, V. S. (2006). *Euler Through Time: A New Look at Old Themes.* American Mathematical Society. ISBN 978-0-8218-3580-7.

- Vardi, Ilan (April 1998). "Archimedes' Cattle Problem" (PDF). *American Mathematical Monthly* **105** (4): 305–319. doi:10.2307/2589706.

- van der Waerden, Bartel L.; Dresden, Arnold (trans) (1961). *Science Awakening.* Vol. 1 or Vol 2. New York: Oxford University Press.

- Weil, André (1984). *Number Theory: an Approach Through History – from Hammurapi to Legendre.* Boston: Birkhäuser. ISBN 978-0-8176-3141-3.

This article incorporates material from the Citizendium article "Number theory", which is licensed under the Creative Commons Attribution-ShareAlike 3.0 Unported License but not under the GFDL.

8.11 External links

- Hazewinkel, Michiel, ed. (2001), "Number theory", *Encyclopedia of Mathematics*, Springer, ISBN 978-1-55608-010-4

- Quotations related to Number theory at Wikiquote

- Number Theory Web

A Lehmer sieve, which is a primitive digital computer once used for finding primes and solving simple Diophantine equations.

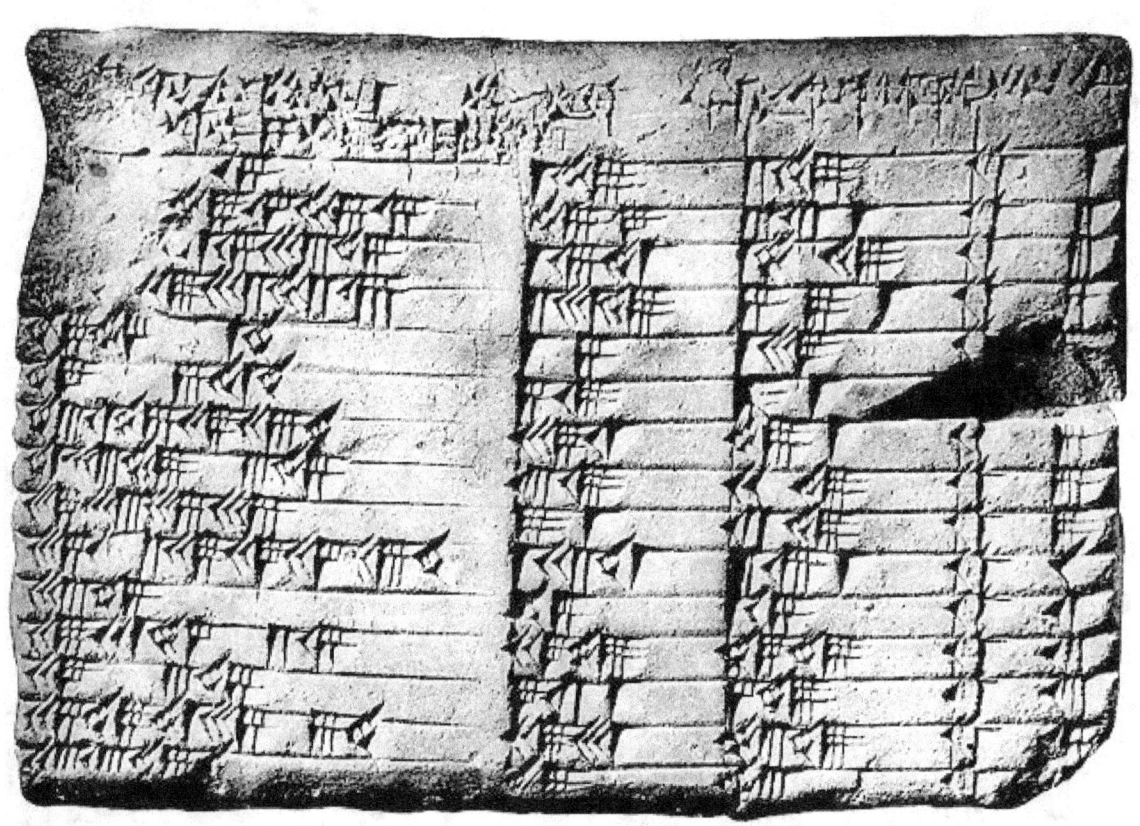

The Plimpton 322 tablet

DIOPHANTI
ALEXANDRINI
ARITHMETICORVM
LIBRI SEX,
ET DE NVMERIS MVLTANGVLIS,
LIBER VNVS.

Nunc primùm Græcè & Latinè editi, atque absolutissimis
Commentariis illustrati.

AVCTORE CLAVDIO GASPARE BACHETO
MEZIRIACO SEBVSIANO, V.C.

LVTETIAE PARISIORVM,
Sumptibus Sebastiani Cramoisy, via
Iacobæa, sub Ciconiis.

M. DC. XXI.
CVM PRIVILEGIO REGIS.

Leonhard Euler

DISQVISITIONES

ARITHMETICAE

AVCTORE

D. CAROLO FRIDERICO GAVSS

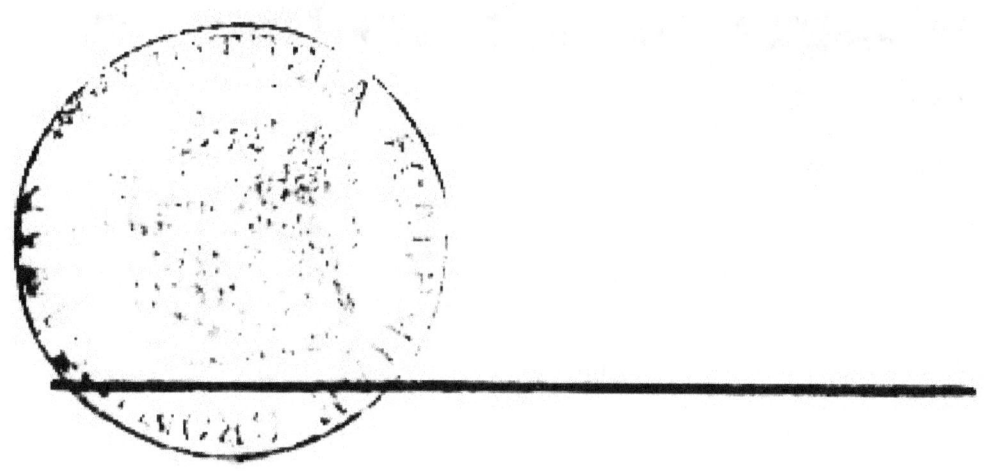

LIPSIAE

IN COMMISSIS APVD GERH. FLEISCHER, JUN.

1801.

Carl Friedrich Gauss

Peter Gustav Lejeune Dirichlet

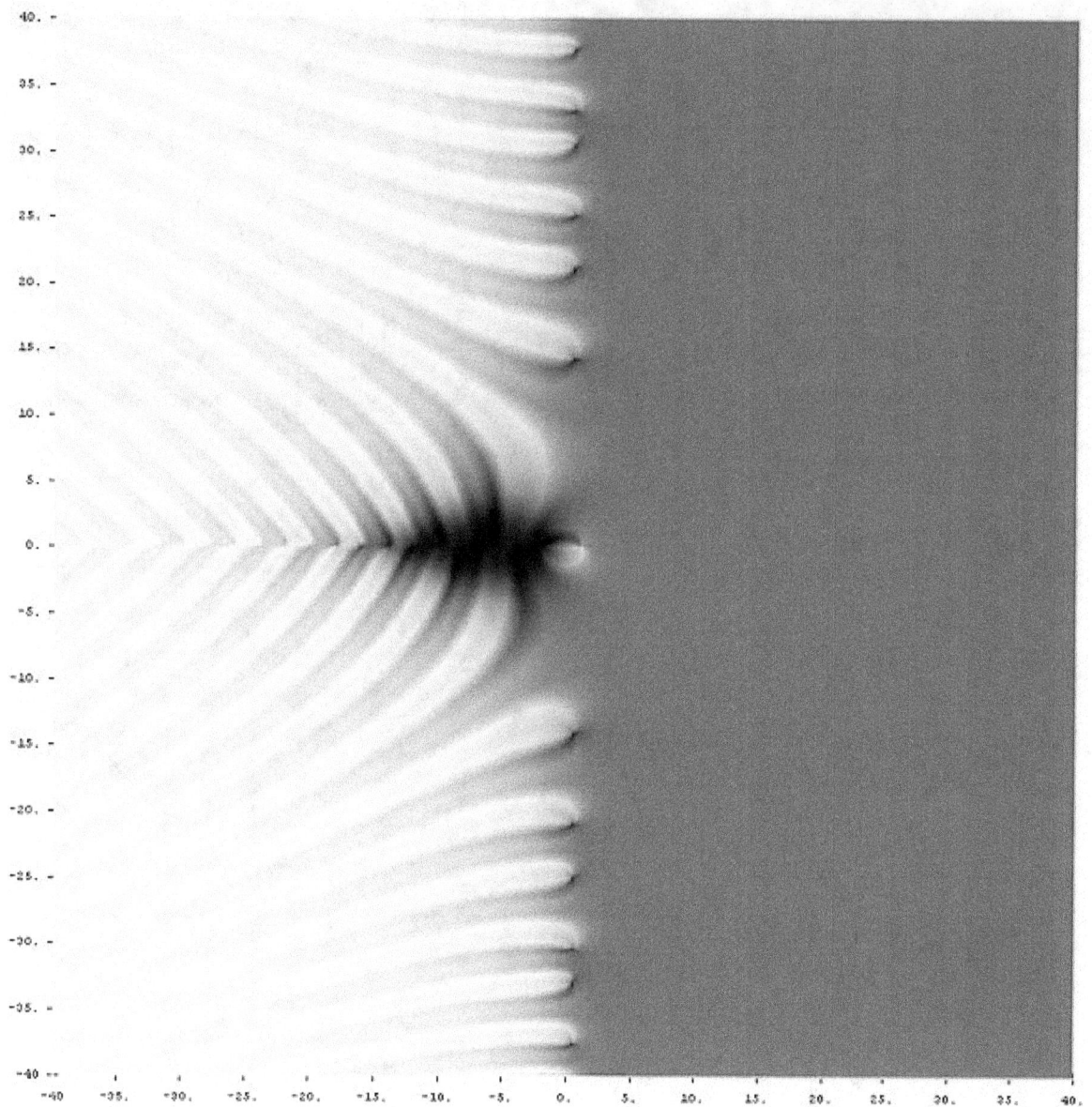

Riemann zeta function ζ(s) in the complex plane. The color of a point s *gives the value of* ζ(s): *dark colors denote values close to zero and hue gives the value's argument.*

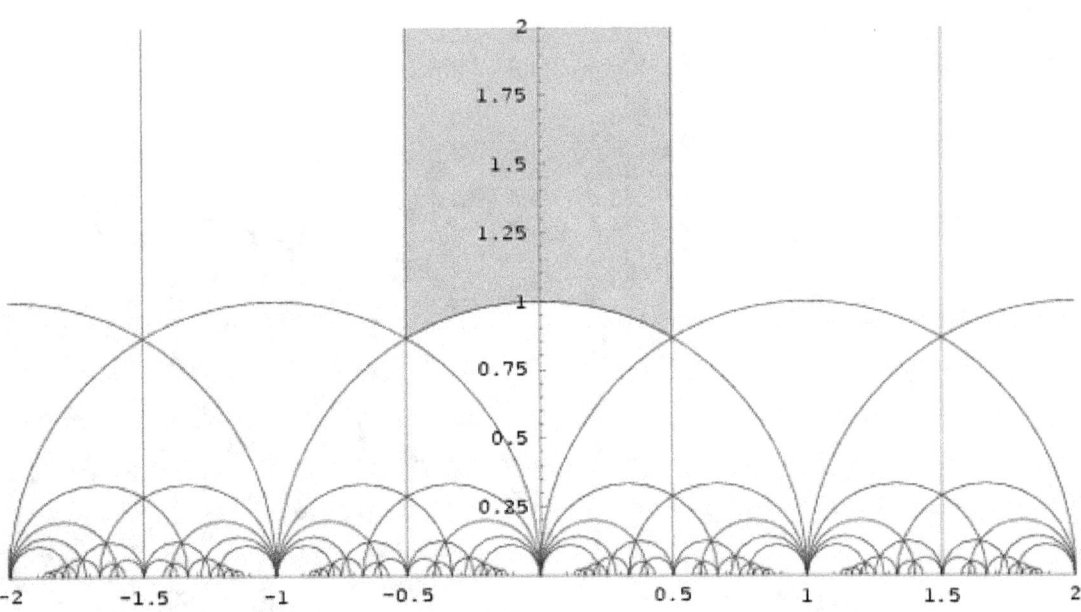

The action of the modular group on the upper half plane. The region in grey is the standard fundamental domain.

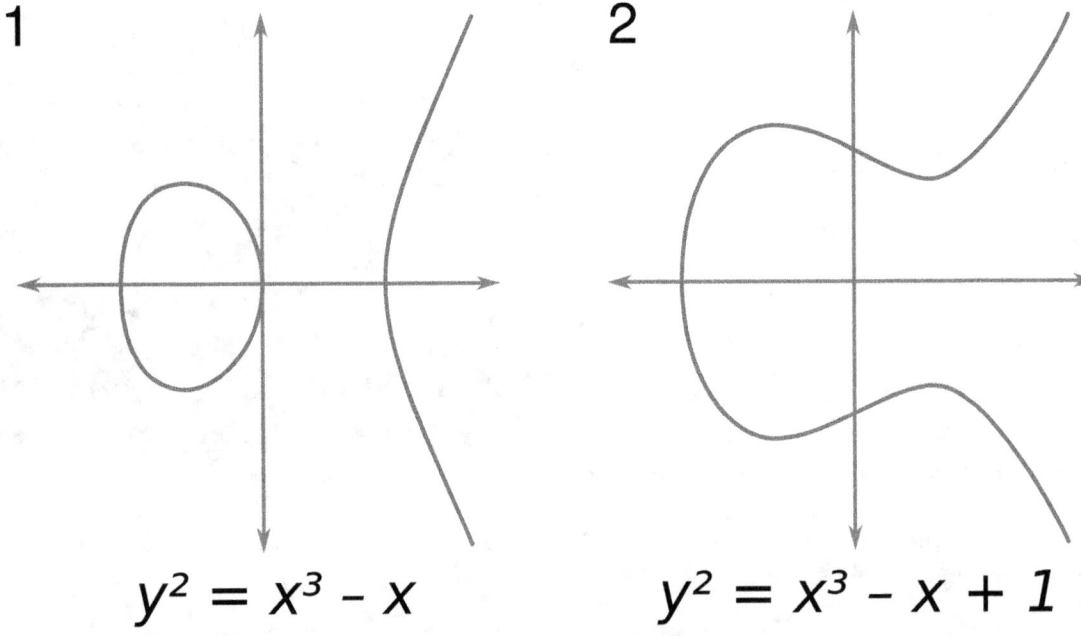

Two examples of an elliptic curve, i.e., a curve of genus 1 having at least one rational point. (Either graph can be seen as a slice of a torus in four-dimensional space.)

Chapter 9

History of statistics

The **History of statistics** can be said to start around 1749 although, over time, there have been changes to the inter-pretation of the word *statistics*. In early times, the meaning was restricted to information about states. This was later extended to include all collections of information of all types, and later still it was extended to include the analysis and interpretation of such data. In modern terms, "statistics" means both sets of collected information, as in national accounts and temperature records, and analytical work which requires statistical inference.

Statistical activities are often associated with models expressed using probabilities, and require probability theory for them to be put on a firm theoretical basis: see History of probability.

A number of statistical concepts have had an important impact on a wide range of sciences. These include the design of experiments and approaches to statistical inference such as Bayesian inference, each of which can be considered to have their own sequence in the development of the ideas underlying modern statistics.

9.1 Introduction

By the 18th century, the term "statistics" designated the systematic collection of demographic and economic data by states. For at least two millennia, these data were mainly tabulations of human and material resources that might be taxed or put to military use. In the early 19th century, collection intensified, and the meaning of "statistics" broadened to include the discipline concerned with the collection, summary, and analysis of data. Today, data are collected and statistics are computed and widely distributed in government, business, most of the sciences and sports, and even for many pastimes. Electronic computers have expedited more elaborate statistical computation even as they have facilitated the collection and aggregation of data. A single data analyst may have available a set of data-files with millions of records, each with dozens or hundreds of separate measurements. These were collected over time from computer activity (for example, a stock exchange) or from computerized sensors, point-of-sale registers, and so on. Computers then produce simple, accurate summaries, and allow more tedious analyses, such as those that require inverting a large matrix or perform hundreds of steps of iteration, that would never be attempted by hand. Faster computing has allowed statisticians to develop "computer-intensive" methods which may look at all permutations, or use randomization to look at 10,000 permutations of a problem, to estimate answers that are not easy to quantify by theory alone.

The term "mathematical statistics" designates the mathematical theories of probability and statistical inference, which are used in statistical practice. The relation between statistics and probability theory developed rather late, however. In the 19th century, statistics increasingly used probability theory, whose initial results were found in the 17th and 18th centuries, particularly in the analysis of games of chance (gambling). By 1800, astronomy used probability models and statistical theories, particularly the method of least squares. Early probability theory and statistics was systematized in the 19th century and statistical reasoning and probability models were used by social scientists to advance the new sciences of experimental psychology and sociology, and by physical scientists in thermodynamics and statistical mechanics. The development of statistical reasoning was closely associated with the development of inductive logic and the scientific method, which are concerns that move statisticians away from the narrower area of mathematical statistics. Much of the theoretical work was readily available by the time computers were available to exploit them. By the 1970s, Johnson and Kotz produced a four-volume Compendium on Statistical Distributions (First Edition 1969-1972), which is still an invaluable resource.

Applied statistics can be regarded as not a field of mathematics but an autonomous mathematical science, like computer science and operations research. Unlike mathematics, statistics had its origins in public administration. Applications arose early in demography and economics; large areas of micro- and macro-economics today are "statistics" with an emphasis on time-series analyses. With its emphasis on learning from data and making best predictions, statistics also has been shaped by areas of academic research including psychological testing, medicine and epidemiology. The ideas of statistical testing have considerable overlap with decision science. With its concerns with searching and effectively presenting data, statistics has overlap with information science and computer science.

9.2 Etymology

*Look up **statistics** in wiktionary, the free dictionary.*

The term *statistics* is ultimately derived from the New Latin *statisticum collegium* ("council of state") and the Italian word *statista* ("statesman" or "politician"). The German *Statistik*, first introduced by Gottfried Achenwall (1749), originally designated the analysis of data about the state, signifying the "science of state" (then called *political arithmetic* in English). It acquired the meaning of the collection and classification of data generally in the early 19th century. It was introduced into English in 1791 by Sir John Sinclair when he published the first of 21 volumes titled *Statistical Account of Scotland*.[1]

Thus, the original principal purpose of *Statistik* was data to be used by governmental and (often centralized) administrative bodies. The collection of data about states and localities continues, largely through national and international statistical services. In particular, censuses provide frequently updated information about the population.

The first book to have 'statistics' in its title was "Contributions to Vital Statistics" (1845) by Francis GP Neison, actuary to the Medical Invalid and General Life Office.

9.3 Origins in probability theory

Main article: History of probability
See also: Timeline of probability and statistics

Basic forms of statistics have been used since the beginning of civilization. Early empires often collated censuses of the population or recorded the trade in various commodities. The Roman Empire was one of the first states to extensively gather data on the size of the empire's population, geographical area and wealth.

The use of statistical methods dates back to least to the 5th century BCE. The historian Thucydides in his *History of the Peloponnesian War* [2] describes how the Athenians calculated the height of the wall of Platea by counting the number of bricks in an unplastered section of the wall sufficiently near them to be able to count them. The count was repeated several times by a number of soldiers. The most frequent value (in modern terminology - the mode) so determined was taken to be the most likely value of the number of bricks. Multiplying this value by the height of the bricks used in the wall allowed the Athenians to determine the height of the ladders necessary to scale the walls.

In the Indian epic - the Mahabharata (Book 3: The Story of Nala) - King Rtuparna estimated the number of fruit and leaves (2095 fruit and 50,000,000 - five crores - leaves) on two great branches of a Vibhitaka tree by counting them on a single twig. This number was then multiplied by the number of twigs on the branches. This estimate was later checked and found to be very close to the actual number. With knowledge of this method Nala was subsequently able to regain his kingdom.

The earliest writing on statistics was found in a 9th-century book entitled: "Manuscript on Deciphering Cryptographic Messages", written by Al-Kindi (801–873 CE). In his book, Al-Kindi gave a detailed description of how to use statistics and frequency analysis to decipher encrypted messages. This text arguably gave rise to the birth of both statistics and cryptanalysis.[3][4]

The Trial of the Pyx is a test of the purity of the coinage of the Royal Mint which has been held on a regular basis since the 12th century. The Trial itself is based on statistical sampling methods. After minting a series of coins - originally from ten pounds of silver - a single coin was placed in the Pyx - a box in Westminster Abbey. After a given period - now once a year - the coins are removed and weighed. A sample of coins removed from the box are then tested for purity.

The *Nuova Cronica*, a 14th-century history of Florence by the Florentine banker and official Giovanni Villani, includes much statistical information on population, ordinances, commerce and trade, education, and religious facilities and has been described as the first introduction of statistics as a positive element in history,[5] though neither the term nor the concept of statistics as a specific field yet existed. But this was proven to be incorrect after the rediscovery of Al-Kindi's book on frequency analysis.[3][4]

The arithmetic mean, although a concept known to the Greeks, was not generalised to more than two values until the 16th century. The invention of the decimal system by Simon Stevin in 1585 seems likely to have facilitated these calculations. This method was first adopted in astronomy by Tycho Brahe who was attempting to reduce the errors in his estimates of the locations of various celestial bodies.

The idea of the median originated in Edward Wright's book on navigation (*Certaine Errors in Navigation*) in 1599 in a section concerning the determination of location with a compass. Wright felt that this value was the most likely to be the correct value in a series of observations.

The birth of statistics is often dated to 1662, when John Graunt, along with William Petty, developed early human statistical and census methods that provided a framework for modern demography. He produced the first life table, giving probabilities of survival to each age. His book *Natural and Political Observations Made upon the Bills of Mortality* used analysis of the mortality rolls to make the first statistically based estimation of the population of London. He knew that there were around 13,000 funerals per year in London and that three people died per eleven families per year. He estimated from the parish records that the average family size was 8 and calculated that the population of London was about 384,000. Laplace in 1802 estimated the population of France with a similar method.

Although the original scope of statistics was limited to data useful for governance, the approach was extended to many fields of a scientific or commercial nature during the 19th century. The mathematical foundations for the subject heavily drew on the new probability theory, pioneered in the 16th century in the correspondence amongst Gerolamo Cardano, Pierre de Fermat and Blaise Pascal. Christiaan Huygens (1657) gave the earliest known scientific treatment of the subject. Jakob Bernoulli's *Ars Conjectandi* (posthumous, 1713) and Abraham de Moivre's *The Doctrine of Chances* (1718) treated the subject as a branch of mathematics. In his book Bernoulli introduced the idea of representing complete certainty as one and probability as a number between zero and one.

The formal study of theory of errors may be traced back to Roger Cotes' *Opera Miscellanea* (posthumous, 1722), but a memoir prepared by Thomas Simpson in 1755 (printed 1756) first applied the theory to the discussion of errors of observation. The reprint (1757) of this memoir lays down the axioms that positive and negative errors are equally probable, and that there are certain assignable limits within which all errors may be supposed to fall; continuous errors are discussed and a probability curve is given. Simpson discussed several possible distributions of error. He first considered the uniform distribution and then the discrete symmetric triangular distribution followed by the continuous symmetric triangle distribution. Tobias Mayer, in his study of the libration of the moon (*Kosmographische Nachrichten*, Nuremberg, 1750), invented the first formal method for estimating the unknown quantities by generalized the averaging of observations under identical circumstances to the averaging of groups of similar equations.

Ruder Boškovic in 1755 based in his work on the shape of the earth proposed in his book *De Litteraria expeditione per pontificiam ditionem ad dimetiendos duos meridiani gradus a PP. Maire et Boscovicli* that the true value of a series of observations would be that which minimises the sum of absolute errors. In modern terminology this value is the median. The first example of what later became known as the normal curve was studied by Abraham de Moivre who plotted this curve on November 12, 1733.[6] de Moivre was studying the number of heads that occurred when a 'fair' coin was tossed.

In 1761 Thomas Bayes proved Bayes' theorem and in 1765 Joseph Priestley invented the first timeline charts.

Johann Heinrich Lambert in his 1765 book *Anlage zur Architectonic* proposed the semicircle as a distribution of errors:

Sir William Petty, a 17th-century economist who used early statistical methods to analyse demographic data.

$$f(x) = \frac{1}{2}\sqrt{(1 - x^2)}$$

with $-1 < x < 1$.

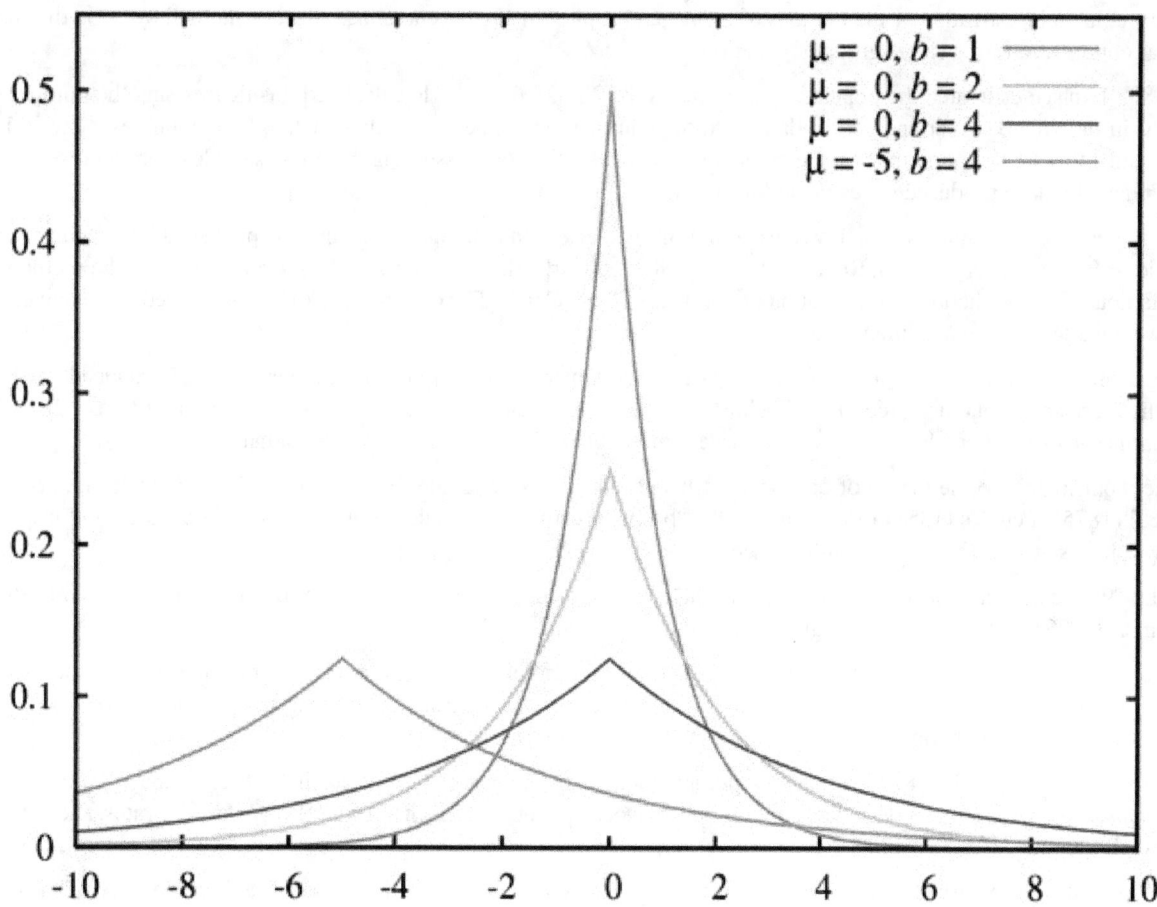

Probability density plots for the Laplace distribution.

Pierre-Simon Laplace (1774) made the first attempt to deduce a rule for the combination of observations from the principles of the theory of probabilities. He represented the law of probability of errors by a curve and deduced a formula for the mean of three observations.

Laplace in 1774 noted that the frequency of an error could be expressed as an exponential function of its magnitude once its sign was disregarded.[7][8] This distribution is now known as the Laplace distribution. Lagrange proposed a parabolic distribution of errors in 1776.

Laplace in 1778 published his second law of errors wherein he noted that the frequency of an error was proportional to the exponential of the square of its magnitude. This was subsequently rediscovered by Gauss (possibly in 1795) and is now best known as the normal distribution which is of central importance in statistics.[9] This distribution was first referred to as the *normal* distribution by Pierce in 1873 who was studying measurement errors when an object was dropped onto a wooden base.[10] He chose the term *normal* because of its frequent occurrence in naturally occurring variables.

Lagrange also suggested in 1781 two other distributions for errors - a Raised cosine distribution and a logarithmic distribution.

Laplace gave (1781) a formula for the law of facility of error (a term due to Joseph Louis Lagrange, 1774), but one which led to unmanageable equations. Daniel Bernoulli (1778) introduced the principle of the maximum product of the probabilities of a system of concurrent errors.

In 1786 William Playfair (1759-1823) introduced the idea of graphical representation into statistics. He invented the line chart, bar chart and histogram and incorporated them into his works on economics, the *Commercial and Political Atlas*. This was followed in 1795 by his invention of the pie chart and circle chart which he used to display the evolution of England's imports and exports. These latter charts came to general attention when he published examples in his *Statistical Breviary* in 1801.

Laplace, in an investigation of the motions of Saturn and Jupiter in 1787, generalized Mayer's method by using different linear combinations of a single group of equations.

In 1802 Laplace estimated the population of France to be 28,328,612.[11] He calculated this figure using the number of births in the previous year and census data for three communities. The census data of these communities showed that they had 2,037,615 persons and that the number of births were 71,866. Assuming that these samples were representative of France, Laplace produced his estimate for the entire population.

The method of least squares, which was used to minimize errors in data measurement, was published independently by Adrien-Marie Legendre (1805), Robert Adrain (1808), and Carl Friedrich Gauss (1809). Gauss had used the method in his famous 1801 prediction of the location of the dwarf planet Ceres. The observations that Gauss based his calculations on were made by the Italian monk Piazzi.

The term *probable error* (*der wahrscheinliche Fehler*) - the median deviation from the mean - was introduced in 1815 by the German astronomer Frederik Wilhelm Bessel. Antoine Augustin Cournot in 1843 was the first to use the term *median* (*valeur médiane*) for the value that divides a probability distribution into two equal halves.

Other contributors to the theory of errors were Ellis (1844), De Morgan (1864), Glaisher (1872), and Giovanni Schiaparelli (1875). Peters's (1856) formula for r , the "probable error" of a single observation was widely used and inspired early robust statistics (resistant to outliers: see Peirce's criterion).

In the 19th century authors on statistical theory included Laplace, S. Lacroix (1816), Littrow (1833), Dedekind (1860), Helmert (1872), Laurent (1873), Liagre, Didion, De Morgan and Boole.

Gustav Theodor Fechner used the median (*Centralwerth*) in sociological and psychological phenomena.[12] It had earlier been used only in astronomy and related fields. Francis Galton used the English term *median* for the first time in 1881 having earlier used the terms *middle-most value* in 1869 and the *medium* in 1880.[13]

Adolphe Quetelet (1796–1874), another important founder of statistics, introduced the notion of the "average man" (*l'homme moyen*) as a means of understanding complex social phenomena such as crime rates, marriage rates, and suicide rates.[14]

The first tests of the normal distribution were invented by the German statistician Wilhelm Lexis in the 1870s. The only data sets available to him that he was able to show were normally distributed were birth rates.

9.3.1   Development of modern statistics

Although the origins of statistical theory lie in the 18th century advances in probability, the modern field of statistics only emerged in the late 19th and early 20th century in three stages. The first wave, at the turn of the century, was led by the work of Sir Francis Galton and Karl Pearson, who transformed statistics into a rigorous mathematical discipline used for analysis, not just in science, but in industry and politics as well. The second wave of the 1910s and 20s was initiated by William Gosset, and reached its culmination in the insights of Sir Ronald Fisher. This involved the development of better experimental models, hypothesis testing and techniques for use with small data samples. The final wave, which mainly saw the refinement and expansion of earlier developments, emerged from the collaborative work between Egon Pearson and Jerzy Neyman in the 1930s.[15] Today, statistical methods are applied in all fields that involve decision making, for making accurate inferences from a collated body of data and for making decisions in the face of uncertainty based on statistical methodology.

The first statistical bodies were established in the early 19th century. The Royal Statistical Society was founded in 1834 and Florence Nightingale, its first female member, pioneered the application of statistical analysis to health problems for the furtherance of epidemiological understanding and public health practice. However, the methods then used would not be considered as modern statistics today.

The Oxford scholar Francis Ysidro Edgeworth's book, *Metretike: or The Method of Measuring Probability and Util-*

Carl Friedrich Gauss, mathematician who developed the method of least squares in 1809.

ity (1887) dealt with probability as the basis of inductive reasoning, and his later works focused on the 'philosophy of chance'.[16] His first paper on statistics (1883) explored the law of error (normal distribution), and his *Methods of Statistics* (1885) introduced an early version of the t distribution, the Edgeworth expansion, the Edgeworth series, the method of variate transformation and the asymptotic theory of maximum likelihood estimates.

The Norwegian Anders Nicolai Kiær introduced the concept of stratified sampling in 1895.[17] Arthur Lyon Bowley introduced new methods of data sampling in 1906 when working on social statistics. Although statistical surveys of social conditions had started with Charles Booth's "Life and Labour of the People in London" (1889-1903) and Seebohm Rowntree's "Poverty, A Study of Town Life" (1901), Bowley's, key innovation consisted of the use of random sampling

The original logo of the Royal Statistical Society, founded in 1834.

techniques. His efforts culminated in his *New Survey of London Life and Labour*.[18]

Sir Francis Galton is credited as one of the principal founders of statistical theory. His contributions to the field included introducing the concepts of standard deviation, correlation, regression and the application of these methods to the study of the variety of human characteristics - height, weight, eyelash length among others. He found that many of these could be fitted to a normal curve distribution.[19]

Galton submitted a paper to *Nature* in 1907 on the usefulness of the median.[20] He examined the accuracy of 787 guesses of the weight of an ox at a country fair. The actual weight was 1208 pounds: the median guess was 1198. The guesses were markedly non-normally distributed.

Galton's publication of *Natural Inheritance* in 1889 sparked the interest of a brilliant mathematician, Karl Pearson,[21] then working at University College London, and he went on to found the discipline of mathematical statistics.[22] He emphasised the statistical foundation of scientific laws and promoted its study and his laboratory attracted students from around the world attracted by his new methods of analysis, including Udny Yule. His work grew to encompass the fields of

Karl Pearson, the founder of mathematical statistics.

biology, epidemiology, anthropometry, medicine and social history. In 1901, with Walter Weldon, founder of biometry, and Galton, he founded the journal *Biometrika* as the first journal of mathematical statistics and biometry.

His work, and that of Galton's, underpins many of the 'classical' statistical methods which are in common use today, including the Correlation coefficient, defined as a product-moment;[23] the method of moments for the fitting of distributions to samples; Pearson's system of continuous curves that forms the basis of the now conventional continuous probability distributions; Chi distance a precursor and special case of the Mahalanobis distance[24] and P-value, defined as the probability measure of the complement of the ball with the hypothesized value as center point and chi distance as radius.[24] He also introduced the term 'standard deviation'.

He also founded the statistical hypothesis testing theory,[24] Pearson's chi-squared test and principal component analysis.[25][26] In 1911 he founded the world's first university statistics department at University College London.

The second wave of mathematical statistics was pioneered by Sir Ronald Fisher who wrote the textbooks that were to define the academic discipline in universities around the world. He also systematized previous results, putting them on a firm mathematical footing. His most important publications were his 1916 seminal paper *The Correlation between Relatives on the Supposition of Mendelian Inheritance* and his classic 1925 work *Statistical Methods for Research Workers*. His paper was the first use to use the statistical term, variance. In 1919, at Rothamsted Experimental Station he started a major study of the extensive collections of data recorded over many years. This resulted in a series of reports under the general title *Studies in Crop Variation*.

Over the next seven years, he pioneered the principles of the design of experiments (see below) and elaborated his studies of analysis of variance. He furthered his studies of the statistics of small samples. Perhaps even more important, he began his systematic approach of the analysis of real data as the springboard for the development of new statistical methods. He developed computational algorithms for analyzing data from his balanced experimental designs. In 1925, this work resulted in the publication of his first book, *Statistical Methods for Research Workers*.[28] This book went through many

Ronald Fisher, "A genius who almost single-handedly created the foundations for modern statistical science", [27]

editions and translations in later years, and it became the standard reference work for scientists in many disciplines. In 1935, this book was followed by *The Design of Experiments*, which was also widely used.

In addition to analysis of variance, Fisher named and promoted the method of maximum likelihood estimation. Fisher also originated the concepts of sufficiency, ancillary statistics, Fisher's linear discriminator and Fisher information. His article *On a distribution yielding the error functions of several well known statistics* (1924) presented Pearson's chi-squared

Anil Kumar Gain, famous Indian statistician

test and William Gosset's t in the same framework as the Gaussian distribution, and his own parameter in the analysis of variance Fisher's z-distribution (more commonly used decades later in the form of the F distribution).[29] The 5% level of significance appears to have been introduced by Fisher in 1925.[30] Fisher stated that deviations exceeding twice the standard deviation are regarded as significant. Before this deviations exceeding three times the probable error were considered significant. For a symmetrical distribution the probable error is half the interquartile range. For a normal distribution the probable error is approximately 2/3 the standard deviation. It appears that Fisher's 5% criterion was rooted in previous practice.

Other important contributions at this time included Charles Spearman's rank correlation coefficient that was a useful extension of the Pearson correlation coefficient. William Sealy Gosset, the English statistician better known under his pseudonym of *Student*, introduced Student's t-distribution, a continuous probability distribution useful in situations where the sample size is small and population standard deviation is unknown.

Egon Pearson (Karl's son) and Jerzy Neyman introduced the concepts of "Type II" error, power of a test and confidence intervals. Jerzy Neyman in 1934 showed that stratified random sampling was in general a better method of estimation than purposive (quota) sampling.[31]

9.4 Design of experiments

In 1747, while serving as surgeon on HM Bark *Salisbury*, James Lind carried out a controlled experiment to develop a cure for scurvy.[32] In this study his subjects' cases "were as similar as I could have them", that is he provided strict entry requirements to reduce extraneous variation. The men were paired, which provided blocking. From a modern perspective, the main thing that is missing is randomized allocation of subjects to treatments.

James Lind is today often described as a one-factor-at-a-time experimenter.[33] Similar one-factor-at-a-time (OFAT) experimentation was performed at the Rothamsted Research Station in the 1840s by Sir John Lawes to determine the optimal inorganic fertilizer for use on wheat.[33]

A theory of statistical inference was developed by Charles S. Peirce in "Illustrations of the Logic of Science" (1877–1878) and "A Theory of Probable Inference" (1883), two publications that emphasized the importance of randomization-based inference in statistics. In another study, Peirce randomly assigned volunteers to a blinded, repeated-measures design to evaluate their ability to discriminate weights.[34][35][36][37]

Peirce's experiment inspired other researchers in psychology and education, which developed a research tradition of randomized experiments in laboratories and specialized textbooks in the 1800s.[34][35][36][37] Peirce also contributed the first English-language publication on an optimal design for regression-models in 1876.[38] A pioneering optimal design for polynomial regression was suggested by Gergonne in 1815. In 1918 Kirstine Smith published optimal designs for polynomials of degree six (and less).[39]

The use of a sequence of experiments, where the design of each may depend on the results of previous experiments, including the possible decision to stop experimenting, was pioneered[40] by Abraham Wald in the context of sequential tests of statistical hypotheses.[41] Surveys are available of optimal sequential designs,[42] and of adaptive designs.[43] One specific type of sequential design is the "two-armed bandit", generalized to the multi-armed bandit, on which early work was done by Herbert Robbins in 1952.[44]

The term "design of experiments" (DOE) derives from early statistical work performed by Sir Ronald Fisher. He was described by Anders Hald as "a genius who almost single-handedly created the foundations for modern statistical science."[45] Fisher initiated the principles of design of experiments and elaborated on his studies of "analysis of variance". Perhaps even more important, Fisher began his systematic approach to the analysis of real data as the springboard for the development of new statistical methods. He began to pay particular attention to the labour involved in the necessary computations performed by hand, and developed methods that were as practical as they were founded in rigour. In 1925, this work culminated in the publication of his first book, *Statistical Methods for Research Workers*.[46] This went into many editions and translations in later years, and became a standard reference work for scientists in many disciplines.[47]

A methodology for designing experiments was proposed by Ronald A. Fisher, in his innovative book *The Design of Experiments* (1935) which also became a standard.[48][49][50][51] As an example, he described how to test the hypothesis that a certain lady could distinguish by flavour alone whether the milk or the tea was first placed in the cup. While this

James Lind carried out the first ever clinical trial in 1747, in an effort to find a treatment for scurvy.

sounds like a frivolous application, it allowed him to illustrate the most important ideas of experimental design: see Lady tasting tea.

Agricultural science advances served to meet the combination of larger city populations and fewer farms. But for crop scientists to take due account of widely differing geographical growing climates and needs, it was important to differentiate local growing conditions. To extrapolate experiments on local crops to a national scale, they had to extend crop sample testing economically to overall populations. As statistical methods advanced (primarily the efficacy of designed experiments instead of one-factor-at-a-time experimentation), representative factorial design of experiments began to enable the meaningful extension, by inference, of experimental sampling results to the population as a whole. But it was hard to decide how representative was the crop sample chosen. Factorial design methodology showed how to estimate and correct for any random variation within the sample and also in the data collection procedures.

9.5 Bayesian statistics

The term *Bayesian* refers to Thomas Bayes (1702–1761), who proved a special case of what is now called Bayes' theorem. However it was Pierre-Simon Laplace (1749–1827) who introduced a general version of the theorem and applied it to celestial mechanics, medical statistics, reliability, and jurisprudence.[52] When insufficient knowledge was available to specify an informed prior, Laplace used uniform priors, according to his "principle of insufficient reason".[52][53] Laplace assumed uniform priors for mathematical simplicity rather than for philosophical reasons.[52] Laplace also introduced primitive versions of conjugate priors and the theorem of von Mises and Bernstein, according to which the posteriors corresponding to initially differing priors ultimately agree, as the number of observations increases.[54] This early Bayesian inference, which used uniform priors following Laplace's principle of insufficient reason, was called "inverse probability" (because it infers backwards from observations to parameters, or from effects to causes[55]).

After the 1920s, inverse probability was largely supplanted by a collection of methods that were developed by Ronald A. Fisher, Jerzy Neyman and Egon Pearson. Their methods came to be called frequentist statistics.[55] Fisher rejected the Bayesian view, writing that "the theory of inverse probability is founded upon an error, and must be wholly rejected".[56] At the end of his life, however, Fisher expressed greater respect for the essay of Bayes, which Fisher believed to have anticipated his own, fiducial approach to probability; Fisher still maintained that Laplace's views on probability were "fallacious rubbish".[56] Neyman started out as a "quasi-Bayesian", but subsequently developed confidence intervals (a key method in frequentist statistics) because "the whole theory would look nicer if it were built from the start without reference to Bayesianism and priors".[57] The word *Bayesian* appeared around 1950, and by the 1960s it became the term preferred by those dissatisfied with the limitations of frequentist statistics.[55][58]

In the 20th century, the ideas of Laplace were further developed in two different directions, giving rise to *objective* and *subjective* currents in Bayesian practice. In the objectivist stream, the statistical analysis depends on only the model assumed and the data analysed.[59] No subjective decisions need to be involved. In contrast, "subjectivist" statisticians deny the possibility of fully objective analysis for the general case.

In the further development of Laplace's ideas, subjective ideas predate objectivist positions. The idea that 'probability' should be interpreted as 'subjective degree of belief in a proposition' was proposed, for example, by John Maynard Keynes in the early 1920s. This idea was taken further by Bruno de Finetti in Italy (*Fondamenti Logici del Ragionamento Probabilistico*, 1930) and Frank Ramsey in Cambridge (*The Foundations of Mathematics*, 1931).[60] The approach was devised to solve problems with the frequentist definition of probability but also with the earlier, objectivist approach of Laplace.[59] The subjective Bayesian methods were further developed and popularized in the 1950s by L.J. Savage.

Objective Bayesian inference was further developed by Harold Jeffreys at the University of Cambridge. His seminal book "Theory of probability" first appeared in 1939 and played an important role in the revival of the Bayesian view of probability.[61][62] In 1957, Edwin Jaynes promoted the concept of maximum entropy for constructing priors, which is an important principle in the formulation of objective methods, mainly for discrete problems. In 1965, Dennis Lindley's 2-volume work "Introduction to Probability and Statistics from a Bayesian Viewpoint" brought Bayesian methods to a wide audience. In 1979, José-Miguel Bernardo introduced reference analysis,[59] which offers a general applicable framework for objective analysis.[63] Other well-known proponents of Bayesian probability theory include I.J. Good, B.O. Koopman, Howard Raiffa, Robert Schlaifer and Alan Turing.

In the 1980s, there was a dramatic growth in research and applications of Bayesian methods, mostly attributed to the discovery of Markov chain Monte Carlo methods, which removed many of the computational problems, and an increasing interest in nonstandard, complex applications.[64] Despite growth of Bayesian research, most undergraduate teaching is still based on frequentist statistics.[65] Nonetheless, Bayesian methods are widely accepted and used, such as for example

Pierre-Simon, marquis de Laplace, one of the main early developers of Bayesian statistics.

in the field of machine learning.[66]

9.6 Important contributors to statistics

See also: List of statisticians and Founders of statistics

9.7 References

[1] Ball, Philip (2004). *Critical Mass*. Farrar, Straus and Giroux. p. 53. ISBN 0-374-53041-6.

[2] Thucydides (1985). *History of the Peloponnesian War*. New York: Penguin Books, Ltd. p. 204.

[3] Singh, Simon (2000). *The code book : the science of secrecy from ancient Egypt to quantum cryptography* (1st Anchor Books ed.). New York: Anchor Books. ISBN 0-385-49532-3.

[4] Ibrahim A. Al-Kadi "The origins of cryptology: The Arab contributions", *Cryptologia*, 16(2) (April 1992) pp. 97–126.

[5] Villani, Giovanni. Encyclopædia Britannica. Encyclopædia Britannica 2006 Ultimate Reference Suite DVD. Retrieved on 2008-03-04.

[6] de Moivre, A. (1738) The doctrine of chances. Woodfall

[7] Laplace, P-S. (1774). "Mémoire sur la probabilité des causes par les évènements". *Mémoires de l'Académie Royale des Sciences Présentés par Divers Savants*, 6, 621–656

[8] Wilson, Edwin Bidwell (1923) "First and second laws of error", *Journal of the American Statistical Association*, 18 (143), 841-851 JSTOR 2965467

[9] Havil J (2003) *Gamma: Exploring Euler's Constant*. Princeton, NJ: Princeton University Press, p. 157

[10] Peirce CS (1873) Theory of errors of observations. Report of the Superintendent US Coast Survey, Washington, Government Printing Office. Appendix no. 21: 200-224

[11] Cochran W.G. (1978) "Laplace's ratio estimators". pp 3-10. In David H.A., (ed). *Contributions to Survey Sampling and Applied Statistics: papers in honor of H. O. Hartley*. Academic Press, New York ISBN122047508,

[12] Keynes, JM (1921) A treatise on probability. Pt II Ch XVII §5 (p 201)

[13] Galton F (1881) Report of the Anthropometric Committee pp 245-260. Report of the 51st Meeting of the British Association for the Advancement of Science

[14] Stigler (1986, Chapter 5: Quetelet's Two Attempts)

[15] Helen Mary Walker (1975). *Studies in the history of statistical method*. Arno Press.

[16] (Stigler 1986, Chapter 9: The Next Generation: Edgeworth)

[17] Bellhouse DR (1988) A brief history of random sampling methods. Handbook of statistics. Vol 6 pp 1-14 Elsevier

[18] Bowley AL (1906) Address to the Economic Science and Statistics Section of the British Association for the Advancement of Science. J Roy Stat Soc 69: 548-557

[19] Galton F (1877) Typical laws of heredity. *Nature* 15: 492-553

[20] Galton F (1907) One Vote, One Value. *Nature* 75: 414

[21] Stigler (1986, Chapter 10: Pearson and Yule)

[22] "The development of modern statistics". Retrieved 2012-12-17.

[23] Stigler, S. M. (1989). "Francis Galton's Account of the Invention of Correlation". *Statistical Science* 4 (2): 73–79. doi:10.1214/ss/1177012580.

[24] Pearson, K. (1900). "On the Criterion that a given System of Deviations from the Probable in the Case of a Correlated System of Variables is such that it can be reasonably supposed to have arisen from Random Sampling". *Philosophical Magazine Series 5* **50** (302): 157–175. doi:10.1080/14786440009463897.

[25] Pearson, K. (1901). "On Lines and Planes of Closest Fit to Systems of Points is Space". *Philosophical Magazine Series 6* **2** (11): 559–572. doi:10.1080/14786440109462720.

[26] Jolliffe, I. T. (2002). *Principal Component Analysis, 2nd ed.* New York: Springer-Verlag.

[27] Hald, Anders (1998). *A History of Mathematical Statistics*. New York: Wiley. ISBN 0-471-17912-4.

[28] Box, *R. A. Fisher*, pp 93–166

[29] Agresti, Alan; David B. Hichcock (2005). "Bayesian Inference for Categorical Data Analysis" (PDF). *Statistical Methods & Applications* **14** (14): 298. doi:10.1007/s10260-005-0121-y.

[30] Fisher RA (1925) Statistical methods for research workers, Edinburgh: Oliver & Boyd

[31] Neyman, J (1934) On the two different aspects of the representative method: The method of stratified sampling and the method of purposive selection. *Journal of the Royal Statistical Society* 97 (4) 557-625 JSTOR 2342192

[32] Dunn, Peter (January 1997). "James Lind (1716-94) of Edinburgh and the treatment of scurvy". *Archive of Disease in Childhood Foetal Neonatal* (United Kingdom: British Medical Journal Publishing Group) **76** (1): 64–65. doi:10.1136/fn.76.1.F64. PMC 1720613. PMID 9059193. Retrieved 2009-01-17.

[33] Klaus Hinkelmann (2012). *Design and Analysis of Experiments, Special Designs and Applications*. John Wiley & Sons. p. xvii.

[34] Charles Sanders Peirce and Joseph Jastrow (1885). "On Small Differences in Sensation". *Memoirs of the National Academy of Sciences* **3**: 73–83.

[35] Hacking, Ian (September 1988). "Telepathy: Origins of Randomization in Experimental Design". *Isis* **79** (A Special Issue on Artifact and Experiment, number 3): 427–451. doi:10.1086/354775. JSTOR 234674. MR 1013489.

[36] Stephen M. Stigler (November 1992). "A Historical View of Statistical Concepts in Psychology and Educational Research". *American Journal of Education* **101** (1): 60–70. doi:10.1086/444032.

[37] Trudy Dehue (December 1997). "Deception, Efficiency, and Random Groups: Psychology and the Gradual Origination of the Random Group Design". *Isis* **88** (4): 653–673. doi:10.1086/383850. PMID 9519574.

[38] Peirce, C. S. (1876). "Note on the Theory of the Economy of Research". *Coast Survey Report*: 197–201., actually published 1879, NOAA PDF Eprint.
Reprinted in *Collected Papers* **7**, paragraphs 139–157, also in *Writings* **4**, pp. 72–78, and in Peirce, C.S. (July–August 1967). "Note on the Theory of the Economy of Research". *Operations Research* **15** (4): 643–648. doi:10.1287/opre.15.4.643. JSTOR 168276.

[39] Smith, Kirstine (1918). "On the Standard Deviations of Adjusted and Interpolated Values of an Observed Polynomial Function and its Constants and the Guidance they give Towards a Proper Choice of the Distribution of Observations". *Biometrika* **12** (1/2): 1–85. doi:10.2307/2331929.

[40] Johnson, N.L. (1961). "Sequential analysis: a survey." *Journal of the Royal Statistical Society*, Series A. Vol. 124 (3), 372–411. (pages 375–376)

[41] Wald, A. (1945) "Sequential Tests of Statistical Hypotheses", Annals of Mathematical Statistics, 16 (2), 117–186.

[42] Chernoff, H. (1972) *Sequential Analysis and Optimal Design*, SIAM Monograph. ISBN 978-0898710069

[43] Zacks, S. (1996) "Adaptive Designs for Parametric Models". In: Ghosh, S. and Rao, C. R., (Eds) (1996). "Design and Analysis of Experiments," *Handbook of Statistics*, Volume 13. North-Holland. ISBN 0-444-82061-2. (pages 151–180)

[44] Robbins, H. (1952). "Some Aspects of the Sequential Design of Experiments". *Bulletin of the American Mathematical Society* **58** (5): 527–535. doi:10.1090/S0002-9904-1952-09620-8.

[45] Hald, Anders (1998) *A History of Mathematical Statistics*. New York: Wiley.

[46] Box, Joan Fisher (1978) *R. A. Fisher: The Life of a Scientist*, Wiley. ISBN 0-471-09300-9 (pp 93–166)

[47] Edwards, A.W.F. (2005). "R. A. Fisher, Statistical Methods for Research Workers, 1925". In Grattan-Guinness, Ivor. *Landmark writings in Western mathematics 1640-1940*. Amsterdam Boston: Elsevier. ISBN 9780444508713.

[48] Stanley, J. C. (1966). "The Influence of Fisher's "The Design of Experiments" on Educational Research Thirty Years Later". *American Educational Research Journal* **3** (3): 223. doi:10.3102/00028312003003223.

[49] Box, JF (February 1980). "R. A. Fisher and the Design of Experiments, 1922-1926". *The American Statistician* **34** (1): 1–7. doi:10.2307/2682986. JSTOR 2682986.

[50] Yates, Frank (June 1964). "Sir Ronald Fisher and the Design of Experiments". *Biometrics* **20** (2): 307–321. doi:10.2307/2528399. JSTOR 2528399.

[51] Stanley, Julian C. (1966). "The Influence of Fisher's "The Design of Experiments" on Educational Research Thirty Years Later". *American Educational Research Journal* **3** (3): 223–229. doi:10.3102/00028312003003223. JSTOR 1161806.

[52] Stigler (1986, Chapter 3: Inverse Probability)

[53] Hald (1998)

[54] Lucien Le Cam (1986) *Asymptotic Methods in Statistical Decision Theory*: Pages 336 and 618–621 (von Mises and Bernstein).

[55] Stephen. E. Fienberg, (2006) When did Bayesian Inference become "Bayesian"? *Bayesian Analysis*, 1 (1), 1–40. See page 5.

[56] Aldrich, A. (2008) "R. A. Fisher on Bayes and Bayes' Theorem", *Bayesian analysis*, 3 (1),161–170

[57] Neyman, J. (1977). "Frequentist probability and frequentist statistics". *Synthese* **36** (1): 97–131. doi:10.1007/BF00485695.

[58] Jeff Miller, "Earliest Known Uses of Some of the Words of Mathematics (B)" "The term Bayesian entered circulation around 1950. R. A. Fisher used it in the notes he wrote to accompany the papers in his Contributions to Mathematical Statistics (1950). Fisher thought Bayes's argument was all but extinct for the only recent work to take it seriously was Harold Jeffreys's Theory of Probability (1939). In 1951 L. J. Savage, reviewing Wald's Statistical Decisions Functions, referred to "modern, or unBayesian, statistical theory" ("The Theory of Statistical Decision," Journal of the American Statistical Association, 46, p. 58.). Soon after, however, Savage changed from being an unBayesian to being a Bayesian."

[59] Bernardo, JM. (2005). "Reference analysis". *Handbook of statistics*. Handbook of Statistics **25**: 17–90. doi:10.1016/S0169-7161(05)25002-2. ISBN 9780444515391.

[60] Gillies, D. (2000), *Philosophical Theories of Probability*. Routledge. ISBN 0-415-18276-X pp 50–1

[61] E. T. Jaynes. *Probability Theory: The Logic of Science* Cambridge University Press, (2003). ISBN 0-521-59271-2

[62] O'Connor, John J.; Robertson, Edmund F., "History of statistics", *MacTutor History of Mathematics archive*, University of St Andrews.

[63] Bernardo, J. M. and Smith, A. F. M. (1994). "Bayesian Theory". Chichester: Wiley.

[64] Wolpert, RL. (2004) "A conversation with James O. Berger", *Statistical Science*, 9, 205–218 doi:10.1214/088342304000000053 MR 2082155

[65] Bernardo, J. M. (2006). "A Bayesian Mathematical Statistics Primer" (PDF). *Proceedings of the Seventh International Conference on Teaching Statistics [CDROM]*. Salvador (Bahia), Brazil: International Association for Statistical Education.

[66] Bishop, C.M. (2007) *Pattern Recognition and Machine Learning*. Springer ISBN 978-0387310732

9.8 Bibliography

- Freedman, D. (1999). "From association to causation: Some remarks on the history of statistics". *Statistical Science* **14** (3): 243–258. doi:10.1214/ss/1009212409. (Revised version, 2002)

- Hald, Anders (2003). *A History of Probability and Statistics and Their Applications before 1750*. Hoboken, NJ: Wiley. ISBN 0-471-47129-1.

- Hald, Anders (1998). *A History of Mathematical Statistics from 1750 to 1930*. New York: Wiley. ISBN 0-471-17912-4.

- Kotz, S., Johnson, N.L. (1992,1992,1997). *Breakthroughs in Statistics*, Vols I,II,III. Springer ISBN 0-387-94037-5, ISBN 0-387-94039-1, ISBN 0-387-94989-5

- Pearson, Egon (1978). *The History of Statistics in the 17th and 18th Centuries against the changing background of intellectual, scientific and religious thought (Lectures by Karl Pearson given at University College London during the academic sessions 1921-1933)*. New York: MacMillan Publishng Co., Inc. p. 744. ISBN 0-02-850120-9.

- Salsburg, David (2001). *The Lady Tasting Tea: How Statistics Revolutionized Science in the Twentieth Century*. ISBN 0-7167-4106-7

- Stigler, Stephen M. (1986). *The History of Statistics: The Measurement of Uncertainty before 1900*. Belknap Press/Harvard University Press. ISBN 0-674-40341-X.

- Stigler, Stephen M. (1999) *Statistics on the Table: The History of Statistical Concepts and Methods*. Harvard University Press. ISBN 0-674-83601-4

- David, H. A. (1995). "First (?) Occurrence of Common Terms in Mathematical Statistics". *The American Statistician* **49** (2): 121–133. doi:10.2307/2684625. JSTOR 2684625.

9.9 External links

- JEHPS: Recent publications in the history of probability and statistics

- Electronic Journ@l for History of Probability and Statistics/Journ@l Electronique d'Histoire des Probabilités et de la Statistique

- Figures from the History of Probability and Statistics (Univ. of Southampton)

- Materials for the History of Statistics (Univ. of York)

- Probability and Statistics on the Earliest Uses Pages (Univ. of Southampton)

- Earliest Uses of Symbols in Probability and Statistics on Earliest Uses of Various Mathematical Symbols

Chapter 10

History of trigonometry

Early study of triangles can be traced to the 2nd millennium BC, in Egyptian mathematics (Rhind Mathematical Papyrus) and Babylonian mathematics. Systematic study of trigonometric functions began in Hellenistic mathematics, reaching India as part of Hellenistic astronomy. In Indian astronomy, the study of trigonometric functions flowered in the Gupta period, especially due to Aryabhata (6th century CE). During the Middle Ages, the study of trigonometry continued in Islamic mathematics, hence it was adopted as a separate subject in the Latin West beginning in the Renaissance with Regiomontanus. The development of modern trigonometry shifted during the western Age of Enlightenment, beginning with 17th-century mathematics (Isaac Newton and James Stirling) and reaching its modern form with Leonhard Euler (1748).

10.1 Etymology

The term "trigonometry" was derived from the Greek "τριγωνομετρία" ("*trigonometria*"), meaning "triangle measuring", from "τρίγωνο" (triangle) + "μετρεῖν" (to measure).

Our modern word "sine" is derived from the Latin word *sinus*, which means "bay", "bosom" or "fold", translating Arabic *jayb*. The Arabic term is in origin a corruption of Sanskrit *jīvā*, or "chord". Sanskrit *jīvā* in learned usage was a synonym of *jyā* "chord", originally the term for "bow-string". Sanskrit *jīvā* was loaned into Arabic as *jiba*.[1][2] This term was then transformed[2] into the genuine Arabic word *jayb*, meaning "bosom, fold, bay", either by the Arabs or by a mistake[1] of the European translators such as Robert of Chester (perhaps because the words were written without vowels[1]), who translated *jayb* into Latin as *sinus*. The words "minute" and "second" are derived from the Latin phrases *partes minutae primae* and *partes minutae secundae*.[3] Particularly Fibonacci's *sinus rectus arcus* proved influential in establishing the term *sinus*.[4][5] These roughly translate to "first small parts" and "second small parts".

10.2 Development

10.2.1 Early trigonometry

The ancient Egyptians and Babylonians had known of theorems on the ratios of the sides of similar triangles for many centuries. However, as pre-Hellenic societies lacked the concept of an angle measure, they were limited to studying the sides of triangles instead.[6]

The Babylonian astronomers kept detailed records on the rising and setting of stars, the motion of the planets, and the solar and lunar eclipses, all of which required familiarity with angular distances measured on the celestial sphere.[2] Based on one interpretation of the Plimpton 322 cuneiform tablet (c. 1900 BC), some have even asserted that the ancient Babylonians had a table of secants.[7] There is, however, much debate as to whether it is a table of Pythagorean triples, a solution of quadratic equations, or a trigonometric table.

The Egyptians, on the other hand, used a primitive form of trigonometry for building pyramids in the 2nd millennium BC.[2] The Rhind Mathematical Papyrus, written by the Egyptian scribe Ahmes (c. 1680–1620 BC), contains the following problem related to trigonometry:[2]

"If a pyramid is 250 cubits high and the side of its base 360 cubits long, what is its *seked*?"

Ahmes' solution to the problem is the ratio of half the side of the base of the pyramid to its height, or the run-to-rise ratio of its face. In other words, the quantity he found for the *seked* is the cotangent of the angle to the base of the pyramid and its face.[2]

10.2.2 Greek mathematics

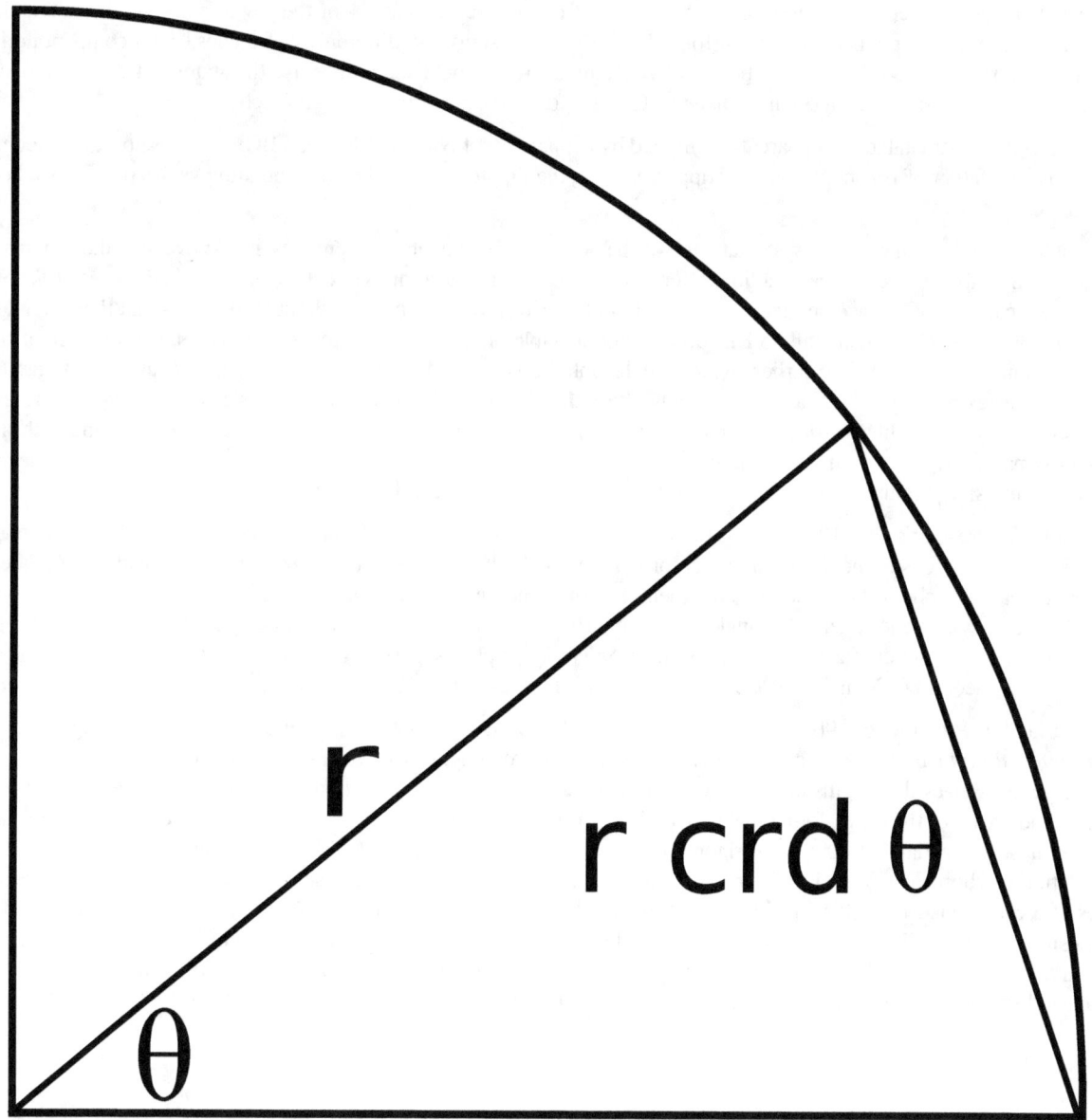

The chord of an angle subtends the arc of the angle.

Ancient Greek and Hellenistic mathematicians made use of the chord. Given a circle and an arc on the circle, the chord is the line that subtends the arc. A chord's perpendicular bisector passes through the center of the circle and bisects the angle. One half of the bisected chord is the sine of one half the bisected angle, that is,

$$\text{chord } \theta = 2 \sin \frac{\theta}{2},$$

and consequently the sine function is also known as the "half-chord". Due to this relationship, a number of trigonometric identities and theorems that are known today were also known to Hellenistic mathematicians, but in their equivalent chord form.[8]

Although there is no trigonometry in the works of Euclid and Archimedes, in the strict sense of the word, there are theorems presented in a geometric way (rather than a trigonometric way) that are equivalent to specific trigonometric laws or formulas.[6] For instance, propositions twelve and thirteen of book two of the *Elements* are the laws of cosines for obtuse and acute angles, respectively. Theorems on the lengths of chords are applications of the law of sines. And Archimedes' theorem on broken chords is equivalent to formulas for sines of sums and differences of angles.[6] To compensate for the lack of a table of chords, mathematicians of Aristarchus' time would sometimes use the statement that, in modern notation, $\sin \alpha / \sin \beta < \alpha / \beta < \tan \alpha / \tan \beta$ whenever $0° < \beta < \alpha < 90°$, now known as Aristarchus' inequality.[9]

The first trigonometric table was apparently compiled by Hipparchus of Nicaea (180 – 125 BCE), who is now consequently known as "the father of trigonometry."[10] Hipparchus was the first to tabulate the corresponding values of arc and chord for a series of angles.[4][10]

Although it is not known when the systematic use of the 360° circle came into mathematics, it is known that the systematic introduction of the 360° circle came a little after Aristarchus of Samos composed *On the Sizes and Distances of the Sun and Moon* (ca. 260 BC), since he measured an angle in terms of a fraction of a quadrant.[9] It seems that the systematic use of the 360° circle is largely due to Hipparchus and his table of chords. Hipparchus may have taken the idea of this division from Hypsicles who had earlier divided the day into 360 parts, a division of the day that may have been suggested by Babylonian astronomy.[11] In ancient astronomy, the zodiac had been divided into twelve "signs" or thirty-six "decans". A seasonal cycle of roughly 360 days could have corresponded to the signs and decans of the zodiac by dividing each sign into thirty parts and each decan into ten parts.[5] It is due to the Babylonian sexagesimal numeral system that each degree is divided into sixty minutes and each minute is divided into sixty seconds.[5]

Menelaus of Alexandria (ca. 100 AD) wrote in three books his *Sphaerica*. In Book I, he established a basis for spherical triangles analogous to the Euclidean basis for plane triangles.[8] He establishes a theorem that is without Euclidean analogue, that two spherical triangles are congruent if corresponding angles are equal, but he did not distinguish between congruent and symmetric spherical triangles.[8] Another theorem that he establishes is that the sum of the angles of a spherical triangle is greater than 180°.[8] Book II of *Sphaerica* applies spherical geometry to astronomy. And Book III contains the "theorem of Menelaus".[8] He further gave his famous "rule of six quantities".[12]

Later, Claudius Ptolemy (ca. 90 – ca. 168 AD) expanded upon Hipparchus' *Chords in a Circle* in his *Almagest*, or the *Mathematical Syntaxis*. The Almagest is primarily a work on astronomy, and astronomy relies on trigonometry. Ptolemy's table of chords gives the lengths of chords of a circle of diameter 120 as a function of the number of degrees n in the corresponding arc of the circle, for n ranging from 1/2 to 180 by increments of 1/2.[13] The thirteen books of the *Almagest* are the most influential and significant trigonometric work of all antiquity.[14] A theorem that was central to Ptolemy's calculation of chords was what is still known today as Ptolemy's theorem, that the sum of the products of the opposite sides of a cyclic quadrilateral is equal to the product of the diagonals. A special case of Ptolemy's theorem appeared as proposition 93 in Euclid's *Data*. Ptolemy's theorem leads to the equivalent of the four sum-and-difference formulas for sine and cosine that are today known as Ptolemy's formulas, although Ptolemy himself used chords instead of sine and cosine.[14] Ptolemy further derived the equivalent of the half-angle formula

$$\sin^2 \left(\tfrac{x}{2} \right) = \tfrac{1 - \cos(x)}{2}. \text{ [14]}$$

Ptolemy used these results to create his trigonometric tables, but whether these tables were derived from Hipparchus' work cannot be determined.[14]

Neither the tables of Hipparchus nor those of Ptolemy have survived to the present day, although descriptions by other ancient authors leave little doubt that they once existed.[15]

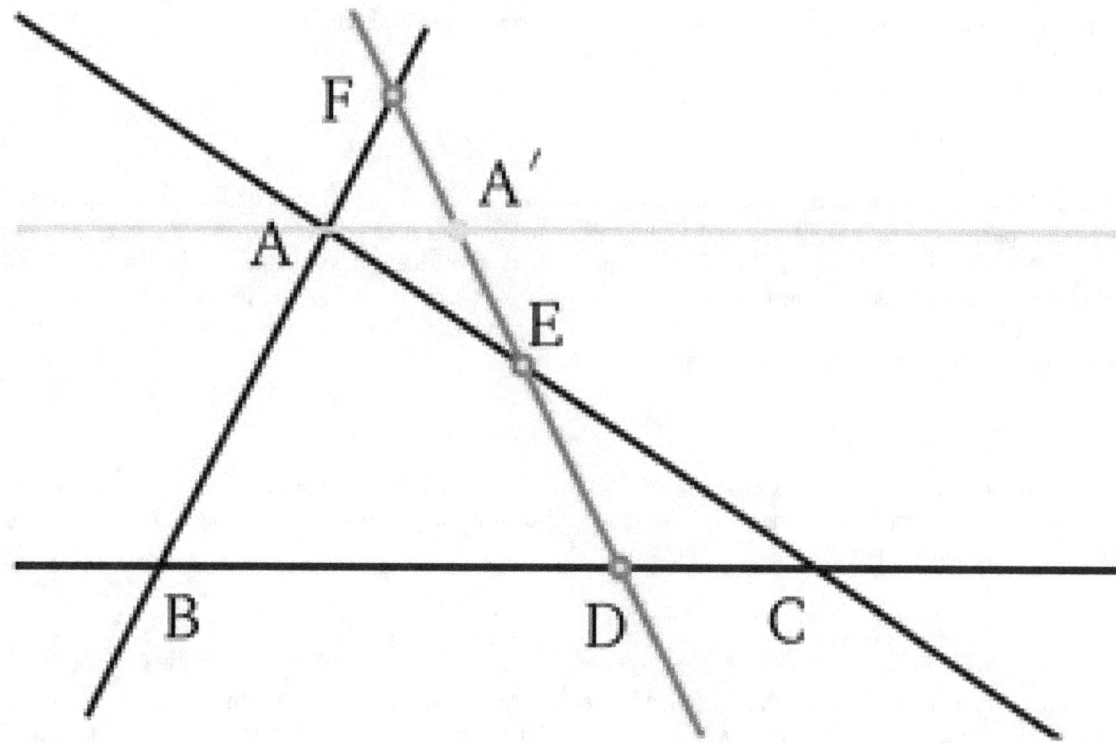

Menelaus' theorem

10.2.3 Indian mathematics

See also: Indian Mathematics and Indian Astronomy

Some of the early and very significant developments of trigonometry were in India. Influential works from the 4th–5th century, known as the Siddhantas (of which there were five, the most complvor of which is the Surya Siddhanta[16]) first defined the sine as the modern relationship between half an angle and half a chord, while also defining the cosine, versine, and inverse sine.[17] Soon afterwards, another Indian mathematician and astronomer, Aryabhata (476–550 AD), collected and expanded upon the developments of the Siddhantas in an important work called the *Aryabhatiya*.[18] The *Siddhantas* and the *Aryabhatiya* contain the earliest surviving tables of sine values and versine (1 − cosine) values, in 3.75° intervals from 0° to 90°, to an accuracy of 4 decimal places.[19] They used the words *jya* for sine, *kojya* for cosine, *utkrama-jya* for versine, and *otkram jya* for inverse sine. The words *jya* and *kojya* eventually became *sine* and *cosine* respectively after a mistranslation described above.

In the 7th century, Bhaskara I produced a formula for calculating the sine of an acute angle without the use of a table. He also gave the following approximation formula for sin(x), which had a relative error of less than 1.9%:

$$\sin x \approx \frac{16x(\pi - x)}{5\pi^2 - 4x(\pi - x)}, \qquad (0 \le x \le \pi).$$

Later in the 7th century, Brahmagupta redeveloped the formula

$$1 - \sin^2(x) = \cos^2(x) = \sin^2\left(\frac{\pi}{2} - x\right)$$

(also derived earlier, as mentioned above) and the Brahmagupta interpolation formula for computing sine values.[20]

Another later Indian author on trigonometry was Bhaskara II in the 12th century. Bhaskara II developed spherical trigonometry, and discovered many trigonometric results.

Bhaskara II was the first to discover $\sin(a+b)$ and $\sin(a-b)$ trigonometric results like:

- $\sin(a+b) = \sin a \cos b + \cos a \sin b$
- $\sin(a-b) = \sin a \cos b - \cos a \sin b$

Madhava (c. 1400) made early strides in the analysis of trigonometric functions and their infinite series expansions. He developed the concepts of the power series and Taylor series, and produced the power series expansions of sine, cosine, tangent, and arctangent.[21][22] Using the Taylor series approximations of sine and cosine, he produced a sine table to 12 decimal places of accuracy and a cosine table to 9 decimal places of accuracy. He also gave the power series of π and the θ, radius, diameter, and circumference of a circle in terms of trigonometric functions. His works were expanded by his followers at the Kerala School up to the 16th century.[21][22]

The Indian text the Yuktibhāṣā contains proof for the expansion of the sine and cosine functions and the derivation and proof of the power series for inverse tangent, discovered by Madhava. The Yuktibhāṣā also contains rules for finding the sines and the cosines of the sum and difference of two angles.

10.2.4 Islamic mathematics

The Indian works were later translated and expanded in the medieval Islamic world by Muslim mathematicians of mostly Persian and Arab descent, who enunciated a large number of theorems which freed the subject of trigonometry from dependence upon the complete quadrilateral, as was the case in Hellenistic mathematics due to the application of Menelaus' theorem. According to E. S. Kennedy, it was after this development in Islamic mathematics that "the first real trigonometry emerged, in the sense that only then did the object of study become the spherical or plane triangle, its sides and angles."[24]

In addition to Indian works, Hellenistic methods dealing with spherical triangles were also known, particularly the method of Menelaus of Alexandria, who developed "Menelaus' theorem" to deal with spherical problems.[8][25] However, E. S. Kennedy points out that while it was possible in pre-Islamic mathematics to compute the magnitudes of a spherical figure, in principle, by use of the table of chords and Menelaus' theorem, the application of the theorem to spherical problems was very difficult in practice.[26] In order to observe holy days on the Islamic calendar in which timings were determined by phases of the moon, astronomers initially used Menelaus' method to calculate the place of the moon and stars, though this method proved to be clumsy and difficult. It involved setting up two intersecting right triangles; by applying Menelaus' theorem it was possible to solve one of the six sides, but only if the other five sides were known. To tell the time from the sun's altitude, for instance, repeated applications of Menelaus' theorem were required. For medieval Islamic astronomers, there was an obvious challenge to find a simpler trigonometric method.[27]

In the early 9th century AD, Muhammad ibn Mūsā al-Khwārizmī produced accurate sine and cosine tables, and the first table of tangents. He was also a pioneer in spherical trigonometry. In 830 AD, Habash al-Hasib al-Marwazi produced the first table of cotangents.[28][29] Muhammad ibn Jābir al-Harrānī al-Battānī (Albatenius) (853-929 AD) discovered the reciprocal functions of secant and cosecant, and produced the first table of cosecants for each degree from 1° to 90°.[29]

By the 10th century AD, in the work of Abū al-Wafā' al-Būzjānī, Muslim mathematicians were using all six trigonometric functions.[30] Abu al-Wafa had sine tables in 0.25° increments, to 8 decimal places of accuracy, and accurate tables of tangent values.[30] He also developed the following trigonometric formula:[31]

$$\sin(2x) = 2\sin(x)\cos(x)$$

In his original text, Abū al-Wafā' states: "If we want that, we multiply the given sine by the cosine minutes, and the result is half the sine of the double".[31] Abū al-Wafā also established the angle addition and difference identities presented with complete proofs:[31]

$$\sin(\alpha \pm \beta) = \sqrt{\sin^2 \alpha - (\sin \alpha \sin \beta)^2} \pm \sqrt{\sin^2 \beta - (\sin \alpha \sin \beta)^2}$$

$$\sin(\alpha \pm \beta) = \sin \alpha \cos \beta \pm \cos \alpha \sin \beta$$

For the second one, the text states: "We multiply the sine of each of the two arcs by the cosine of the other *minutes*. If we want the sine of the sum, we add the products, if we want the sine of the difference, we take their difference".[31]

He also discovered the law of sines for spherical trigonometry:[28]

$$\frac{\sin A}{\sin a} = \frac{\sin B}{\sin b} = \frac{\sin C}{\sin c}.$$

Also in the late 10th and early 11th centuries AD, the Egyptian astronomer Ibn Yunus performed many careful trigono-metric calculations and demonstrated the following trigonometric identity:[32]

$$\cos a \cos b = \frac{\cos(a+b) + \cos(a-b)}{2}$$

Al-Jayyani (989–1079) of al-Andalus wrote *The book of unknown arcs of a sphere*, which is considered "the first treatise on spherical trigonometry" in its modern form.[33] It "contains formulae for right-handed triangles, the general law of sines, and the solution of a spherical triangle by means of the polar triangle." This treatise later had a "strong influence on European mathematics", and his "definition of ratios as numbers" and "method of solving a spherical triangle when all sides are unknown" are likely to have influenced Regiomontanus.[33]

The method of triangulation was first developed by Muslim mathematicians, who applied it to practical uses such as surveying[34] and Islamic geography, as described by Abu Rayhan Biruni in the early 11th century. Biruni himself in-troduced triangulation techniques to measure the size of the Earth and the distances between various places.[35] In the late 11th century, Omar Khayyám (1048–1131) solved cubic equations using approximate numerical solutions found by interpolation in trigonometric tables. In the 13th century, Nasīr al-Dīn al-Tūsī was the first to treat trigonometry as a mathematical discipline independent from astronomy, and he developed spherical trigonometry into its present form.[29] He listed the six distinct cases of a right-angled triangle in spherical trigonometry, and in his *On the Sector Figure*, he stated the law of sines for plane and spherical triangles, discovered the law of tangents for spherical triangles, and provided proofs for both these laws.[36]

In the 15th century, Jamshīd al-Kāshī provided the first explicit statement of the law of cosines in a form suitable for triangulation. In France, the law of cosines is still referred to as the *theorem of Al-Kashi*. He also gave trigonometric tables of values of the sine function to four sexagesimal digits (equivalent to 8 decimal places) for each 1° of argument with differences to be added for each 1/60 of 1°. Ulugh Beg also gives accurate tables of sines and tangents correct to 8 decimal places around the same time.

10.2.5   Chinese mathematics

In China, Aryabhata's table of sines were translated into the Chinese mathematical book of the *Kaiyuan Zhanjing*, com-piled in 718 AD during the Tang Dynasty.[37] Although the Chinese excelled in other fields of mathematics such as solid geometry, binomial theorem, and complex algebraic formulas, early forms of trigonometry were not as widely appre-ciated as in the earlier Greek, Hellenistic, Indian and Islamic worlds.[38] Instead, the early Chinese used an empirical substitute known as *chong cha*, while practical use of plane trigonometry in using the sine, the tangent, and the secant were known.[37] However, this embryonic state of trigonometry in China slowly began to change and advance during the Song Dynasty (960–1279), where Chinese mathematicians began to express greater emphasis for the need of spherical trigonometry in calendrical science and astronomical calculations.[37] The polymath Chinese scientist, mathematician and official Shen Kuo (1031–1095) used trigonometric functions to solve mathematical problems of chords and arcs.[37] Vic-tor J. Katz writes that in Shen's formula "technique of intersecting circles", he created an approximation of the arc s of a circle given the diameter d, sagitta v, and length c of the chord subtending the arc, the length of which he approximated as[39]

$$s = c + \frac{2v^2}{d}.$$

Sal Restivo writes that Shen's work in the lengths of arcs of circles provided the basis for spherical trigonometry developed in the 13th century by the mathematician and astronomer Guo Shoujing (1231–1316).[40] As the historians L. Gauchet and Joseph Needham state, Guo Shoujing used spherical trigonometry in his calculations to improve the calendar system and Chinese astronomy.[37][41] Along with a later 17th-century Chinese illustration of Guo's mathematical proofs, Needham states that:

> Guo used a quadrangular spherical pyramid, the basal quadrilateral of which consisted of one equatorial and one ecliptic arc, together with two meridian arcs, one of which passed through the summer solstice point...By such methods he was able to obtain the du lü (degrees of equator corresponding to degrees of ecliptic), the ji cha (values of chords for given ecliptic arcs), and the cha lü (difference between chords of arcs differing by 1 degree).[42]

Despite the achievements of Shen and Guo's work in trigonometry, another substantial work in Chinese trigonometry would not be published again until 1607, with the dual publication of *Euclid's Elements* by Chinese official and astronomer Xu Guangqi (1562–1633) and the Italian Jesuit Matteo Ricci (1552–1610).[43]

10.2.6 European mathematics

In 1342, Levi ben Gershon, known as Gersonides, wrote *On Sines, Chords and Arcs*, in particular proving the sine law for plane triangles and giving five-figure sine tables.[44]

A simplified trigonometric table, the "*toleta de marteloio*", was used by sailors in the Mediterranean Sea during the 14th-15th Centuries to calculate navigation courses. It is described by Ramon Llull of Majorca in 1295, and laid out in the 1436 atlas of Venetian captain Andrea Bianco.

Regiomontanus was perhaps the first mathematician in Europe to treat trigonometry as a distinct mathematical discipline,[45] in his *De triangulis omnimodus* written in 1464, as well as his later *Tabulae directionum* which included the tangent function, unnamed.

The *Opus palatinum de triangulis* of Georg Joachim Rheticus, a student of Copernicus, was probably the first in Europe to define trigonometric functions directly in terms of right triangles instead of circles, with tables for all six trigonometric functions; this work was finished by Rheticus' student Valentin Otho in 1596.

In the 17th century, Isaac Newton and James Stirling developed the general Newton–Stirling interpolation formula for trigonometric functions.

In the 18th century, Leonhard Euler's *Introductio in analysin infinitorum* (1748) was mostly responsible for establishing the analytic treatment of trigonometric functions in Europe, deriving their infinite series and presenting "Euler's formula" $e^{ix} = \cos x + i \sin x$. Euler used the near-modern abbreviations *sin.*, *cos.*, *tang.*, *cot.*, *sec.*, and *cosec.* Prior to this, Roger Cotes had computed the derivative of sine in his *Harmonia Mensurarum* (1722).[46] Also in the 18th century, Brook Taylor defined the general Taylor series and gave the series expansions and approximations for all six trigonometric functions. The works of James Gregory in the 17th century and Colin Maclaurin in the 18th century were also very influential in the development of trigonometric series.

10.3 See also

- Greek mathematics
- History of mathematics
- Trigonometric functions
- Trigonometry
- Ptolemy's table of chords
- Aryabhata's sine table

10.4   Citations and footnotes

[1] Boyer (1991), page 252: *It was Robert of Chester's translation from the Arabic that resulted in our word "sine." The Hindus had given the name jiva to the half-chord in trigonometry, and the Arabs had taken this over as jiba. In the Arabic language there is also the word jaib meaning "bay" or "inlet." When Robert of Chester came to translate the technical word jiba, he seems to have confused this with the word jaib (perhaps because vowels were omitted); hence, he used the word sinus, the Latin word for "bay" or "inlet."*

[2] Maor, Eli (1998). *Trigonometric Delights.* Princeton University Press. p. 20. ISBN 0-691-09541-8.

[3] Boyer (1991), page 252: "It was Robert of Chester's translation from the Arabic that resulted in our word "sine." The Hindus had given the name jiva to the half-chord in trigonometry, and the Arabs had taken this over as jiba. In the Arabic language there is also the word jaib meaning 'bay' or 'inlet.' When Robert of Chester came to translate the technical word jiba, he seems to have confused this with the word jaib (perhaps because vowels were omitted); hence, he used the word sinus, the Latin word for 'bay' or 'inlet.'"

[4] O'Connor (1996).

[5] Boyer (1991). "Greek Trigonometry and Mensuration". pp. 166–167. It should be recalled that form the days of Hipparchus until modern times there were no such things as trigonometric *ratios*. The Greeks, and after them the Hindus and the Arabs, used trigonometric *lines*. These at first took the form, as we have seen, of chords in a circle, and it became incumbent upon Ptolemy to associate numerical values (or approximations) with the chords. [...] It is not unlikely that the 260-degree measure was carried over from astronomy, where the zodiac had been divided into twelve "signs" or 36 "decans." A cycle of the seasons of roughly 360 days could readily be made to correspond to the system of zodiacal signs and decans by subdividing each sign into thirty parts and each decan into ten parts. Our common system of angle measure may stem from this correspondence. Moreover since the Babylonian position system for fractions was so obviously superior to the Egyptians unit fractions and the Greek common fractions, it was natural for Ptolemy to subdivide his degrees into sixty *partes minutae primae*, each of these latter into sixty *partes minutae secundae*, and so on. It is from the Latin phrases that translators used in this connection that our words "minute" and "second" have been derived. It undoubtedly was the sexagesimal system that led Ptolemy to subdivide the diameter of his trigonometric circle into 120 parts; each of these he further subdivided into sixty minutes and each minute of length sixty seconds. Missing or empty |title= (help)

[6] Boyer (1991). "Greek Trigonometry and Mensuration". pp. 158–159. Trigonometry, like other branches of mathematics, was not the work of any one man, or nation. Theorems on ratios of the sides of similar triangles had been known to, and used by, the ancient Egyptians and Babylonians. In view of the pre-Hellenic lack of the concept of angle measure, such a study might better be called "trilaterometry," or the measure of three sided polygons (trilaterals), than "trigonometry," the measure of parts of a triangle. With the Greeks we first find a systematic study of relationships between angles (or arcs) in a circle and the lengths of chords subtending these. Properties of chords, as measures of central and inscribed angles in circles, were familiar to the Greeks of Hippocrates' day, and it is likely that Eudoxus had used ratios and angle measures in determining the size of the earth and the relative distances of the sun and the moon. In the works of Euclid there is no trigonometry in the strict sense of the word, but there are theorems equivalent to specific trigonometric laws or formulas. Propositions II.12 and 13 of the *Elements*, for example, are the laws of cosines for obtuse and acute angles respectively, stated in geometric rather than trigonometric language and proved by a method similar to that used by Euclid in connection with the Pythagorean theorem. Theorems on the lengths of chords are essentially applications of the modern law of sines. We have seen that Archimedes' theorem on the broken chord can readily be translated into trigonometric language analogous to formulas for sines of sums and differences of angles. Missing or empty |title= (help)

[7] Joseph (2000b, pp.383–84).

[8] Boyer (1991). "Greek Trigonometry and Mensuration". p. 163. In Book I of this treatise Menelaus establishes a basis for spherical triangles analogous to that of Euclid I for plane triangles. Included is a theorem without Euclidean analogue – that two spherical triangles are congruent if corresponding angles are equal (Menelaus did not distinguish between congruent and symmetric spherical triangles); and the theorem $A + B + C > 180°$ is established. The second book of the *Sphaerica* describes the application of spherical geometry to astronomical phenomena and is of little mathematical interest. Book III, the last, contains the well known "theorem of Menelaus" as part of what is essentially spherical trigonometry in the typical Greek form – a geometry or trigonometry of chords in a circle. In the circle in Fig. 10.4 we should write that chord AB is twice the sine of half the central angle AOB (multiplied by the radius of the circle). Menelaus and his Greek successors instead referred to AB simply as the chord corresponding to the arc AB. If BOB' is a diameter of the circle, then chord A' is twice the cosine of half the angle AOB (multiplied by the radius of the circle). Missing or empty |title= (help)

[9] Boyer (1991). "Greek Trigonometry and Mensuration". p. 159. Instead we have an Aristarchan treatise, perhaps composed earlier (ca. 260 BC), *On the Sizes and Distances of the Sun and Moon*, which assumes a geocentric universe. In this work Aristarchus made the observation that when the moon is just half-full, the angle between the lines of sight to the sun and the moon is less than a right angle by one thirtieth of a quadrant. (The systematic introduction of the 360° circle came a little later. In trigonometric language of today this would mean that the ratio of the distance of the moon to that of the sun (the ration ME to SE in Fig. 10.1) is sin 3°. Trigonometric tables not having been developed yet, Aristarchus fell back upon a well-known geometric theorem of the time which now would be expressed in the inequalities $\sin \alpha / \sin \beta < \alpha/\beta < \tan \alpha / \tan \beta$, for $0° < \beta < \alpha < 90°$.) Missing or empty |title= (help)

[10] Boyer (1991). "Greek Trigonometry and Mensuration". p. 162. For some two and a half centuries, from Hippocrates to Eratosthenes, Greek mathematicians had studied relationships between lines and circles and had applied these in a variety of astronomical problems, but no systematic trigonometry had resulted. Then, presumably during the second half of the 2nd century BC, the first trigonometric table apparently was compiled by the astronomer Hipparchus of Nicaea (ca. 180–ca. 125 BC), who thus earned the right to be known as "the father of trigonometry." Aristarchus had known that in a given circle the ratio of arc to chord decreases as the arc decreases from 180° to 0°, tending toward a limit of 1. However, it appears that not until Hipparchus undertook the task had anyone tabulated corresponding values of arc and chord for a whole series of angles. Missing or empty |title= (help)

[11] Boyer (1991). "Greek Trigonometry and Mensuration". p. 162. It is not known just when the systematic use of the 360° circle came into mathematics, but it seems to be due largely to Hipparchus in connection with his table of chords. It is possible that he took over from Hypsicles, who earlier had divided the day into parts, a subdivision that may have been suggested by Babylonian astronomy. Missing or empty |title= (help)

[12] Needham, Volume 3, 108.

[13] Toomer, G. J. (1998), *Ptolemy's Almagest*, Princeton University Press, ISBN 0-691-00260-6

[14] Boyer (1991). "Greek Trigonometry and Mensuration". pp. 164–166. The theorem of Menelaus played a fundamental role in spherical trigonometry and astronomy, but by far the most influential and significant trigonometric work of all antiquity was composed by Ptolemy of Alexandria about half a century after Menelaus. [...] Of the life of the author we are as little informed as we are of that of the author of the Elements. We do not know when or where Euclid and Ptolemy were born. We know that Ptolemy made observations at Alexandria from AD. 127 to 151 and, therefore, assume that he was born at the end of the 1st century. Suidas, a writer who lived in the 10th century, reported that Ptolemy was alive under Marcus Aurelius (emperor from AD 161 to 180).
Ptolemy's *Almagest* is presumed to be heavily indebted for its methods to the *Chords in a Circle* of Hipparchus, but the extent of the indebtedness cannot be reliably assessed. It is clear that in astronomy Ptolemy made use of the catalog of star positions bequeathed by Hipparchus, but whether or not Ptolemy's trigonometric tables were derived in large part from his distinguished predecessor cannot be determined. [...] Central to the calculation of Ptolemy's chords was a geometric proposition still known as "Ptolemy's theorem": [...] that is, the sum of the products of the opposite sides of a cyclic quadrilateral is equal to the product of the diagonals. [...] A special case of Ptolemy's theorem had appeared in Euclid's *Data* (Proposition 93): [...] Ptolemy's theorem, therefore, leads to the result $\sin(\alpha - \beta) = \sin \alpha \cos \beta - \cos \alpha \sin B$. Similar reasoning leads to the formula [...] These four sum-and-difference formulas consequently are often known today as Ptolemy's formulas.
It was the formula for sine of the difference – or, more accurately, chord of the difference – that Ptolemy found especially useful in building up his tables. Another formula that served him effectively was the equivalent of our half-angle formula. Missing or empty |title= (help)

[15] Boyer, pp. 158–168.

[16] Boyer (1991), p. 208.

[17] Boyer (1991), p. 209.

[18] Boyer (1991), p. 210.

[19] Boyer (1991), p. 215.

[20] Joseph (2000a, pp.285–86).

[21] O'Connor and Robertson (2000).

[22] Pearce (2002).

[23] Charles Henry Edwards (1994). *The historical development of the calculus.* Springer Study Edition Series (3 ed.). Springer. p. 205. ISBN 978-0-387-94313-8.

[24] Kennedy, E. S. (1969). "The History of Trigonometry". *31st Yearbook* (National Council of Teachers of Mathematics, Washington DC). (cf. Haq, Syed Nomanul. "The Indian and Persian background". pp. 60–3., in Seyyed Hossein Nasr, Oliver Leaman (1996). *History of Islamic Philosophy*. Routledge. pp. 52–70. ISBN 0-415-13159-6.)

[25] O'Connor, John J.; Robertson, Edmund F., "Menelaus of Alexandria", *MacTutor History of Mathematics archive*, University of St Andrews. "Book 3 deals with spherical trigonometry and includes Menelaus's theorem."

[26] Kennedy, E. S. (1969). "The History of Trigonometry". *31st Yearbook* (National Council of Teachers of Mathematics, Washington DC): 337. (cf. Haq, Syed Nomanul. "The Indian and Persian background". p. 68., in Seyyed Hossein Nasr, Oliver Leaman (1996). *History of Islamic Philosophy*. Routledge. pp. 52–70. ISBN 0-415-13159-6.)

[27] Gingerich, Owen (April 1986). "Islamic astronomy". *Scientific American* **254** (10): 74. doi:10.1038/scientificamerican0486-74. Retrieved 2008-05-18.

[28] Jacques Sesiano, "Islamic mathematics", p. 157, in Selin, Helaine; D'Ambrosio, Ubiratan, eds. (2000). *Mathematics Across Cultures: The History of Non-western Mathematics*. Springer. ISBN 1-4020-0260-2.

[29] "trigonometry". *Encyclopædia Britannica*. Retrieved 2008-07-21.

[30] Boyer (1991) p. 238.

[31] Moussa, Ali (2011). "Mathematical Methods in Abū al-Wafā"'s Almagest and the Qibla Determinations". *Arabic Sciences and Philosophy* (Cambridge University Press) **21** (1): 1–56. doi:10.1017/S095742391000007X.

[32] William Charles Brice, 'An Historical atlas of Islam', p.413

[33] O'Connor, John J.; Robertson, Edmund F., "Abu Abd Allah Muhammad ibn Muadh Al-Jayyani", *MacTutor History of Mathematics archive*, University of St Andrews.

[34] Donald Routledge Hill (1996), "Engineering", in Roshdi Rashed, *Encyclopedia of the History of Arabic Science*, Vol. 3, p. 751–795 [769].

[35] O'Connor, John J.; Robertson, Edmund F., "Abu Arrayhan Muhammad ibn Ahmad al-Biruni", *MacTutor History of Mathematics archive*, University of St Andrews.

[36] Berggren, J. Lennart (2007). "Mathematics in Medieval Islam". *The Mathematics of Egypt, Mesopotamia, China, India, and Islam: A Sourcebook*. Princeton University Press. p. 518. ISBN 978-0-691-11485-9.

[37] Needham, Volume 3, 109.

[38] Needham, Volume 3, 108–109.

[39] Katz, 308.

[40] Restivo, 32.

[41] Gauchet, 151.

[42] Needham, Volume 3, 109–110.

[43] Needham, Volume 3, 110.

[44] Simonson, Shai. "The Mathematics of Levi ben Gershon, the Ralbag" (PDF). Retrieved 2009-06-22.

[45] Boyer, p. 274

[46] "The calculus of the trigonometric functions", Historia Mathematica Volume 14, Issue 4, November 1987, Pages 311–324, by Victor J. Katz doi 10.1016/0315-0860(87)90064-4, the proof of Cotes is mentioned on p. 315.

10.5 References

- Boyer, Carl B. (1991). *A History of Mathematics* (Second ed.). John Wiley & Sons, Inc. ISBN 0-471-54397-7.

- Gauchet, L. (1917). *Note Sur La Trigonométrie Sphérique de Kouo Cheou-King.*

- Joseph, George G. (2000). *The Crest of the Peacock.* Princeton, NJ: Princeton University Press. ISBN 0-691-00659-8.

- Joseph, George G. (2000). *The Crest of the Peacock: Non-European Roots of Mathematics* (2nd ed.). London: Penguin Books. ISBN 0-691-00659-8.

- Katz, Victor J. (November 1987). "The calculus of the trigonometric functions". *Historia Mathematica* **14** (4): 311-—324. doi:10.1016/0315-0860(87)90064-4. Retrieved 1 September 2014.

- Katz, Victor J. (2007). The *Mathematics of Egypt, Mesopotamia, China, India, and Islam: A Sourcebook.* Princeton: Princeton University Press. ISBN 0-691-11485-4.

- Needham, Joseph (1986). *Science and Civilization in China: Volume 3, Mathematics and the Sciences of the Heavens and the Earth.* Taipei: Caves Books, Ltd.

- O'Connor, J.J., and E.F. Robertson, "Trigonometric functions", *MacTutor History of Mathematics Archive.* (1996).

- O'Connor, J.J., and E.F. Robertson, "Madhava of Sangamagramma", *MacTutor History of Mathematics Archive.* (2000).

- Pearce, Ian G., "Madhava of Sangamagramma", *MacTutor History of Mathematics Archive.* (2002).

- Restivo, Sal. (1992). *Mathematics in Society and History: Sociological Inquiries.* Dordrecht: Kluwer Academic Publishers. ISBN 1-4020-0039-1.

Guo Shoujing (1231–1316)

Isaac Newton in a 1702 portrait by Godfrey Kneller.

Chapter 11

Numeral system

This article is about different methods of expressing numbers with symbols. For the classification of numbers in mathematics, see Number. For how numbers are expressed using words, see Numeral (linguistics).

A **numeral system** (or **system of numeration**) is a writing system for expressing numbers, that is, a mathematical notation for representing numbers of a given set, using digits or other symbols in a consistent manner. It can be seen as the context that allows the symbols "11" to be interpreted as the binary symbol for *three*, the decimal symbol for *eleven*, or a symbol for other numbers in different bases.

The number the numeral represents is called its value.

Ideally, a numeral system will:

- Represent a useful set of numbers (e.g. all integers, or rational numbers)
- Give every number represented a unique representation (or at least a standard representation)
- Reflect the algebraic and arithmetic structure of the numbers.

For example, the usual decimal representation of whole numbers gives every non zero whole number a unique representation as a finite sequence of digits, beginning by a non-zero digit. However, when decimal representation is used for the rational or real numbers, such numbers in general have an infinite number of representations, for example 2.31 can also be written as 2.310, 2.3100000, 2.309999999..., etc., all of which have the same meaning except for some scientific and other contexts where greater precision is implied by a larger number of figures shown.

Numeral systems are sometimes called *number systems*, but that name is ambiguous, as it could refer to different systems of numbers, such as the system of real numbers, the system of complex numbers, the system of *p*-adic numbers, etc. Such systems are, however, not the topic of this article.

11.1 Main numeral systems

The most commonly used system of numerals is the Hindu–Arabic numeral system. Two Indian mathematicians are credited with developing it. Aryabhata of Kusumapura developed the place-value notation in the 5th century and a century later Brahmagupta introduced the symbol for zero.[1] The numeral system and the zero concept, developed by the Hindus in India, slowly spread to other surrounding countries due to their commercial and military activities with India. The Arabs adopted and modified it. Even today, the Arabs call the numerals they use "Rakam Al-Hind" or the Hindu numeral system. The Arabs translated Hindu texts on numerology and spread them to the western world due to their trade links with them. The Western world modified them and called them the Arabic numerals, as they learned from them. Hence the current western numeral system is the modified version of the Hindu numeral system developed in India. It also exhibits a great similarity to the Sanskrit–Devanagari notation, which is still used in India.

The simplest numeral system is the unary numeral system, in which every natural number is represented by a corresponding number of symbols. If the symbol / is chosen, for example, then the number seven would be represented by ///////. Tally marks represent one such system still in common use. The unary system is only useful for small numbers, although it plays an important role in theoretical computer science. Elias gamma coding, which is commonly used in data compression, expresses arbitrary-sized numbers by using unary to indicate the length of a binary numeral.

The unary notation can be abbreviated by introducing different symbols for certain new values. Very commonly, these values are powers of 10; so for instance, if / stands for one, − for ten and + for 100, then the number 304 can be compactly represented as +++ //// and the number 123 as + − − /// without any need for zero. This is called sign-value notation. The ancient Egyptian numeral system was of this type, and the Roman numeral system was a modification of this idea.

More useful still are systems which employ special abbreviations for repetitions of symbols; for example, using the first nine letters of the alphabet for these abbreviations, with A standing for "one occurrence", B "two occurrences", and so on, one could then write C+ D/ for the number 304. This system is used when writing Chinese numerals and other East Asian numerals based on Chinese. The number system of the English language is of this type ("three hundred [and] four"), as are those of other spoken languages, regardless of what written systems they have adopted. However, many languages use mixtures of bases, and other features, for instance 79 in French is *soixante dix-neuf* (60 + 10 + 9) and in Welsh is *pedwar ar bymtheg a thrigain* (4 + (5 + 10) + (3 × 20)) or (somewhat archaic) *pedwar ugain namyn un* (4 × 20 − 1). In English, one could say "four score less one", as in the famous Gettysburg Address representing "87 years ago" as "four score and seven years ago".

More elegant is a *positional system*, also known as place-value notation. Again working in base 10, ten different digits 0, ..., 9 are used and the position of a digit is used to signify the power of ten that the digit is to be multiplied with, as in $304 = 3 \times 100 + 0 \times 10 + 4 \times 1$ or more precisely $3 \times 10^2 + 0 \times 10^1 + 4 \times 10^0$. Note that zero, which is not needed in the other systems, is of crucial importance here, in order to be able to "skip" a power. The Hindu–Arabic numeral system, which originated in India and is now used throughout the world, is a positional base-10 system.

Arithmetic is much easier in positional systems than in the earlier additive ones; furthermore, additive systems need a large number of different symbols for the different powers of 10; a positional system needs only ten different symbols (assuming that it uses base 10).

Positional decimal system is presently universally used in human writing. The base 1000 is also used, by grouping the digits and considering a sequence of three decimal digits as a single digit. This is the meaning of the common notation 1,000,234,567 used for very large numbers.

In computers, the main numeral systems are based on the positional system in base 2 (binary numeral system), with two binary digits, 0 and 1. Positional systems obtained by grouping binary digits by three (octal numeral system) or four (hexadecimal numeral system) are commonly used. For very large integers, bases 2^{32} or 2^{64} (grouping binary digits by 32 or 64, the length of the machine word) are used, as, for example, in GMP.

The numerals used when writing numbers with digits or symbols can be divided into two types that might be called the arithmetic numerals 0,1,2,3,4,5,6,7,8,9 and the geometric numerals 1, 10, 100, 1000, 10000 ..., respectively. The sign-value systems use only the geometric numerals and the positional systems use only the arithmetic numerals. A sign-value system does not need arithmetic numerals because they are made by repetition (except for the Ionic system), and a positional system does not need geometric numerals because they are made by position. However, the spoken language uses *both* arithmetic and geometric numerals.

In certain areas of computer science, a modified base-k positional system is used, called bijective numeration, with digits 1, 2, ..., k ($k \geq 1$), and zero being represented by an empty string. This establishes a bijection between the set of all such digit-strings and the set of non-negative integers, avoiding the non-uniqueness caused by leading zeros. Bijective base-k numeration is also called k-adic notation, not to be confused with p-adic numbers. Bijective base-1 is the same as unary.

11.2   Positional systems in detail

See also: Positional notation

In a positional base-b numeral system (with b a natural number greater than 1 known as the radix), b basic symbols (or digits) corresponding to the first b natural numbers including zero are used. To generate the rest of the numerals, the position of the symbol in the figure is used. The symbol in the last position has its own value, and as it moves to the left its value is multiplied by b.

For example, in the decimal system (base 10), the numeral 4327 means $(4 \times 10^3) + (3 \times 10^2) + (2 \times 10^1) + (7 \times 10^0)$, noting that $10^0 = 1$.

In general, if b is the base, one writes a number in the numeral system of base b by expressing it in the form $a_n b^n + a_{n-1} b^{n-1} + a_{n-2} b^{n-2} + \ldots + a_0 b^0$ and writing the enumerated digits $a_n a_{n-1} a_{n-2} \ldots a_0$ in descending order. The digits are natural numbers between 0 and $b - 1$, inclusive.

If a text (such as this one) discusses multiple bases, and if ambiguity exists, the base (itself represented in base 10) is added in subscript to the right of the number, like this: number$_{\text{base}}$. Unless specified by context, numbers without subscript are considered to be decimal.

By using a dot to divide the digits into two groups, one can also write fractions in the positional system. For example, the base-2 numeral 10.11 denotes $1 \times 2^1 + 0 \times 2^0 + 1 \times 2^{-1} + 1 \times 2^{-2} = 2.75$.

In general, numbers in the base b system are of the form:

$$(a_n a_{n-1} \cdots a_1 a_0 . c_1 c_2 c_3 \cdots)_b = \sum_{k=0}^{n} a_k b^k + \sum_{k=1}^{\infty} c_k b^{-k}.$$

The numbers b^k and b^{-k} are the weights of the corresponding digits. The position k is the logarithm of the corresponding weight w, that is $k = \log_b w = \log_b b^k$. The highest used position is close to the order of magnitude of the number.

The number of tally marks required in the unary numeral system for *describing the weight* would have been **w**. In the positional system, the number of digits required to describe it is only $k + 1 = \log_b w + 1$, for $k \geq 0$. For example, to describe the weight 1000 then four digits are needed because $\log_{10} 1000 + 1 = 3 + 1$. The number of digits required to *describe the position* is $\log_b k + 1 = \log_b \log_b w + 1$ (in positions 1, 10, 100,... only for simplicity in the decimal example).

Note that a number has a terminating or repeating expansion if and only if it is rational; this does not depend on the base. A number that terminates in one base may repeat in another (thus $0.3_{10} = 0.0100110011001\ldots_2$). An irrational number stays aperiodic (with an infinite number of non-repeating digits) in all integral bases. Thus, for example in base 2, $\pi = 3.1415926\ldots_{10}$ can be written as the aperiodic $11.001001000011111\ldots_2$.

Putting overscores, n, or dots, $\dot{n}$, above the common digits is a convention used to represent repeating rational expansions. Thus:

$$14/11 = 1.272727272727\ldots = 1.\overline{27} \text{ or } 321.3217878787878\ldots = 321.32\overline{178}.$$

If $b = p$ is a prime number, one can define base-p numerals whose expansion to the left never stops; these are called the p-adic numbers.

11.3 Generalized variable-length integers

More general is using a mixed radix notation (here written little-endian) like $a_0 a_1 a_2$ for $a_0 + a_1 b_1 + a_2 b_1 b_2$, etc.

This is used in punycode, one aspect of which is the representation of a sequence of non-negative integers of arbitrary size in the form of a sequence without delimiters, of "digits" from a collection of 36: a–z and 0–9, representing 0–25 and 26–35 respectively. A digit lower than a threshold value marks that it is the most-significant digit, hence the end of the number. The threshold value depends on the position in the number. For example, if the threshold value for the first digit is b (i.e. 1) then a (i.e. 0) marks the end of the number (it has just one digit), so in numbers of more than one digit the range is only b–9 (1–35), therefore the weight b_1 is 35 instead of 36. Suppose the threshold values for the second and third digits are c (2), then the third digit has a weight $34 \times 35 = 1190$ and we have the following sequence:

a (0), ba (1), ca (2), .., 9a (35), bb (36), cb (37), .., 9b (70), bca (71), .., 99a (1260), bcb (1261), etc.

Unlike a regular based numeral system, there are numbers like 9b where 9 and b each represents 35; yet the representation is unique because ac and aca are not allowed – the a would terminate the number.

The flexibility in choosing threshold values allows optimization depending on the frequency of occurrence of numbers of various sizes.

The case with all threshold values equal to 1 corresponds to bijective numeration, where the zeros correspond to separators of numbers with digits which are non-zero.

11.4   See also

- List of numeral systems
- Computer numbering formats
- Golden ratio base
- List of numeral system topics
- *n*-ary
- Number names
- Quater-imaginary base
- Quipu
- Recurring decimal
- Residue numeral system
- Short and long scales
- Subtractive notation
- -yllion
- Numerical cognition
- Number system

11.5   References

[1] David Eugene Smith; Louis Charles Karpinski (1911). *The Hindu-Arabic numerals*. Ginn and Company.

11.6   Sources

- Georges Ifrah. *The Universal History of Numbers : From Prehistory to the Invention of the Computer*, Wiley, 1999. ISBN 0-471-37568-3.

- D. Knuth. *The Art of Computer Programming*. Volume 2, 3rd Ed. Addison–Wesley. pp. 194–213, "Positional Number Systems".

- A.L. Kroeber (Alfred Louis Kroeber) (1876–1960), Handbook of the Indians of California, Bulletin 78 of the Bureau of American Ethnology of the Smithsonian Institution (1919)

- J.P. Mallory and D.Q. Adams, *Encyclopedia of Indo-European Culture*, Fitzroy Dearborn Publishers, London and Chicago, 1997.

- Hans J. Nissen; Peter Damerow; Robert K. Englund (1993). *Archaic Bookkeeping: Early Writing and Techniques of Economic Administration in the Ancient Near East.* University Of Chicago Press. ISBN 978-0-226-58659-5.

- Schmandt-Besserat, Denis (1996). *How Writing Came About.* University of Texas Press. ISBN 978-0-292-77704-0.

- Zaslavsky, Claudia (1999). *Africa counts: number and pattern in African cultures.* Chicago Review Press. ISBN 978-1-55652-350-2.

11.7 External links

- Numerical Mechanisms and Children's Concept of Numbers

- Software for converting from one numeral system to another

- Online conversion of fractional numbers between numeral systems

- Open source numeral systems converter

- Open source numeral systems calculator

- Online multi numeral system converter

Chapter 12

Kenneth O. May Prize

Kenneth O. May Prize and **Medal** in history of mathematics is an award of the International Commission on the History of Mathematics (ICHM) "for the encouragement and promotion of the history of mathematics internationally". It was established in 1989 and is named in honor of Kenneth O. May, the founder of ICHM. Since then, the award is given every four years, at the ICHM congress.

12.1 Kenneth O. May Prize winners

Source: (1989-2005) A Brief History of the Kenneth O. May Prize

- 2013: Menso Folkerts and Jens Høyrup [1]

- 2009: Ivor Grattan-Guinness and Radha Charan Gupta [2]

- 2005: Henk J. M. Bos

- 2001: Ubiratàn D'Ambrosio and Lam Lay Yong

- 1997: René Taton

- 1993: Christoph Scriba and Hans Wussing

- 1989: Dirk Struik and Adolph P. Yushkevich

12.2 References

[1] Craig Fraser – ICHM: Awarding of the May Prizes for 2013

[2] BLC Newsletter August 2009

- A Brief History of the Kenneth O. May Prize in the History of Mathematics

- BLC Newsletter August 2009

Chapter 13

List of important publications in mathematics

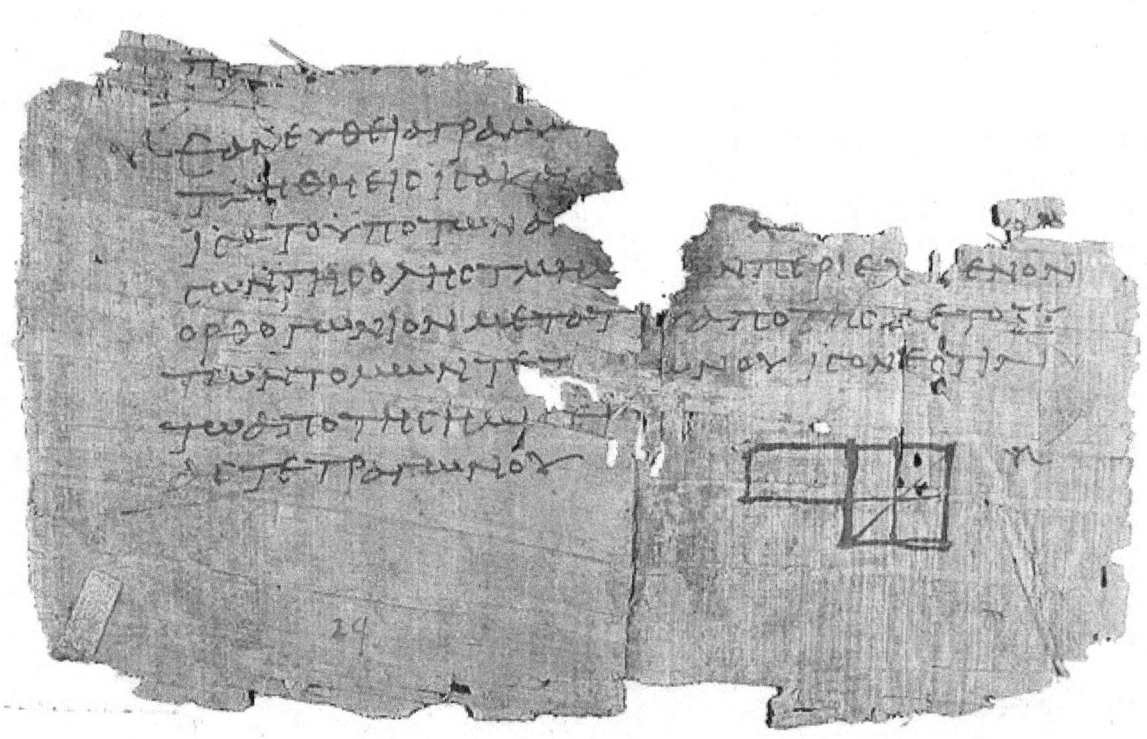

One of the oldest surviving fragments of Euclid's Elements, *found at Oxyrhynchus and dated to circa AD 100. The diagram accompanies Book II, Proposition 5.*[1]

This is a list of **important publications** in mathematics, organized by field.

Some reasons why a particular publication might be regarded as important:

- **Topic creator** – A publication that created a new topic

- **Breakthrough** – A publication that changed scientific knowledge significantly

- **Influence** – A publication which has significantly influenced the world or has had a massive impact on the teaching of mathematics.

Among published compilations of important publications in mathematics are *Landmark writings in Western mathematics 1640–1940* by Ivor Grattan-Guinness[2] and *A Source Book in Mathematics* by David Eugene Smith.[3]

13.1   Algebra

13.1.1   Theory of equations

Baudhayana Sulba Sutra

- Baudhayana (8th century BC)

Description: Believed to have been written around the 8th century BC, this is one of the oldest mathematical texts. It laid the foundations of Indian mathematics and was influential in South Asia and its surrounding regions, and perhaps even Greece. Though this was primarily a geometrical text, it also contained some important algebraic developments, including the earliest list of Pythagorean triples discovered algebraically, geometric solutions of linear equations, the earliest use of quadratic equations of the forms $ax^2 = c$ and $ax^2 + bx = c$, and integral solutions of simultaneous Diophantine equations with up to four unknowns.

The Nine Chapters on the Mathematical Art

- *The Nine Chapters on the Mathematical Art* from the 10th–2nd century BCE.

Description:Contains the earliest description of Gaussian elimination for solving system of linear equations, it also contains method for finding square root and cubic root.

The Sea Island Mathematical Manual

- Liu Hui (220-280)

Description, contains the application of right angle triangles for survey of depth or height of distant objects.

The Mathematical Classic of Sun Zi

- Sunzi (5th century)

Description: Contains the earlist description of Chinese remainder theorem.

Aryabhatiya

- Aryabhata (499 CE)

Description: Aryabhatia introduced the method known as "Modus Indorum" or the method of the Indians that has become our algebra today. This algebra came along with the Hindu Number system to Arabia and then migrated to Europe. The text contains 33 verses covering mensuration (kṣetra vyāvahāra), arithmetic and geometric progressions, gnomon / shadows (shanku-chhAyA), simple, quadratic, simultaneous, and indeterminate equations. It also gave the modern standard algorithm for solving first-order diophantine equations.

Jigu Suanjing

Jigu Suanjing (626AD)

Description: This book by Tang dynasty mathematician Wang Xiaotong Contains the world's earliest third order equation.

Brāhmasphuṭasiddhānta

- Brahmagupta (628 AD)

Description: Contained rules for manipulating both negative and positive numbers, a method for computing square roots, and general methods of solving linear and some quadratic equations. [4] [5] [6] [7]

Al-Kitāb al-mukhtaṣar fī hīsāb al-ğabr wa'l-muqābala

- Muhammad ibn Mūsā al-Khwārizmī (820)

Description: The first book on the systematic algebraic solutions of linear and quadratic equations by the Persian scholar Muhammad ibn Mūsā al-Khwārizmī. The book is considered to be the foundation of modern algebra and Islamic mathematics. The word "algebra" itself is derived from the *al-Jabr* in the title of the book.[8]

Yigu yanduan

- Liu Yi (12th century)

Contains the earliest invention of 4th order polynomial equation.

Mathematical Treatise in Nine Sections

- Qin Jiushao (1247)

Description: This 13th century book contains the earliest complete solution of 19th century Horner's method of solving high order polynomial equations (up to 10th order). It also contains a complete solution of Chinese remainder theorem, which predates Euler and Gauss by several centuries.

Ceyuan haijing

- Li Zhi (1248)

Description: Contains the application of high order polynomial equation in solving complex geometry problems.

Jade Mirror of the Four Unknowns

- Zhu Shijie (1303)

Description Contains the method of establishing system of high order polynomial equations of up to four unknowns.

Ars Magna

- Gerolamo Cardano (1545)

Description: Otherwise known as *The Great Art*, provided the first published methods for solving cubic and quartic equations (due to Scipione del Ferro, Niccolò Fontana Tartaglia, and Lodovico Ferrari), and exhibited the first published calculations involving non-real complex numbers.[9][10]

Vollständige Anleitung zur Algebra

- Leonhard Euler (1770)

Description: Also known as Elements of Algebra, Euler's textbook on elementary algebra is one of the first to set out algebra in the modern form we would recognize today. The first volume deals with determinate equations, while the second part deals with Diophantine equations. The last section contains a proof of Fermat's Last Theorem for the case $n = 3$, making some valid assumptions regarding $\mathbf{Q}(\sqrt{-3})$ that Euler did not prove.[11]

Demonstratio nova theorematis omnem functionem algebraicam rationalem integram unius variabilis in factores reales primi vel secundi gradus resolvi posse

- Carl Friedrich Gauss (1799)

Description: Gauss' doctoral dissertation,[12] which contained a widely accepted (at the time) but incomplete proof[13] of the fundamental theorem of algebra.

13.1.2   Abstract algebra

Group theory

Réflexions sur la résolution algébrique des équations

- Joseph Louis Lagrange (1770)

Description: The title means "Reflections on the algebraic solutions of equations". Made the prescient observation that the roots of the Lagrange resolvent of a polynomial equation are tied to permutations of the roots of the original equation, laying a more general foundation for what had previously been an ad hoc analysis and helping motivate the later development of the theory of permutation groups, group theory, and Galois theory. The Lagrange resolvent also introduced the discrete Fourier transform of order 3.

Articles Publiés par Galois dans les Annales de Mathématiques

- Journal de Mathematiques pures et Appliquées, II (1846)

Description: Posthumous publication of the mathematical manuscripts of Évariste Galois by Joseph Liouville. Included are Galois' papers *Mémoire sur les conditions de résolubilité des équations par radicaux* and *Des équations primitives qui sont solubles par radicaux*.

Traité des substitutions et des équations algébriques

- Camille Jordan (1870)

Online version: Online version

Description: Traité des substitutions et des équations algébriques (Treatise on Substitutions and Algebraic Equations). The first book on group theory, giving a then-comprehensive study of permutation groups and Galois theory. In this book, Jordan introduced the notion of a simple group and epimorphism (which he called *l'isomorphisme mériédrique*),[14] proved part of the Jordan–Hölder theorem, and discussed matrix groups over finite fields as well as the Jordan normal form.[15]

Theorie der Transformationsgruppen

- Sophus Lie, Friedrich Engel (1888–1893).

Publication data: 3 volumes, B.G. Teubner, Verlagsgesellschaft, mbH, Leipzig, 1888–1893. Volume 1, Volume 2, Volume 3.

Description: The first comprehensive work on transformation groups, serving as the foundation for the modern theory of Lie groups.

Solvability of groups of odd order

- Walter Feit and John Thompson (1960)

Description: Gave a complete proof of the solvability of finite groups of odd order, establishing the long-standing Burnside conjecture that all finite non-abelian simple groups are of even order. Many of the original techniques used in this paper were used in the eventual classification of finite simple groups.

Homological algebra

Homological Algebra

- Henri Cartan and Samuel Eilenberg (1956)

Description: Provided the first fully worked out treatment of abstract homological algebra, unifying previously disparate presentations of homology and cohomology for associative algebras, Lie algebras, and groups into a single theory.

Sur Quelques Points d'Algèbre Homologique

- Alexander Grothendieck (1957)

Description: Revolutionized homological algebra by introducing abelian categories and providing a general framework for Cartan and Eilenberg's notion of derived functors.

13.2 Algebraic geometry

13.2.1 Theorie der Abelschen Functionen

- Bernhard Riemann (1857)

Publication data: *Journal für die Reine und Angewandte Mathematik*

Description: Developed the concept of Riemann surfaces and their topological properties beyond Riemann's 1851 thesis work, proved an index theorem for the genus (the original formulation of the Riemann–Hurwitz formula), proved the Riemann inequality for the dimension of the space of meromorphic functions with prescribed poles (the original formulation of the Riemann–Roch theorem), discussed birational transformations of a given curve and the dimension of the corresponding moduli space of inequivalent curves of a given genus, and solved more general inversion problems than those investigated by Abel and Jacobi. André Weil once wrote that this paper "*is one of the greatest pieces of mathematics that has ever been written; there is not a single word in it that is not of consequence.*" [16]

13.2.2 *Faisceaux Algébriques Cohérents*

- Jean-Pierre Serre

Publication data: *Annals of Mathematics*, 1955

Description: *FAC*, as it is usually called, was foundational for the use of sheaves in algebraic geometry, extending beyond the case of complex manifolds. Serre introduced Čech cohomology of sheaves in this paper, and, despite some technical deficiencies, revolutionized formulations of algebraic geometry. For example, the long exact sequence in sheaf cohomology allows one to show that some surjective maps of sheaves induce surjective maps on sections; specifically, these are the maps whose kernel (as a sheaf) has a vanishing first cohomology group. The dimension of a vector space of sections of a coherent sheaf is finite, in projective geometry, and such dimensions include many discrete invariants of varieties, for example Hodge numbers. While Grothendieck's derived functor cohomology has replaced Čech cohomology for technical reasons, actual calculations, such as of the cohomology of projective space, are usually carried out by Čech techniques, and for this reason Serre's paper remains important.

13.2.3 *Géométrie Algébrique et Géométrie Analytique*

- Jean-Pierre Serre (1956)

Description: In mathematics, algebraic geometry and analytic geometry are closely related subjects, where *analytic geometry* is the theory of complex manifolds and the more general analytic spaces defined locally by the vanishing of analytic functions of several complex variables. A (mathematical) theory of the relationship between the two was put in place during the early part of the 1950s, as part of the business of laying the foundations of algebraic geometry to include, for example, techniques from Hodge theory. (*NB* While analytic geometry as use of Cartesian coordinates is also in a sense included in the scope of algebraic geometry, that is not the topic being discussed in this article.) The major paper consolidating the theory was *Géométrie Algébrique et Géométrie Analytique* by Serre, now usually referred to as *GAGA*. A *GAGA-style result* would now mean any theorem of comparison, allowing passage between a category of objects from algebraic geometry, and their morphisms, and a well-defined subcategory of analytic geometry objects and holomorphic mappings.

13.2.4 Le théorème de Riemann–Roch, d'après A. Grothendieck

- Armand Borel, Jean-Pierre Serre (1958)

Description: Borel and Serre's exposition of Grothendieck's version of the Riemann–Roch theorem, published after Grothendieck made it clear that he was not interested in writing up his own result. Grothendieck reinterpreted both sides of the formula that Hirzebruch proved in 1953 in the framework of morphisms between varieties, resulting in a sweeping generalization.[17] In his proof, Grothendieck broke new ground with his concept of Grothendieck groups, which led to the development of K-theory.[18]

13.2.5 *Éléments de géométrie algébrique*

- Alexander Grothendieck (1960–1967)

Description: Written with the assistance of Jean Dieudonné, this is Grothendieck's exposition of his reworking of the foundations of algebraic geometry. It has become the most important foundational work in modern algebraic geometry. The approach expounded in EGA, as these books are known, transformed the field and led to monumental advances.

13.2.6 *Séminaire de géométrie algébrique*

- Alexander Grothendieck et al.

Description: These seminar notes on Grothendieck's reworking of the foundations of algebraic geometry report on work done at IHÉS starting in the 1960s. SGA 1 dates from the seminars of 1960–1961, and the last in the series, SGA 7, dates from 1967 to 1969. In contrast to EGA, which is intended to set foundations, SGA describes ongoing research as it unfolded in Grothendieck's seminar; as a result, it is quite difficult to read, since many of the more elementary and foundational results were relegated to EGA. One of the major results building on the results in SGA is Pierre Deligne's proof of the last of the open Weil conjectures in the early 1970s. Other authors who worked on one or several volumes of SGA include Michel Raynaud, Michael Artin, Jean-Pierre Serre, Jean-Louis Verdier, Pierre Deligne, and Nicholas Katz.

13.3 Number theory

13.3.1 *Brāhmasphuṭasiddhānta*

- Brahmagupta (628)

Description: Brahmagupta's Brāhmasphuṭasiddhānta is the first book that mentions zero as a number, hence Brahmagupta is considered the first to formulate the concept of zero. The current system of the four fundamental operations (addition, subtraction, multiplication and division) based on the Hindu-Arabic number system also first appeared in Brahmasphutasiddhanta. It was also one of the first texts to provide concrete ideas on positive and negative numbers.

13.3.2 *De fractionibus continuis dissertatio*

- Leonhard Euler (1744)

Description: First presented in 1737, this paper [19] provided the first then-comprehensive account of the properties of continued fractions. It also contains the first proof that the number e is irrational.[20]

13.3.3 *Recherches d'Arithmétique*

- Joseph Louis Lagrange (1775)

Description: Developed a general theory of binary quadratic forms to handle the general problem of when an integer is representable by the form $ax^2 + by^2 + cxy$. This included a reduction theory for binary quadratic forms, where he proved that every form is equivalent to a certain canonically chosen reduced form.[21][22]

13.3.4   *Disquisitiones Arithmeticae*

- Carl Friedrich Gauss (1801)

Description: The *Disquisitiones Arithmeticae* is a profound and masterful book on number theory written by German mathematician Carl Friedrich Gauss and first published in 1801 when Gauss was 24. In this book Gauss brings together results in number theory obtained by mathematicians such as Fermat, Euler, Lagrange and Legendre and adds many important new results of his own. Among his contributions was the first complete proof known of the Fundamental theorem of arithmetic, the first two published proofs of the law of quadratic reciprocity, a deep investigation of binary quadratic forms going beyond Lagrange's work in Recherches d'Arithmétique, a first appearance of Gauss sums, cyclotomy, and the theory of constructible polygons with a particular application to the constructibility of the regular 17-gon. Of note, in section V, article 303 of Disquisitiones, Gauss summarized his calculations of class numbers of imaginary quadratic number fields, and in fact found all imaginary quadratic number fields of class numbers 1, 2, and 3 (confirmed in 1986) as he had conjectured.[23] In section VII, article 358, Gauss proved what can be interpreted as the first non-trivial case of the Riemann Hypothesis for curves over finite fields (the Hasse–Weil theorem).[24]

13.3.5   *Beweis des Satzes, daß jede unbegrenzte arithmetische Progression, deren erstes Glied und Differenz ganze Zahlen ohne gemeinschaftlichen Factor sind, unendlich viele Primzahlen enthält*

- Peter Gustav Lejeune Dirichlet (1837)

Description: Pioneering paper in analytic number theory, which introduced Dirichlet characters and their L-functions to establish Dirichlet's theorem on arithmetic progressions.[25] In subsequent publications, Dirichlet used these tools to determine, among other things, the class number for quadratic forms.

13.3.6   *Über die Anzahl der Primzahlen unter einer gegebenen Grösse*

- Bernhard Riemann (1859)

Description: *Über die Anzahl der Primzahlen unter einer gegebenen Grösse* (or *On the Number of Primes Less Than a Given Magnitude*) is a seminal 8-page paper by Bernhard Riemann published in the November 1859 edition of the *Monthly Reports of the Berlin Academy*. Although it is the only paper he ever published on number theory, it contains ideas which influenced dozens of researchers during the late 19th century and up to the present day. The paper consists primarily of definitions, heuristic arguments, sketches of proofs, and the application of powerful analytic methods; all of these have become essential concepts and tools of modern analytic number theory. It also contains the famous Riemann Hypothesis, one of the most important open problems in mathematics.[26]

13.3.7   *Vorlesungen über Zahlentheorie*

- Peter Gustav Lejeune Dirichlet and Richard Dedekind

Description: *Vorlesungen über Zahlentheorie* (*Lectures on Number Theory*) is a textbook of number theory written by German mathematicians P. G. Lejeune Dirichlet and R. Dedekind, and published in 1863. The *Vorlesungen* can be seen as a watershed between the classical number theory of Fermat, Jacobi and Gauss, and the modern number theory of Dedekind, Riemann and Hilbert. Dirichlet does not explicitly recognise the concept of the group that is central to modern algebra, but many of his proofs show an implicit understanding of group theory

13.3.8 *Zahlbericht*

Main article: Zahlbericht

- David Hilbert (1897)

Description: Unified and made accessible many of the developments in algebraic number theory made during the nineteenth century. Although criticized by André Weil (who stated "*more than half of his famous Zahlbericht is little more than an account of Kummer's number-theoretical work, with inessential improvements*")[27] and Emmy Noether,[28] it was highly influential for many years following its publication.

13.3.9 Fourier Analysis in Number Fields and Hecke's Zeta-Functions

- John Tate (1950)

Description: Generally referred to simply as *Tate's Thesis*, Tate's Princeton Ph.D. thesis, under Emil Artin, is a reworking of Erich Hecke's theory of zeta- and L-functions in terms of Fourier analysis on the adeles. The introduction of these methods into number theory made it possible to formulate extensions of Hecke's results to more general L-functions such as those arising from automorphic forms.

13.3.10 Automorphic Forms on GL(2)

Main article: Automorphic Forms on GL(2)

- Hervé Jacquet and Robert Langlands (1970)

Description: This publication offers evidence towards Langlands' conjectures by reworking and expanding the classical theory of modular forms and their L-functions through the introduction of representation theory.

13.3.11 La conjecture de Weil. I.

- Pierre Deligne (1974)

Description: Proved the Riemann hypothesis for varieties over finite fields, settling the last of the open Weil conjectures.

13.3.12 Endlichkeitssätze für abelsche Varietäten über Zahlkörpern

- Gerd Faltings (1983)

Description: Faltings proves a collection of important results in this paper, the most famous of which is the first proof of the Mordell conjecture (a conjecture dating back to 1922). Other theorems proved in this paper include an instance of the Tate conjecture (relating the homomorphisms between two abelian varieties over a number field to the homomorphisms between their Tate modules) and some finiteness results concerning abelian varieties over number fields with certain properties.

13.3.13   Modular Elliptic Curves and Fermat's Last Theorem

- Andrew Wiles (1995)

Description: This article proceeds to prove a special case of the Shimura–Taniyama conjecture through the study of the deformation theory of Galois representations. This in turn implies the famed Fermat's Last Theorem. The proof's method of identification of a deformation ring with a Hecke algebra (now referred to as an $R=T$ theorem) to prove modularity lifting theorems has been an influential development in algebraic number theory.

13.3.14   The geometry and cohomology of some simple Shimura varieties

- Michael Harris and Richard Taylor (2001)

Description: Harris and Taylor provide the first proof of the local Langlands conjecture for GL(n). As part of the proof, this monograph also makes an in depth study of the geometry and cohomology of certain Shimura varieties at primes of bad reduction.

13.3.15   Le lemme fondamental pour les algèbres de Lie

- Ngô Bảo Châu

Description: Ngô Bảo Châu proved a long standing unsolved problem in the classical Langlands program, using methods from the Geometric Langlands program.

13.4   Analysis

13.4.1   *Introductio in analysin infinitorum*

- Leonhard Euler (1748)

Description: The eminent historian of mathematics Carl Boyer once called Euler's *Introductio in analysin infinitorum* the greatest modern textbook in mathematics.[29] Published in two volumes,[30][31] this book more than any other work succeeded in establishing analysis as a major branch of mathematics, with a focus and approach distinct from that used in geometry and algebra.[32] Notably, Euler identified functions rather than curves to be the central focus in his book.[33] Logarithmic, exponential, trigonometric, and transcendental functions were covered, as were expansions into partial fractions, evaluations of $\zeta(2k)$ for k a positive integer between 1 and 13, infinite series-infinite product formulas,[29] continued fractions, and partitions of integers.[34] In this work, Euler proved that every rational number can be written as a finite continued fraction, that the continued fraction of an irrational number is infinite, and derived continued fraction expansions for e and $\sqrt{e}$.[30] This work also contains a statement of Euler's formula and a statement of the pentagonal number theorem, which he had discovered earlier and would publish a proof for in 1751.

13.4.2   Calculus

Yuktibhāṣā

- Jyeshtadeva (1501)

Description: Written in India in 1501, this was the world's first calculus text. "This work laid the foundation for a complete system of fluxions"[35] and served as a summary of the Kerala School's achievements in calculus, trigonometry and mathematical analysis, most of which were earlier discovered by the 14th century mathematician Madhava. It is possible that this text influenced the later development of calculus in Europe. Some of its important developments in calculus include: the fundamental ideas of differentiation and integration, the derivative, differential equations, term by term integration, numerical integration by means of infinite series, the relationship between the area of a curve and its integral, and the mean value theorem.

Nova methodus pro maximis et minimis, itemque tangentibus, quae nec fractas nec irrationales quantitates moratur, et singulare pro illi calculi genus

- Gottfried Leibniz (1684)

Description: Leibniz's first publication on differential calculus, containing the now familiar notation for differentials as well as rules for computing the derivatives of powers, products and quotients.

Philosophiae Naturalis Principia Mathematica

- Isaac Newton

Description: The *Philosophiae Naturalis Principia Mathematica* (Latin: "mathematical principles of natural philosophy", often *Principia* or *Principia Mathematica* for short) is a three-volume work by Isaac Newton published on 5 July 1687. Perhaps the most influential scientific book ever published, it contains the statement of Newton's laws of motion forming the foundation of classical mechanics as well as his law of universal gravitation, and derives Kepler's laws for the motion of the planets (which were first obtained empirically). Here was born the practice, now so standard we identify it with science, of explaining nature by postulating mathematical axioms and demonstrating that their conclusion are observable phenomena. In formulating his physical theories, Newton freely used his unpublished work on calculus. When he submitted Principia for publication, however, Newton chose to recast the majority of his proofs as geometric arguments.[36]

Institutiones calculi differentialis cum eius usu in analysi finitorum ac doctrina serierum

- Leonhard Euler (1755)

Description: Published in two books,[37] Euler's textbook on differential calculus presented the subject in terms of the function concept, which he had introduced in his 1748 *Introductio in analysin infinitorum*. This work opens with a study of the calculus of finite differences and makes a thorough investigation of how differentiation behaves under substitutions.[38] Also included is a systematic study of Bernoulli polynomials and the Bernoulli numbers (naming them as such), a demonstration of how the Bernoulli numbers are related to the coefficients in the Euler–Maclaurin formula and the values of $\zeta(2n)$,[39] a further study of Euler's constant (including its connection to the gamma function), and an application of partial fractions to differentiation.[40]

Über die Darstellbarkeit einer Function durch eine trigonometrische Reihe

- Bernhard Riemann (1867)

Description: Written in 1853, Riemann's work on trigonometric series was published posthumously. In it, he extended Cauchy's definition of the integral to that of the Riemann integral, allowing some functions with dense subsets of discontinuities on an interval to be integrated (which he demonstrated by an example).[41] He also stated the Riemann series theorem,[41] proved the Riemann-Lebesgue lemma for the case of bounded Riemann integrable functions,[42] and developed the Riemann localization principle.[43]

Intégrale, longueur, aire

- Henri Lebesgue (1901)

Description: Lebesgue's doctoral dissertation, summarizing and extending his research to date regarding his development of measure theory and the Lebesgue integral.

13.4.3   Complex analysis

Grundlagen für eine allgemeine Theorie der Functionen einer veränderlichen complexen Grösse

- Bernhard Riemann (1851)

Description: Riemann's doctoral dissertation introduced the notion of a Riemann surface, conformal mapping, simple connectivity, the Riemann sphere, the Laurent series expansion for functions having poles and branch points, and the Riemann mapping theorem.

13.4.4   Functional analysis

Théorie des opérations linéaires

- Stefan Banach (1932; originally published 1931 in Polish under the title *Teorja operacyj.*)

Description: The first mathematical monograph on the subject of linear metric spaces, bringing the abstract study of functional analysis to the wider mathematical community. The book introduced the ideas of a normed space and the notion of a so-called *B*-space, a complete normed space. The *B*-spaces are now called Banach spaces and are one of the basic objects of study in all areas of modern mathematical analysis. Banach also gave proofs of versions of the open mapping theorem, closed graph theorem, and Hahn–Banach theorem.

13.4.5   Fourier analysis

Mémoire sur la propagation de la chaleur dans les corps solides

- Joseph Fourier (1807)[44]

Description: Introduced Fourier analysis, specifically Fourier series. Key contribution was to not simply use trigonometric series, but to model *all* functions by trigonometric series.

When Fourier submitted his paper in 1807, the committee (which included Lagrange, Laplace, Malus and Legendre, among others) concluded: *...the manner in which the author arrives at these equations is not exempt of difficulties and [...] his analysis to integrate them still leaves something to be desired on the score of generality and even rigour*. Making Fourier series rigorous, which in detail took over a century, led directly to a number of developments in analysis, notably the rigorous statement of the integral via the Dirichlet integral and later the Lebesgue integral.

Sur la convergence des séries trigonométriques qui servent à représenter une fonction arbitraire entre des limites données

- Peter Gustav Lejeune Dirichlet (1829, expanded German edition in 1837)

Description: In his habilitation thesis on Fourier series, Riemann characterized this work of Dirichlet as "*the first profound paper about the subject*".[45] This paper gave the first rigorous proof of the convergence of Fourier series under fairly general conditions (piecewise continuity and monotonicity) by considering partial sums, which Dirichlet transformed into a particular Dirichlet integral involving what is now called the Dirichlet kernel. This paper introduced the nowhere continuous Dirichlet function and an early version of the Riemann–Lebesgue lemma.[46]

On convergence and growth of partial sums of Fourier series

- Lennart Carleson (1966)

Description: Settled Lusin's conjecture that the Fourier expansion of any L^2 function converges almost everywhere.

13.5 Geometry

See also: List of books in computational geometry and List of books about polyhedra

13.5.1 *Baudhayana Sulba Sutra*

- Baudhayana

Description: Written around the 8th century BC, this is one of the oldest geometrical texts. It laid the foundations of Indian mathematics and was influential in South Asia and its surrounding regions, and perhaps even Greece. Among the important geometrical discoveries included in this text are: the earliest list of Pythagorean triples discovered algebraically, the earliest statement of the Pythagorean theorem, geometric solutions of linear equations, several approximations of π, the first use of irrational numbers, and an accurate computation of the square root of 2, correct to a remarkable five decimal places. Though this was primarily a geometrical text, it also contained some important algebraic developments, including the earliest use of quadratic equations of the forms $ax^2 = c$ and $ax^2 + bx = c$, and integral solutions of simultaneous Diophantine equations with up to four unknowns.

13.5.2 *Euclid's Elements*

- Euclid

Publication data: c. 300 BC

Online version: Interactive Java version

Description: This is often regarded as not only the most important work in geometry but one of the most important works in mathematics. It contains many important results in geometry, number theory and the first algorithm as well. More than any specific result in the publication, it seems that the major achievement of this publication is the popularization of logic and mathematical proof as a method of solving problems.

13.5.3 *The Nine Chapters on the Mathematical Art*

- Unknown author

Description: This was a Chinese mathematics book, mostly geometric, composed during the Han Dynasty, perhaps as early as 200 BC. It remained the most important textbook in China and East Asia for over a thousand years, similar to the position of Euclid's *Elements* in Europe. Among its contents: Linear problems solved using the principle known later in the West as the *rule of false position*. Problems with several unknowns, solved by a principle similar to Gaussian elimination. Problems involving the principle known in the West as the Pythagorean theorem. The earliest solution of a matrix using a method equivalent to the modern method.

13.5.4   *The Conics*

- Apollonius of Perga

Description: The Conics was written by Apollonius of Perga, a Greek mathematician. His innovative methodology and terminology, especially in the field of conics, influenced many later scholars including Ptolemy, Francesco Maurolico, Isaac Newton, and René Descartes. It was Apollonius who gave the ellipse, the parabola, and the hyperbola the names by which we know them.

13.5.5   *Surya Siddhanta*

- Unknown (400 CE)

Description: Contains the roots of modern trigonometry. It describes the archeo-astronomy theories, principles and methods of the ancient Hindus. This siddhanta is supposed to be the knowledge that the Sun god gave to an Asura called Maya. It uses sine (jya), cosine (kojya or "perpendicular sine") and inverse sine (otkram jya) for the first time, and also contains the earliest use of the tangent and secant. Later Indian mathematicians such as Aryabhata made references to this text, while later Arabic and Latin translations were very influential in Europe and the Middle East.

13.5.6   *Aryabhatiya*

- Aryabhata (499 CE)

Description: This was a highly influential text during the Golden Age of mathematics in India. The text was highly concise and therefore elaborated upon in commentaries by later mathematicians. It made significant contributions to geometry and astronomy, including introduction of sine/ cosine, determination of the approximate value of pi and accurate calculation of the earth's circumference.

13.5.7   *La Géométrie*

- René Descartes

Description: La Géométrie was published in 1637 and written by René Descartes. The book was influential in developing the Cartesian coordinate system and specifically discussed the representation of points of a plane, via real numbers; and the representation of curves, via equations.

13.5.8   *Grundlagen der Geometrie*

- David Hilbert

Online version: English

Publication data: Hilbert, David (1899). *Grundlagen der Geometrie*. Teubner-Verlag Leipzig. ISBN 1-4020-2777-X.

Description: Hilbert's axiomatization of geometry, whose primary influence was in its pioneering approach to meta-mathematical questions including the use of models to prove axiom independence and the importance of establishing the consistency and completeness of an axiomatic system.

13.5.9 *Regular Polytopes*

- H.S.M. Coxeter

Description: *Regular Polytopes* is a comprehensive survey of the geometry of regular polytopes, the generalisation of regular polygons and regular polyhedra to higher dimensions. Originating with an essay entitled *Dimensional Analogy* written in 1923, the first edition of the book took Coxeter 24 years to complete. Originally written in 1947, the book was updated and republished in 1963 and 1973.

13.5.10 Differential geometry

Recherches sur la courbure des surfaces

- Leonhard Euler (1760)

Publication data: Mémoires de l'académie des sciences de Berlin **16** (1760) pp. 119–143; published 1767. (Full text and an English translation available from the Dartmouth Euler archive.)

Description: Established the theory of surfaces, and introduced the idea of principal curvatures, laying the foundation for subsequent developments in the differential geometry of surfaces.

Disquisitiones generales circa superficies curvas

- Carl Friedrich Gauss (1827)

Publication data: "Disquisitiones generales circa superficies curvas", *Commentationes Societatis Regiae Scientiarum Gottingesis Recentiores* Vol. **VI** (1827), pp. 99–146; "General Investigations of Curved Surfaces" (published 1965) Raven Press, New York, translated by A.M.Hiltebeitel and J.C.Morehead.

Description: Groundbreaking work in differential geometry, introducing the notion of Gaussian curvature and Gauss' celebrated Theorema Egregium.

Über die Hypothesen, welche der Geometrie zu Grunde Liegen

- Bernhard Riemann (1854)

Publication data: "Über die Hypothesen, welche der Geometrie zu Grunde Liegen", *Abhandlungen der Königlichen Gesellschaft der Wissenschaften zu Göttingen*, Vol. 13, 1867.English translate

Description: Riemann's famous Habiltationsvortrag, in which he introduced the notions of a manifold, Riemannian metric, and curvature tensor.

Leçons sur la théorie génerale des surfaces et les applications géométriques du calcul infinitésimal

- Gaston Darboux

Publication data: Darboux, Gaston (1887,1889,1896). *Leçons sur la théorie génerale des surfaces: Volume I, Volume II, Volume III, Volume IV*. Gauthier-Villars.

Description: Leçons sur la théorie génerale des surfaces et les applications géométriques du calcul infinitésimal (on the General Theory of Surfaces and the Geometric Applications of Infinitesimal Calculus). A treatise covering virtually every aspect of the 19th century differential geometry of surfaces.

13.6   Topology

13.6.1 *Analysis situs*

- Henri Poincaré (1895, 1899–1905)

Description: Poincaré's Analysis Situs and his Compléments à l'Analysis Situs laid the general foundations for algebraic topology. In these papers, Poincaré introduced the notions of homology and the fundamental group, provided an early formulation of Poincaré duality, gave the Euler–Poincaré characteristic for chain complexes, and mentioned several important conjectures including the Poincaré conjecture.

13.6.2 *L'anneau d'homologie d'une représentation, Structure de l'anneau d'homologie d'une représentation*

- Jean Leray (1946)

Description: These two Comptes Rendus notes of Leray from 1946 introduced the novel concepts of sheafs, sheaf cohomology, and spectral sequences, which he had developed during his years of captivity as a prisoner of war. Leray's announcements and applications (published in other Comptes Rendus notes from 1946) drew immediate attention from other mathematicians. Subsequent clarification, development, and generalization by Henri Cartan, Jean-Louis Koszul, Armand Borel, Jean-Pierre Serre, and Leray himself allowed these concepts to be understood and applied to many other areas of mathematics.[47] Dieudonné would later write that these notions created by Leray *"undoubtedly rank at the same level in the history of mathematics as the methods invented by Poincaré and Brouwer"*.[48]

13.6.3 Quelques propriétés globales des variétés differentiables

- René Thom (1954)

Description: In this paper, Thom proved the Thom transversality theorem, introduced the notions of oriented and unoriented cobordism, and demonstrated that cobordism groups could be computed as the homotopy groups of certain Thom spaces. Thom completely characterized the unoriented cobordism ring and achieved strong results for several problems, including Steenrod's problem on the realization of cycles.[49][50]

13.7   Category theory

13.7.1 General theory of natural equivalences

- Samuel Eilenberg and Saunders Mac Lane (1945)

Description: The first paper on category theory. Mac Lane later wrote in *Categories for the Working Mathematician* that he and Eilenberg introduced categories so that they could introduce functors, and they introduced functors so that they could introduce natural equivalences. Prior to this paper, "natural" was used in an informal and imprecise way to designate constructions that could be made without making any choices. Afterwards, "natural" had a precise meaning which occurred in a wide variety of contexts and had powerful and important consequences.

13.7.2 Categories for the Working Mathematician

- Saunders Mac Lane (1971, second edition 1998)

Description: Saunders Mac Lane, one of the founders of category theory, wrote this exposition to bring categories to the masses. Mac Lane brings to the fore the important concepts that make category theory useful, such as adjoint functors and universal properties.

13.7.3 Higher Topos Theory

- Jacob Lurie (2010)

Description: *This purpose of this book is twofold: to provide a general introduction to higher category theory (using the formalism of "quasicategories" or "weak Kan complexes"), and to apply this theory to the study of higher versions of Grothendieck topoi. A few applications to classical topology are included.* (see arXiv.)

13.8 Set theory

13.8.1 Über eine Eigenschaft des Inbegriffes aller reellen algebraischen Zahlen

- Georg Cantor (1874)

Online version: Online version

Description: Contains the first proof that the set of all real numbers is uncountable; also contains a proof that the set of algebraic numbers is denumerable. (For history and controversies about this article, see Cantor's first uncountability proof.)

13.8.2 Grundzüge der Mengenlehre

- Felix Hausdorff

Description: First published in 1914, this was the first comprehensive introduction to set theory. Besides the systematic treatment of known results in set theory, the book also contains chapters on measure theory and topology, which were then still considered parts of set theory. Here Hausdorff presents and develops highly original material which was later to become the basis for those areas.

13.8.3 The consistency of the axiom of choice and of the generalized continuum-hypothesis with the axioms of set theory

- Kurt Gödel (1938)

Description: Gödel proves the results of the title. Also, in the process, introduces the class L of constructible sets, a major influence in the development of axiomatic set theory.

13.8.4 The Independence of the Continuum Hypothesis

- Paul J. Cohen (1963, 1964)

Description: Cohen's breakthrough work proved the independence of the continuum hypothesis and axiom of choice with respect to Zermelo–Fraenkel set theory. In proving this Cohen introduced the concept of *forcing* which led to many other major results in axiomatic set theory.

13.9   Logic

13.9.1   The Laws of Thought

- George Boole (1854)

Description: Published in 1854, The Laws of Thought was the first book to provide a mathematical foundation for logic. Its aim was a complete re-expression and extension of Aristotle's logic in the language of mathematics. Boole's work founded the discipline of algebraic logic and would later be central for Claude Shannon in the development of digital logic.

13.9.2   Begriffsschrift

- Gottlob Frege (1879)

Description: Published in 1879, the title **Begriffsschrift** is usually translated as *concept writing* or *concept notation*; the full title of the book identifies it as "*a formula language, modelled on that of arithmetic, of pure thought*". Frege's motivation for developing his formal logical system was similar to Leibniz's desire for a *calculus ratiocinator*. Frege defines a logical calculus to support his research in the foundations of mathematics. **Begriffsschrift** is both the name of the book and the calculus defined therein. It was arguably the most significant publication in logic since Aristotle.

13.9.3   Formulario mathematico

- Giuseppe Peano (1895)

Description: First published in 1895, the **Formulario mathematico** was the first mathematical book written entirely in a formalized language. It contained a description of mathematical logic and many important theorems in other branches of mathematics. Many of the notations introduced in the book are now in common use.

13.9.4   Principia Mathematica

- Bertrand Russell and Alfred North Whitehead (1910–1913)

Description: The *Principia Mathematica* is a three-volume work on the foundations of mathematics, written by Bertrand Russell and Alfred North Whitehead and published in 1910–1913. It is an attempt to derive all mathematical truths from a well-defined set of axioms and inference rules in symbolic logic. The questions remained whether a contradiction could be derived from the Principia's axioms, and whether there exists a mathematical statement which could neither be proven nor disproven in the system. These questions were settled, in a rather surprising way, by Gödel's incompleteness theorem in 1931.

13.9.5   Systems of Logic Based on Ordinals

- Alan Turing's Ph.D. thesis

13.9.6   Über formal unentscheidbare Sätze der Principia Mathematica und verwandter Systeme, I

(On Formally Undecidable Propositions of Principia Mathematica and Related Systems)

- Kurt Gödel (1931)

Online version: Online version

Description: In mathematical logic, **Gödel's incompleteness theorems** are two celebrated theorems proved by Kurt Gödel in 1931. The first incompleteness theorem states:

> For any formal system such that (1) it is ω -consistent (omega-consistent), (2) it has a recursively definable set of axioms and rules of derivation, and (3) every recursive relation of natural numbers is definable in it, there exists a formula of the system such that, according to the intended interpretation of the system, it expresses a truth about natural numbers and yet it is not a theorem of the system.

13.10 Combinatorics

13.10.1 On sets of integers containing no k elements in arithmetic progression

- Endre Szemerédi (1975)

Description: Settled a conjecture of Paul Erdős and Pál Turán (now known as Szemerédi's theorem) that if a sequence of natural numbers has positive upper density then it contains arbitrarily long arithmetic progressions. Szemerédi's solution has been described as a "masterpiece of combinatorics"[51] and it introduced new ideas and tools to the field including a weak form of the Szemerédi regularity lemma.[52]

13.10.2 Graph theory

Solutio problematis ad geometriam situs pertinentis

- Leonhard Euler (1741)

- Euler's original publication (in Latin)

Description: Euler's solution of the Königsberg bridge problem in *Solutio problematis ad geometriam situs pertinentis* (*The solution of a problem relating to the geometry of position*) is considered to be the first theorem of graph theory.

On the evolution of random graphs

- Paul Erdős and Alfréd Rényi (1960)

Description: Provides a detailed discussion of sparse random graphs, including distribution of components, occurrence of small subgraphs, and phase transitions.[53]

Network Flows and General Matchings

- Ford, L., & Fulkerson, D.

- Flows in Networks. Prentice-Hall, 1962.

Description: Presents the Ford-Fulkerson algorithm for solving the maximum flow problem, along with many ideas on flow-based models.

13.11   Computational complexity theory

See List of important publications in theoretical computer science.

13.12   Probability theory

See list of important publications in statistics.

13.13   Game theory

13.13.1   Zur Theorie der Gesellschaftsspiele

- John von Neumann (1928)

Description: Went well beyond Émile Borel's initial investigations into strategic two-person game theory by proving the minimax theorem for two-person, zero-sum games.

13.13.2   *Theory of Games and Economic Behavior*

- Oskar Morgenstern, John von Neumann (1944)

Description: This book led to the investigation of modern game theory as a prominent branch of mathematics. This profound work contained the method for finding optimal solutions for two-person zero-sum games.

13.13.3   *Equilibrium Points in N-person Games*

- John Forbes Nash
- *Proceedings of the National Academy of Sciences* 36 (1950), 48–49. MR 0031701
- "Equilibrium Points in N-person Games"

Description: Nash equilibrium

13.13.4   *On Numbers and Games*

- John Horton Conway

Description: The book is in two, {0,1|}, parts. The zeroth part is about numbers, the first part about games – both the values of games and also some real games that can be played such as Nim, Hackenbush, Col and Snort amongst the many described.

13.13.5   *Winning Ways for your Mathematical Plays*

- Elwyn Berlekamp, John Conway and Richard K. Guy

Description: A compendium of information on mathematical games. It was first published in 1982 in two volumes, one focusing on Combinatorial game theory and surreal numbers, and the other concentrating on a number of specific games.

13.14 Fractals

13.14.1 *How Long Is the Coast of Britain? Statistical Self-Similarity and Fractional Dimension*

- Benoît Mandelbrot

Description: A discussion of self-similar curves that have fractional dimensions between 1 and 2. These curves are examples of fractals, although Mandelbrot does not use this term in the paper, as he did not coin it until 1975. Shows Mandelbrot's early thinking on fractals, and is an example of the linking of mathematical objects with natural forms that was a theme of much of his later work.

13.15 Numerical analysis

13.15.1 Optimization

Method of Fluxions

- Isaac Newton

Description: *Method of Fluxions* was a book written by Isaac Newton. The book was completed in 1671, and published in 1736. Within this book, Newton describes a method (the Newton–Raphson method) for finding the real zeroes of a function.

Essai d'une nouvelle méthode pour déterminer les maxima et les minima des formules intégrales indéfinies

- Joseph Louis Lagrange (1761)

Description: Major early work on the calculus of variations, building upon some of Lagrange's prior investigations as well as those of Euler. Contains investigations of minimal surface determination as well as the initial appearance of Lagrange multipliers.

Математические методы организации и планирования производства

- Leonid Kantorovich (1939) "[The Mathematical Method of Production Planning and Organization]" (in Russian).

Description: Kantorovich wrote the first paper on production planning, which used Linear Programs as the model. He received the Nobel prize for this work in 1975.

Decomposition Principle for Linear Programs

- George Dantzig and P. Wolfe
- Operations Research 8:101–111, 1960.

Description: Dantzig's is considered the father of linear programming in the western world. He independently invented the simplex algorithm. Dantzig and Wolfe worked on decomposition algorithms for large-scale linear programs in factory and production planning.

How good is the simplex algorithm?

- Victor Klee and George J. Minty

- Klee, Victor; Minty, George J. (1972). "How good is the simplex algorithm?". In Shisha, Oved. *Inequalities III (Proceedings of the Third Symposium on Inequalities held at the University of California, Los Angeles, Calif., September 1–9, 1969, dedicated to the memory of Theodore S. Motzkin)*. New York-London: Academic Press. pp. 159–175. MR 332165.

Description: Klee and Minty gave an example showing that the simplex algorithm can take exponentially many steps to solve a linear program.

Полиномиальный алгоритм в линейном программировании

- Khachiyan, Leonid Genrikhovich (1979). Полиномиальный алгоритм в линейном программировании [A polynomial algorithm for linear programming]. *Doklady Akademii Nauk SSSR* (in Russian) **244**: 1093–1096..

Description: Khachiyan's work on Ellipsoid method. This was the first polynomial time algorithm for linear programming.

13.16　Early manuscripts

These are publications that are not necessarily relevant to a mathematician nowadays, but are nonetheless important publications in the history of mathematics.

13.16.1　*Rhind Mathematical Papyrus*

- Ahmes (scribe)

Description: It is one of the oldest mathematical texts, dating to the Second Intermediate Period of ancient Egypt. It was copied by the scribe Ahmes (properly *Ahmose*) from an older Middle Kingdom papyrus. It laid the foundations of Egyptian mathematics and in turn, later influenced Greek and Hellenistic mathematics. Besides describing how to obtain an approximation of π only missing the mark by less than one per cent, it is describes one of the earliest attempts at squaring the circle and in the process provides persuasive evidence against the theory that the Egyptians deliberately built their pyramids to enshrine the value of π in the proportions. Even though it would be a strong overstatement to suggest that the papyrus represents even rudimentary attempts at analytical geometry, Ahmes did make use of a kind of an analogue of the cotangent.

13.16.2　*Archimedes Palimpsest*

- Archimedes of Syracuse

Description: Although the only mathematical tools at its author's disposal were what we might now consider secondary-school geometry, he used those methods with rare brilliance, explicitly using infinitesimals to solve problems that would now be treated by integral calculus. Among those problems were that of the center of gravity of a solid hemisphere, that of the center of gravity of a frustum of a circular paraboloid, and that of the area of a region bounded by a parabola and one of its secant lines. For explicit details of the method used, see Archimedes' use of infinitesimals.

13.16.3 *The Sand Reckoner*

- Archimedes of Syracuse

Online version: Online version

Description: The first known (European) system of number-naming that can be expanded beyond the needs of everyday life.

13.17 Textbooks

13.17.1 *Synopsis of Pure Mathematics*

- G. S. Carr

Description: Contains over 6000 theorems of mathematics, assembled by George Shoobridge Carr for the purpose of training students in the art of mathematics, studied extensively by Ramanujan. (first half here) It was one of the few books that attempts to summarize the entirety of known mathematics.

13.17.2 *Arithmetick: or, The Grounde of Arts*

- Robert Recorde

Description: Written in 1542, it was the first really popular arithmetic book written in the English Language.

13.17.3 *Cocker's Arithmetick*

- Edward Cocker (authorship disputed)

Description: Textbook of arithmetic published in 1678 by John Hawkins, who claimed to have edited manuscripts left by Edward Cocker, who had died in 1676. This influential mathematics textbook used to teach arithmetic in schools in the United Kingdom for over 150 years.

13.17.4 *The Schoolmaster's Assistant, Being a Compendium of Arithmetic both Practical and Theoretical*

- Thomas Dilworth

Description: An early and popular English arithmetic textbook published in America in the 18th century. The book reached from the introductory topics to the advanced in five sections.

13.17.5 *Geometry*

- Andrei Kiselyov

Publication data: 1892

Description: The most widely used and influential textbook in Russian mathematics. (See Kiselyov page and MAA review.)

13.17.6   *A Course of Pure Mathematics*

- G. H. Hardy

Description: A classic textbook in introductory mathematical analysis, written by G. H. Hardy. It was first published in 1908, and went through many editions. It was intended to help reform mathematics teaching in the UK, and more specifically in the University of Cambridge, and in schools preparing pupils to study mathematics at Cambridge. As such, it was aimed directly at "scholarship level" students — the top 10% to 20% by ability. The book contains a large number of difficult problems. The content covers introductory calculus and the theory of infinite series.

13.17.7   Moderne Algebra

- B. L. van der Waerden

Description: The first introductory textbook (graduate level) expounding the abstract approach to algebra developed by Emil Artin and Emmy Noether. First published in German in 1931 by Springer Verlag. A later English translation was published in 1949 by Frederick Ungar Publishing Company.

13.17.8   Algebra

- Saunders Mac Lane and Garrett Birkhoff

Description: A definitive introductory text for abstract algebra using a category theoretic approach. Both a rigorous introduction from first principles, and a reasonably comprehensive survey of the field.

13.17.9   *Calculus, Vol. 1*

- Tom M. Apostol

13.17.10   *Algebraic Geometry*

- Robin Hartshorne

Description: The first comprehensive introductory (graduate level) text in algebraic geometry that used the language of schemes and cohomology. Published in 1977, it lacks aspects of the scheme language which are nowadays considered central, like the functor of points.

13.17.11   *Naive Set Theory*

- Paul Halmos

Description: An undergraduate introduction to not-very-naive set theory which has lasted for decades. It is still considered by many to be the best introduction to set theory for beginners. While the title states that it is naive, which is usually taken to mean without axioms, the book does introduce all the axioms of Zermelo–Fraenkel set theory and gives correct and rigorous definitions for basic objects. Where it differs from a "true" axiomatic set theory book is its character: There are no long-winded discussions of axiomatic minutiae, and there is next to nothing about topics like large cardinals. Instead it aims, and succeeds, in being intelligible to someone who has never thought about set theory before.

13.17.12 *Cardinal and Ordinal Numbers*

- Wacław Sierpiński

Description: The *nec plus ultra* reference for basic facts about cardinal and ordinal numbers. If you have a question about the cardinality of sets occurring in everyday mathematics, the first place to look is this book, first published in the early 1950s but based on the author's lectures on the subject over the preceding 40 years.

13.17.13 *Set Theory: An Introduction to Independence Proofs*

- Kenneth Kunen

Description: This book is not really for beginners, but graduate students with some minimal experience in set theory and formal logic will find it a valuable self-teaching tool, particularly in regard to forcing. It is far easier to read than a true reference work such as Jech, *Set Theory*. It may be the best textbook from which to learn forcing, though it has the disadvantage that the exposition of forcing relies somewhat on the earlier presentation of Martin's axiom.

13.17.14 **Topologie**

- Pavel Sergeevich Alexandrov
- Heinz Hopf

Description: First published round 1935, this text was a pioneering "reference" text book in topology, already incorporating many modern concepts from set-theoretic topology, homological algebra and homotopy theory.

13.17.15 **General Topology**

- John L. Kelley

Description: First published in 1955,for many years the only introductory graduate level textbook in the U.S.A. teaching the basics of point set, as opposed to algebraic, topology. Prior to this the material, essential for advanced study in many fields, was only available in bits and pieces from texts on other topics or journal articles.

13.17.16 **Topology from the Differentiable Viewpoint**

- John Milnor

Description: This short book introduces the main concepts of differential topology in Milnor's lucid and concise style. While the book does not cover very much, its topics are explained beautifully in a way that illuminates all their details.

13.17.17 *Number Theory, An approach through history from Hammurapi to Legendre*

- André Weil

Description: An historical study of number theory, written by one of the 20th century's greatest researchers in the field. The book covers some thirty six centuries of arithmetical work but the bulk of it is devoted to a detailed study and exposition of the work of Fermat, Euler, Lagrange, and Legendre. The author wishes to take the reader into the workshop of his subjects to share their successes and failures. A rare opportunity to see the historical development of a subject through the mind of one of its greatest practitioners.

13.17.18   An Introduction to the Theory of Numbers

- G. H. Hardy and E. M. Wright

Description: *An Introduction to the Theory of Numbers* was first published in 1938, and is still in print, with the latest edition being the 6th (2008). It is likely that almost every serious student and researcher into number theory has consulted this book, and probably has it on their bookshelf. It was not intended to be a textbook, and is rather an introduction to a wide range of differing areas of number theory which would now almost certainly be covered in separate volumes. The writing style has long been regarded as exemplary, and the approach gives insight into a variety of areas without requiring much more than a good grounding in algebra, calculus and complex numbers.

13.17.19   *Foundations of Differential Geometry*

- Shoshichi Kobayashi and Katsumi Nomizu

13.17.20   *Hodge Theory and Complex Algebraic Geometry I*

13.17.21   *Hodge Theory and Complex Algebraic Geometry II*

- Claire Voisin

13.18   Popular writing

13.18.1   Gödel, Escher, Bach

- Douglas Hofstadter

Description: *Gödel, Escher, Bach: an Eternal Golden Braid* is a Pulitzer Prize-winning book, first published in 1979 by Basic Books. It is a book about how the creative achievements of logician Kurt Gödel, artist M. C. Escher and composer Johann Sebastian Bach interweave. As the author states: "I realized that to me, Gödel and Escher and Bach were only shadows cast in different directions by some central solid essence. I tried to reconstruct the central object, and came up with this book."

13.18.2   The World of Mathematics

- James R. Newman

Description: *The World of Mathematics* was specially designed to make mathematics more accessible to the inexperienced. It comprises nontechnical essays on every aspect of the vast subject, including articles by and about scores of eminent mathematicians, as well as literary figures, economists, biologists, and many other eminent thinkers. Includes the work of Archimedes, Galileo, Descartes, Newton, Gregor Mendel, Edmund Halley, Jonathan Swift, John Maynard Keynes, Henri Poincaré, Lewis Carroll, George Boole, Bertrand Russell, Alfred North Whitehead, John von Neumann, and many others. In addition, an informative commentary by distinguished scholar James R. Newman precedes each essay or group of essays, explaining their relevance and context in the history and development of mathematics. Originally published in 1956, it does not include many of the exciting discoveries of the later years of the 20th century but it has no equal as a general historical survey of important topics and applications.

13.19 References

[1] Bill Casselman. "One of the Oldest Extant Diagrams from Euclid". University of British Columbia. Retrieved 2008-09-26.

[2] Ivor Grattan-Guinness, *Landmark writings in Western mathematics 1640–1940*, Elsevier Science, 2005

[3] David Eugene Smith, *A Source Book in Mathematics*, Dover Publications, 1984

[4] Shashi S. Sharma. *Mathematics & Astronomers of Ancient India*. Pitambar. p. 29. ISBN 978-81-209-1421-6. Brahmagupta is believed to have composed many important works of mathematics and astronomy. However, two of his most important works are: Brahmasphutasiddhanta (BSS) written in 628 AD, and the Khandakhadyaka...

[5] Miodrag Petković (2009). *Famous puzzles of great mathematicians*. American Mathematical Society. pp. 77, 299. ISBN 978-0-8218-4814-2. many important results from astronomy, arithmetic and algebra", "major work

[6] Helaine Selin, ed. (1997). *Encyclopaedia of the history of science, technology, and medicine in non-western cultures*. Springer. p. 162. ISBN 978-0-7923-4066-9. holds a remarkable place in the history of Eastern civilzation", "most important work", "remarkably modern in outlook", "marvelous piece of pure mathematics", "more remarkable algebraic contributions", "important step towards the integral solutions of [second-order indeterminate] equations", "In geometry, Brahmagupta's achievements were equally praiseworthy.

[7] John Tabak (2004). *Algebra: sets, symbols, and the language of thought*. Infobase Publishing. pp. 38*ff*. ISBN 978-0-8160-4954-7. Brahmagupta's masterpiece", "a great deal of important algebra", "The *Brahma-sphuta-siddhānta* was quickly recognized by Brahmagupta's contemporaries as an important and imaginative work. It inspired numerous commentaries by many generations of mathematicians.

[8] Clark, Allan (1984). *Elements of abstract algebra*. United States: Courier Dover Publications. p. ix. ISBN 978-0-486-64725-8.

[9] O'Connor, J. J.; Robertson, E. F. (1998). "Girolamo Cardano".

[10] Markus Fierz (1983). *Girolamo Cardano: 1501-1576. Physician, Natural Philosopher, Mathematician*. Birkhäuser Boston. ISBN 978-0-8176-3057-7.

[11] Weil, André (1984). *Number Theory: An approach through history From Hammurapi to Legendre*. Birkhäuser. pp. 239–242. ISBN 0-8176-3141-0.

[12] Gauss, J.C.F. (1799). "Demonstratio nova theorematis omnem functionem algebraicam rationalem integram unius variabilis in factores reales primi vel secundi gradus resolvi posse".

[13] O'Connor, J. J.; Robertson, E. F. (1996). "The fundamental theorem of algebra".

[14] ed. by A. N. Kolmogorov... (2001). *Mathematics of the 19th Century: Mathematical Logic, Algebra, Number Theory, and Probability Theory*. Birkhäuser Verlag. pp. 39, 63, 66–68. ISBN 3-7643-6441-6.

[15] O'Connor, J. J.; Robertson, E. F. (2001). "Marie Ennemond Camille Jordan".

[16] Krieger, Martin H. (March 2007). "A 1940 Letter of André Weil on Analogy in Mathematics" (PDF). *Notices of the American Mathematical Society* **52** (3): 338. Retrieved 13 January 2008.

[17] Jackson, Allyn (October 2004). "Comme Appelé du Néant — As If Summoned from the Void: The Life of Alexandre Grothendieck" (PDF). *Notices of the American Mathematical Society* **51** (9): 1045–1046. Retrieved 13 January 2008.

[18] Dieudonné, Jean (1989). *A history of algebraic and differential topology 1900–1960*. Birkhäuser. pp. 598–600. ISBN 0-8176-3388-X.

[19] Euler, L. (1744). "De fractionibus continuis dissertatio" (PDF). Retrieved 23 June 2009.

[20] Sandifer, Ed (February 2006). "How Euler Did It: Who proved e is irrational?" (PDF). *MAA Online*. Retrieved 23 June 2009.

[21] Goldfeld, Dorian (July 1985). "Gauss' Class Number Problem For Imaginary Quadratic Fields" (PDF). *Bulletin of the American Mathematical Society* **13** (1): 24. doi:10.1090/S0273-0979-1985-15352-2.

[22] Weil, André (1984). *Number Theory: An approach through history From Hammurapi to Legendre*. Birkhäuser. pp. 316–322. ISBN 0-8176-3141-0.

[23] Ireland, K.; Rosen, M. (1993). *A Classical Introduction to Modern Number Theory*. New York, New York: Springer-Verlag. pp. 358–361. ISBN 0-387-97329-X.

[24] Silverman, J.; Tate, J. (1992). *Rational Points on Elliptic Curves*. New York, New York: Springer-Verlag. p. 110. ISBN 0-387-97825-9.

[25] Elstrodt, Jürgen (2007). "The Life and Work of Gustav Lejeune Dirichlet (1805–1859)" (PDF). *Clay Mathematics Proceedings*: 21–22.

[26] H. M. Edwards, *Riemann's Zeta Function*, Academic Press, 1974

[27] Lemmermeyer, Franz; Schappacher, Norbert. "Introduction to the English Edition of Hilbert's Zahlbericht" (PDF). p. 3. Retrieved 13 January 2008.

[28] Lemmermeyer, Franz; Schappacher, Norbert. "Introduction to the English Edition of Hilbert's Zahlbericht" (PDF). p. 5. Retrieved 13 January 2008.

[29] Alexanderson, Gerald L. (October 2007). "Euler's Introductio In Analysin Infinitorum" (PDF). *Bulletin of the American Mathematical Society* **44** (4): 635–639. doi:10.1090/S0273-0979-07-01183-4.

[30] Euler, L. "E101 – Introductio in analysin infinitorum, volume 1". Retrieved 16 March 2008.

[31] Euler, L. "E102 – Introductio in analysin infinitorum, volume 2". Retrieved 16 March 2008.

[32] Calinger, Ronald (1982). *Classics of Mathematics*. Oak Park, Illinois: Moore Publishing Company, Inc. pp. 396–397. ISBN 0-935610-13-8.

[33] O'Connor, J. J.; Robertson, E. F. (1995). "The function concept".

[34] Andrews, George E. (October 2007). "Euler's "De Partitio Numerorum"" (PDF). *Bulletin of the American Mathematical Society* **44** (4): 561–573. doi:10.1090/S0273-0979-07-01180-9.

[35] Charles Whish (1834). "On the Hindu Quadrature of the circle and the infinite series of the proportion of the circumference to the diameter exhibited in the four Sastras, the Tantra Sahgraham, Yucti Bhasha, Carana Padhati and Sadratnamala". *Transactions of the Royal Asiatic Society of Great Britain and Ireland* (Royal Asiatic Society of Great Britain and Ireland) **3** (3): 509–523. doi:10.1017/S0950473700001221. JSTOR 25581775.

[36] Gray, Jeremy (2000). "MAA Book Review: Reading the Principia: The Debate on Newton's Mathematical Methods for Natural Philosophy from 1687 to 1736 by Niccolò Guicciardini".

[37] Euler, L. "E212 – Institutiones calculi differentialis cum eius usu in analysi finitorum ac doctrina serierum". Retrieved 21 March 2008.

[38] O'Connor, J. J.; Robertson, E. F. (1998). "Leonhard Euler".

[39] Sandifer, Ed (September 2005). "How Euler Did It: Bernoulli Numbers" (PDF). *MAA Online*. Retrieved 23 June 2009.

[40] Sandifer, Ed (June 2007). "How Euler Did It: Partial Fractions" (PDF). *MAA Online*. Retrieved 23 June 2009.

[41] Bressoud, David (2007). *A Radical Approach to Real Analysis*. Mathematical Association of America. pp. 248–255. ISBN 0-88385-747-2.

[42] Kline, Morris (1990). *Mathematical Thought From Ancient to Modern Times*. Oxford University Press. pp. 1046–1047. ISBN 0-19-506137-3.

[43] Benedetto, John (1997). *Harmonic Analysis and Applications*. CRC Press. pp. 170–171. ISBN 0-8493-7879-6.

[44] *Mémoire sur la propagation de la chaleur dans les corps solides, présenté le 21 décembre 1807 à l'Institut national – Nouveau Bulletin des sciences par la Société philomatique de Paris* **I** (6). Paris: Bernard. March 1808. pp. 112–116. Reprinted in "Mémoire sur la propagation de la chaleur dans les corps solides". *Joseph Fourier – Œuvres complètes, tome 2*. pp. 215–221.

[45] Koch, Helmut (1998). *Mathematics in Berlin: Gustav Peter Lejeune Dirichlet*. Birkhäuser. pp. 33–40. ISBN 3-7643-5943-9.

[46] Elstrodt, Jürgen (2007). "The Life and Work of Gustav Lejeune Dirichlet (1805–1859)" (PDF). *Clay Mathematics Proceedings*: 19–20.

[47] Miller, Haynes (2000). "Leray in Oflag XVIIA: The origins of sheaf theory, sheaf cohomology, and spectral sequences" (PS).

[48] Dieudonné, Jean (1989). *A history of algebraic and differential topology 1900–1960*. Birkhäuser. pp. 123–141. ISBN 0-8176-3388-X.

[49] Dieudonné, Jean (1989). *A history of algebraic and differential topology 1900–1960*. Birkhäuser. pp. 556–575. ISBN 0-8176-3388-X.

[50] Sullivan, Dennis (April 2004). "René Thom's Work On Geometric Homology And Bordism" (PDF). *Bulletin of the American Mathematical Society* **41** (3): 341–350. doi:10.1090/S0273-0979-04-01026-2.

[51] "2008 Steele Prizes; Seminal Contribution to Research: Endre Szemerédi" (PDF). *Notices of the American Mathematical Society* **55** (4): 488. April 2008. Retrieved 19 July 2008.

[52] "Interview with Endre Szemerédi" (PDF). *Notices of the American Mathematical Society* **60** (2): 226. April 2013. doi:10.1090/noti948. Retrieved 27 January 2013.

[53] Bollobás, Béla (2002). *Modern Graph Theory*. Springer. p. 252. ISBN 978-0-387-98488-9.

INSTITVTIONVM
CALCVLI INTEGRALIS
VOLVMEN PRIMVM

IN QVO METHODVS INTEGRANDI A PRIMIS PRIN-
CIPIIS VSQVE AD INTEGRATIONEM AEQVATIONVM DIFFE-
RENTIALIVM PRIMI GRADVS PERTRACTATVR.

AVCTORE

LEONHARDO EVLERO

ACAD. SCIENT. BORVSSIAE DIRECTORE VICENNALI ET SOCIO
ACAD. PETROP. PARISIN. ET LONDIN.

PETROPOLI

Impenfis Academiae Imperialis Scientiarum

1768.

v

Institutiones calculi differentialis

Chapter 14

Lists of mathematicians

This is a list of lists of **mathematicians.**

Lists by nationality, ethnicity or religion

- List of African-American mathematicians
- List of Chinese mathematicians
- List of Greek mathematicians
- List of Hungarian mathematicians
- List of Indian mathematicians
- List of Italian mathematicians
- List of Jewish American mathematicians
- List of Jewish mathematicians
- List of Polish mathematicians
- List of Russian mathematicians
- List of Slovenian mathematicians
- List of Ukrainian mathematicians
- List of Welsh mathematicians

Lists by profession

- List of actuaries
- List of game theorists
- List of geometers
- List of logicians
- List of mathematical probabilists
- List of statisticians

271

Other lists of mathematicians

- List of Cambridge mathematicians

- List of amateur mathematicians

- List of mathematicians born in the 19th century

- List of centenarians (scientists and mathematicians)

- List of female mathematicians

- List of films about mathematicians

- List of mathematicians who studied chess

- List of mathematicians, physicians, and scientists educated at Jesus College, Oxford

- Senior Wrangler (University of Cambridge)

14.1   See also

- The Mathematics Genealogy Project – database for the academic genealogy of mathematicians

- List of mathematical artists

14.2   External links

- The MacTutor History of Mathematics archive – Extensive list of detailed biographies

- The Oberwolfach Photo Collection – Photographs of mathematicians from all over the world

- Photos of mathematicians – A collection of photos of mathematicians (and computer scientists) made by Andrej Bauer.

- Calendar of mathematicians' birthdays and death anniversaries

- Periodic Table of Mathematicians

Chapter 15

List of mathematics history topics

This is a **list of mathematics history topics**, by Wikipedia page. See also list of mathematicians, timeline of mathematics, history of mathematics, list of publications in mathematics.

- 1729 (anecdote)

- Adequality

- Archimedes Palimpsest

- Archimedes' use of infinitesimals

- Arithmetization of analysis

- Brachistochrone curve

- Chinese mathematics

- Cours d'Analyse

- Edinburgh Mathematical Society

- Erlangen programme

- Fermat's last theorem

- Greek mathematics

- Thomas Little Heath

- Hilbert's problems

- History of topos theory

- Hyperbolic quaternion

- Indian mathematics

- Islamic mathematics

- Italian school of algebraic geometry

- Kraków School of Mathematics

- Law of Continuity

- Lwów School of Mathematics

- Nicolas Bourbaki

- Non-Euclidean geometry

- Scottish Café

- Seven bridges of Königsberg

- Spectral theory

- Synthetic geometry

- Tautochrone curve

- Unifying theories in mathematics

- Waring's problem

- Warsaw School of Mathematics

15.1   Academic positions

- Lowndean Professor of Astronomy and Geometry

- Lucasian professor

- Rouse Ball Professor of Mathematics

- Sadleirian Chair

Chapter 16

Timeline of mathematics

This is a timeline of pure and applied mathematics history.

16.1 Rhetorical stage

16.1.1 Before 1000 BC

- ca. 70,000 BC — South Africa, ochre rocks adorned with scratched geometric patterns.[1]

- ca. 35,000 BC to 20,000 BC — Africa and France, earliest known prehistoric attempts to quantify time.[2][3][4]

- c. 20,000 BC — Nile Valley, Ishango Bone: possibly the earliest reference to prime numbers and Egyptian multiplication.

- c. 3400 BC — Mesopotamia, the Sumerians invent the first numeral system, and a system of weights and measures.

- c. 3100 BC — Egypt, earliest known decimal system allows indefinite counting by way of introducing new symbols.[5]

- c. 2800 BC — Indus Valley Civilization on the Indian subcontinent, earliest use of decimal ratios in a uniform system of ancient weights and measures, the smallest unit of measurement used is 1.704 millimetres and the smallest unit of mass used is 28 grams.

- 2700 BC — Egypt, precision surveying.

- 2400 BC — Egypt, precise astronomical calendar, used even in the Middle Ages for its mathematical regularity.

- c. 2000 BC — Mesopotamia, the Babylonians use a base-60 positional numeral system, and compute the first known approximate value of π at 3.125.

- c. 2000 BC — Scotland, Carved Stone Balls exhibit a variety of symmetries including all of the symmetries of Platonic solids.

- 1800 BC — Egypt, Moscow Mathematical Papyrus, findings volume of a frustum.

- c. 1800 BC — Berlin Papyrus 6619 (Egypt, 19th dynasty) contains a quadratic equation and its solution.[5]

- 1650 BC — Rhind Mathematical Papyrus, copy of a lost scroll from around 1850 BC, the scribe Ahmes presents one of the first known approximate values of π at 3.16, the first attempt at squaring the circle, earliest known use of a sort of cotangent, and knowledge of solving first order linear equations.

- 1046 BC to 256 BC — China, *Chou Pei Suan Ching*, arithmetic and geometric algorithms and proofs.

16.2   Syncopated stage

16.2.1   1st millennium BC

- c. 1000 BC — Vulgar fractions used by the Egyptians. However, only unit fractions are used (i.e., those with 1 as the numerator) and interpolation tables are used to approximate the values of the other fractions.[6]

- first half of 1st millennium BC — Vedic India — Yajnavalkya, in his Shatapatha Brahmana, describes the motions of the sun and the moon, and advances a 95-year cycle to synchronize the motions of the sun and the moon.

- c. 8th century BC — the Yajur Veda, one of the four Hindu Vedas, contains the earliest concept of infinity, and states "if you remove a part from infinity or add a part to infinity, still what remains is infinity."

- 800 BC — Baudhayana, author of the Baudhayana Sulba Sutra, a Vedic Sanskrit geometric text, contains quadratic equations, and calculates the square root of two correctly to five decimal places.

- 624 BC – 546 BC — Thales of Miletus has various theorems attributed to him.

- c. 600 BC — the other Vedic "Sulba Sutras" ("rule of chords" in Sanskrit) use Pythagorean triples, contain of a number of geometrical proofs, and approximate π at 3.16.

- second half of 1st millennium BC — The Lo Shu Square, the unique normal magic square of order three, was discovered in China.

- 530 BC — Pythagoras studies propositional geometry and vibrating lyre strings; his group also discovers the irrationality of the square root of two.

- c. 510 BC — Anaxagoras

- c. 500 BC — Indian grammarian Pānini writes the *Astadhyayi*, which contains the use of metarules, transformations and recursions, originally for the purpose of systematizing the grammar of Sanskrit.

- c. 500 BC Oenopides of Chios

- 470 BC – 410 BC — Hippocrates of Chios utilizes lunes in an attempt to square the circle.

- 5th century BC — Apastamba, author of the Apastamba Sulba Sutra, another Vedic Sanskrit geometric text, makes an attempt at squaring the circle and also calculates the square root of 2 correct to five decimal places.

- 490 BC – 430 BC Zeno of Elea *Zeno's paradoxes*

- 5th c. BC Theodorus of Cyrene

- 460 BC – 370 BC Democritus

- 460 BC – 399 BC Hippias

- 428 BC Archytas

- 423 BC – 347 BC Plato

- 417 BC – 317 BC Theaetetus (mathematician)

- c. 400 BC — Jaina mathematicians in India write the *Surya Prajinapti*, a mathematical text classifing all numbers into three sets: enumerable, innumerable and infinite. It also recognises five different types of infinity: infinite in one and two directions, infinite in area, infinite everywhere, and infinite perpetually.

- 408 BC –355 BC Eudoxus of Cnidus

- 5th century Antiphon the Sophist

- 5th century (late) Bryson of Heraclea

- 400 BC – 350 BC Thymaridas

- 395 BC 313 BC Xenocrates

- 4th century BC — Indian texts use the Sanskrit word "Shunya" to refer to the concept of "void" (zero).

- 390 BC- 320 BC Dinostratus

- 380- 290 Autolycus of Pitane

- 370 BC — Eudoxus states the method of exhaustion for area determination.

- 370 BC – 300 BC Aristaeus the Elder

- 370 BC – 300 BC Callippus

- 350 BC — Aristotle discusses logical reasoning in *Organon*.

- 330 BC — the earliest known work on Chinese geometry, the *Mo Jing*, is compiled.

- 310 BC – 230 BC Aristarchus of Samos

- 390 BC – 310 BC Heraclides of Pontus

- 380 BC – 320 BC Menaechmus

- 300 BC — Jain mathematicians in India write the *Bhagabati Sutra*, which contains the earliest information on combinations.

- 300 BC — Euclid in his *Elements* studies geometry as an axiomatic system, proves the infinitude of prime numbers and presents the Euclidean algorithm; he states the law of reflection in *Catoptrics*, and he proves the fundamental theorem of arithmetic.

- c. 300 BC — Brahmi numerals (ancestor of the common modern base 10 numeral system) are conceived in India.

- 370 – 300 — Eudemus of Rhodes works on histories of arithmetic, geometry and astronomy now lost.[7]

- 300 BC — Mesopotamia, the Babylonians invent the earliest calculator, the abacus.

- c. 300 BC — Indian mathematician Pingala writes the *Chhandah-shastra*, which contains the first Indian use of zero as a digit (indicated by a dot) and also presents a description of a binary numeral system, along with the first use of Fibonacci numbers and Pascal's triangle.

- c. 3rd century BC – Kātyāyana

- 280 BC – 210 BC Nicomedes (mathematician)

- 280 BC – 220BC Philon of Byzantium

- 279 BC – 206 BC Chrysippus

- 280 BC – 220 BC Conon of Samos

- 250 BC – 190 BC Dionysodorus

- 202 BC to 186 BC — *Book on Numbers and Computation*, a mathematical treatise, is written in Han Dynasty China.

- 262 −198 BC Apollonius of Perga

- 260 BC — Archimedes proved that the value of π lies between 3 + 1/7 (approx. 3.1429) and 3 + 10/71 (approx. 3.1408), that the area of a circle was equal to π multiplied by the square of the radius of the circle and that the area enclosed by a parabola and a straight line is 4/3 multiplied by the area of a triangle with equal base and height. He also gave a very accurate estimate of the value of the square root of 3.

- c. 250 BC — late Olmecs had already begun to use a true zero (a shell glyph) several centuries before Ptolemy in the New World. See 0 (number).

- 240 BC — Eratosthenes uses his sieve algorithm to quickly isolate prime numbers.

- 240 BC 190 BC Diocles (mathematician)

- 225 BC — Apollonius of Perga writes *On Conic Sections* and names the ellipse, parabola, and hyperbola.

- 206 BC to 8 AD — Counting rods are invented in China.

- 200 BC – 140 BC Zenodorus (mathematician)

- 150 BC — Jain mathematicians in India write the *Sthananga Sutra*, which contains work on the theory of numbers, arithmetical operations, geometry, operations with fractions, simple equations, cubic equations, quartic equations, and permutations and combinations.

- c. 150 BC — Perseus (geometer)

- 150 BC — A method of Gaussian elimination appears in the Chinese text *The Nine Chapters on the Mathematical Art*.

- 150 BC — Horner's method appears in the Chinese text *The Nine Chapters on the Mathematical Art*.

- 150 BC — Negative numbers appear in the Chinese text *The Nine Chapters on the Mathematical Art*.

- 150 BC – 75 BC Zeno of Sidon

- 190 BC – 120 BC — Hipparchus develops the bases of trigonometry.

- 190 BC –120 BC Hypsicles

- 160 BC – 100 BC Theodosius of Bithynia

- 135 BC – 51 BC Posidonius

- 78 BC – 37 BC Jing Fang

- 50 BC — Indian numerals, a descendant of the Brahmi numerals (the first positional notation base-10 numeral system), begins development in India.

- mid 1st century Cleomedes (as late as 400 AD)

- final centuries BC — Indian astronomer Lagadha writes the *Vedanga Jyotisha*, a Vedic text on astronomy that describes rules for tracking the motions of the sun and the moon, and uses geometry and trigonometry for astronomy.

- 1st C. BC Geminus

16.2.2   1st millennium AD

- 1st century — Heron of Alexandria, (Hero) the earliest fleeting reference to square roots of negative numbers.

- c 100 Theon of Smyrna

- 60 – 120 Nicomachus

- 70 – 140 Menelaus of Alexandria Spherical trigonometry

- 78 – 139 Zhang Heng

- 132 – 192 Cai Yong

- c. 3rd century — Ptolemy of Alexandria wrote the *Almagest*.

- 240 – 300 Sporus of Nicaea

- 250 — Diophantus uses symbols for unknown numbers in terms of syncopated algebra, and writes *Arithmetica*, one of the earliest treatises on algebra.

- 263 — Liu Hui computes π using Liu Hui's π algorithm.

- 300 — the earliest known use of zero as a decimal digit is introduced by Indian mathematicians.

- 234 – 305 Porphyry (philosopher)

- 300 – 360 Serenus of Antinouplis

- 300 to 500 — the Chinese remainder theorem is developed by Sun Tzu.

- 300 to 500 — a description of rod calculus is written by Sun Tzu.

- 335 – 405 Theon of Alexandria

- c. 340 — Pappus of Alexandria states his hexagon theorem and his centroid theorem.

- 350 – 415 Hypatia

- c. 400 — the Bakhshali manuscript is written by Jaina mathematicians, which describes a theory of the infinite containing different levels of infinity, shows an understanding of indices, as well as logarithms to base 2, and computes square roots of numbers as large as a million correct to at least 11 decimal places.

- 412 – 485 Proclus

- 420 – 480 Domninus of Larissa

- b 440 Marinus of Neapolis "I wish everything was mathematics."

- 450 — Zu Chongzhi computes π to seven decimal places. This calculation remains the most accurate calculation for π for close to a thousand years.

- c. 474 – 558 Anthemius of Tralles

- 500 — Aryabhata writes the *Aryabhata-Siddhanta*, which first introduces the trigonometric functions and methods of calculating their approximate numerical values. It defines the concepts of sine and cosine, and also contains the earliest tables of sine and cosine values (in 3.75-degree intervals from 0 to 90 degrees).

- 480 – 540 Eutocius of Ascalon

- 490 – 560 Simplicius of Cilicia

- 6th century — Aryabhata gives accurate calculations for astronomical constants, such as the solar eclipse and lunar eclipse, computes π to four decimal places, and obtains whole number solutions to linear equations by a method equivalent to the modern method.

- 6th century – Yativṛṣabha

- 535 – 566 Zhen Luan

- 550 — Hindu mathematicians give zero a numeral representation in the positional notation Indian numeral system.

- 7th century — Bhaskara I gives a rational approximation of the sine function.

- 7th century — Brahmagupta invents the method of solving indeterminate equations of the second degree and is the first to use algebra to solve astronomical problems. He also develops methods for calculations of the motions and places of various planets, their rising and setting, conjunctions, and the calculation of eclipses of the sun and the moon.

- 628 — Brahmagupta writes the *Brahma-sphuta-siddhanta*, where zero is clearly explained, and where the modern place-value Indian numeral system is fully developed. It also gives rules for manipulating both negative and positive numbers, methods for computing square roots, methods of solving linear and quadratic equations, and rules for summing series, Brahmagupta's identity, and the Brahmagupta theorem.

- 8th century — Virasena gives explicit rules for the Fibonacci sequence, gives the derivation of the volume of a frustum using an infinite procedure, and also deals with the logarithm to base 2 and knows its laws.

- 8th century — Shridhara gives the rule for finding the volume of a sphere and also the formula for solving quadratic equations.

- 773 — Kanka brings Brahmagupta's Brahma-sphuta-siddhanta to Baghdad to explain the Indian system of arithmetic astronomy and the Indian numeral system.

- 773 — Al-Fazaii translates the Brahma-sphuta-siddhanta into Arabic upon the request of King Khalif Abbasid Al Mansoor.

- 9th century — Govindsvamin discovers the Newton-Gauss interpolation formula, and gives the fractional parts of Aryabhata's tabular sines.

- 810 — The House of Wisdom is built in Baghdad for the translation of Greek and Sanskrit mathematical works into Arabic.

- 820 — Al-Khwarizmi — Persian mathematician, father of algebra, writes the *Al-Jabr*, later transliterated as *Algebra*, which introduces systematic algebraic techniques for solving linear and quadratic equations. Translations of his book on arithmetic will introduce the Hindu-Arabic decimal number system to the Western world in the 12th century. The term *algorithm* is also named after him.

- 820 — Al-Mahani conceived the idea of reducing geometrical problems such as doubling the cube to problems in algebra.

- c. 850 — Al-Kindi pioneers cryptanalysis and frequency analysis in his book on cryptography.

- c. 850 — Mahāvīra writes the Gaṇitasārasaṅgraha otherwise known as the Ganita Sara Samgraha which gives systematic rules for expressing a fraction as the sum of unit fractions.

- 895 — Thabit ibn Qurra: the only surviving fragment of his original work contains a chapter on the solution and properties of cubic equations. He also generalized the Pythagorean theorem, and discovered the theorem by which pairs of amicable numbers can be found, (i.e., two numbers such that each is the sum of the proper divisors of the other).

- c. 900 — Abu Kamil of Egypt had begun to understand what we would write in symbols as $x^n \cdot x^m = x^{m+n}$

- 940 — Abu'l-Wafa al-Buzjani extracts roots using the Indian numeral system.

- 953 — The arithmetic of the Hindu-Arabic numeral system at first required the use of a dust board (a sort of handheld blackboard) because "the methods required moving the numbers around in the calculation and rubbing some out as the calculation proceeded." Al-Uqlidisi modified these methods for pen and paper use. Eventually the advances enabled by the decimal system led to its standard use throughout the region and the world.

- 953 — Al-Karaji is the "first person to completely free algebra from geometrical operations and to replace them with the arithmetical type of operations which are at the core of algebra today. He was first to define the monomials x, x^2, x^3, ... and $1/x$, $1/x^2$, $1/x^3$, ... and to give rules for products of any two of these. He started a school of algebra which flourished for several hundreds of years". He also discovered the binomial theorem for integer exponents, which "was a major factor in the development of numerical analysis based on the decimal system".

- 975 — Al-Batani extended the Indian concepts of sine and cosine to other trigonometrical ratios, like tangent, secant and their inverse functions. Derived the formulae: $\sin\alpha = \tan\alpha/\sqrt{1 + \tan^2\alpha}$ and $\cos\alpha = 1/\sqrt{1 + \tan^2\alpha}$

16.3 Symbolic stage

16.3.1 1000–1500

- c. 1000 — Abū Sahl al-Qūhī (Kuhi) solves equations higher than the second degree.

- c. 1000 — Abu-Mahmud al-Khujandi first states a special case of Fermat's Last Theorem.

- c. 1000 — Law of sines is discovered by Muslim mathematicians, but it is uncertain who discovers it first between Abu-Mahmud al-Khujandi, Abu Nasr Mansur, and Abu al-Wafa.

- c. 1000 — Pope Sylvester II introduces the abacus using the Hindu-Arabic numeral system to Europe.

- 1000 — Al-Karaji writes a book containing the first known proofs by mathematical induction. He used it to prove the binomial theorem, Pascal's triangle, and the sum of integral cubes.[8] He was "the first who introduced the theory of algebraic calculus".[9]

- c. 1000 — Ibn Tahir al-Baghdadi studied a slight variant of Thabit ibn Qurra's theorem on amicable numbers, and he also made improvements on the decimal system.

- 1020 — Abul Wáfa gave the formula: $\sin(\alpha + \beta) = \sin \alpha \cos \beta + \sin \beta \cos \alpha$. Also discussed the quadrature of the parabola and the volume of the paraboloid.

- 1021 — Ibn al-Haytham formulated and solved Alhazen's problem geometrically.

- 1030 — Ali Ahmad Nasawi writes a treatise on the decimal and sexagesimal number systems. His arithmetic explains the division of fractions and the extraction of square and cubic roots (square root of 57,342; cubic root of 3, 652, 296) in an almost modern manner.[10]

- 1070 — Omar Khayyám begins to write *Treatise on Demonstration of Problems of Algebra* and classifies cubic equations.

- c. 1100 — Omar Khayyám "gave a complete classification of cubic equations with geometric solutions found by means of intersecting conic sections". He became the first to find general geometric solutions of cubic equations and laid the foundations for the development of analytic geometry and non-Euclidean geometry. He also extracted roots using the decimal system (Hindu-Arabic numeral system).

- 12th century — Indian numerals have been modified by Arab mathematicians to form the modern Hindu-Arabic numeral system (used universally in the modern world).

- 12th century — the Hindu-Arabic numeral system reaches Europe through the Arabs.

- 12th century — Bhaskara Acharya writes the Lilavati, which covers the topics of definitions, arithmetical terms, interest computation, arithmetical and geometrical progressions, plane geometry, solid geometry, the shadow of the gnomon, methods to solve indeterminate equations, and combinations.

- 12th century — Bhāskara II (Bhaskara Acharya) writes the *Bijaganita* (*Algebra*), which is the first text to recognize that a positive number has two square roots.

- 12th century — Bhaskara Acharya conceives differential calculus, and also develops Rolle's theorem, Pell's equation, a proof for the Pythagorean Theorem, proves that division by zero is infinity, computes π to 5 decimal places, and calculates the time taken for the earth to orbit the sun to 9 decimal places.

- 1130 — Al-Samawal gave a definition of algebra: "[it is concerned] with operating on unknowns using all the arithmetical tools, in the same way as the arithmetician operates on the known."[11]

- 1135 — Sharafeddin Tusi followed al-Khayyam's application of algebra to geometry, and wrote a treatise on cubic equations that "represents an essential contribution to another algebra which aimed to study curves by means of equations, thus inaugurating the beginning of algebraic geometry".[11]

- 1202 — Leonardo Fibonacci demonstrates the utility of Hindu-Arabic numerals in his Liber Abaci (*Book of the Abacus*).

- 1247 — Qin Jiushao publishes *Shùshū Jiǔzhāng* (*Mathematical Treatise in Nine Sections*).

- 1260 — Al-Farisi gave a new proof of Thabit ibn Qurra's theorem, introducing important new ideas concerning factorization and combinatorial methods. He also gave the pair of amicable numbers 17296 and 18416 that have also been joint attributed to Fermat as well as Thabit ibn Qurra.[12]

- c. 1250 — Nasir Al-Din Al-Tusi attempts to develop a form of non-Euclidean geometry.

- 1303 — Zhu Shijie publishes *Precious Mirror of the Four Elements*, which contains an ancient method of arranging binomial coefficients in a triangle.

- 14th century — Madhava is considered the father of mathematical analysis, who also worked on the power series for π and for sine and cosine functions, and along with other Kerala school mathematicians, founded the important concepts of calculus.

- 14th century — Parameshvara, a Kerala school mathematician, presents a series form of the sine function that is equivalent to its Taylor series expansion, states the mean value theorem of differential calculus, and is also the first mathematician to give the radius of circle with inscribed cyclic quadrilateral.

- 1400 — Madhava discovers the series expansion for the inverse-tangent function, the infinite series for arctan and sin, and many methods for calculating the circumference of the circle, and uses them to compute π correct to 11 decimal places.

- c. 1400 — Ghiyath al-Kashi "contributed to the development of decimal fractions not only for approximating algebraic numbers, but also for real numbers such as π. His contribution to decimal fractions is so major that for many years he was considered as their inventor. Although not the first to do so, al-Kashi gave an algorithm for calculating nth roots, which is a special case of the methods given many centuries later by [Paolo] Ruffini and [William George] Horner." He is also the first to use the decimal point notation in arithmetic and Arabic numerals. His works include *The Key of arithmetics, Discoveries in mathematics, The Decimal point*, and *The benefits of the zero*. The contents of the *Benefits of the Zero* are an introduction followed by five essays: "On whole number arithmetic", "On fractional arithmetic", "On astrology", "On areas", and "On finding the unknowns [unknown variables]". He also wrote the *Thesis on the sine and the chord* and *Thesis on finding the first degree sine*.

- 15th century — Ibn al-Banna and al-Qalasadi introduced symbolic notation for algebra and for mathematics in general.[11]

- 15th century — Nilakantha Somayaji, a Kerala school mathematician, writes the *Aryabhatiya Bhasya*, which contains work on infinite-series expansions, problems of algebra, and spherical geometry.

- 1424 — Ghiyath al-Kashi computes π to sixteen decimal places using inscribed and circumscribed polygons.

- 1427 — Al-Kashi completes *The Key to Arithmetic* containing work of great depth on decimal fractions. It applies arithmetical and algebraic methods to the solution of various problems, including several geometric ones.

- 1478 — An anonymous author writes the Treviso Arithmetic.

- 1494 — Luca Pacioli writes *Summa de arithmetica, geometria, proportioni et proportionalità*; introduces primitive symbolic algebra using "co" (cosa) for the unknown.

16.3.2 Modern

16th century

- 1501 — Nilakantha Somayaji writes the Tantrasamgraha.

- 1520 — Scipione dal Ferro develops a method for solving "depressed" cubic equations (cubic equations without an x^2 term), but does not publish.

- 1522 — Adam Ries explained the use of Arabic digits and their advantages over Roman numerals.

- 1535 — Niccolò Tartaglia independently develops a method for solving depressed cubic equations but also does not publish.

- 1539 — Gerolamo Cardano learns Tartaglia's method for solving depressed cubics and discovers a method for depressing cubics, thereby creating a method for solving all cubics.

- 1540 — Lodovico Ferrari solves the quartic equation.

- 1544 — Michael Stifel publishes *Arithmetica integra*.

- 1550 — Jyeshtadeva, a Kerala school mathematician, writes the *Yuktibhāṣā*, the world's first calculus text, which gives detailed derivations of many calculus theorems and formulae.

- 1572 — Rafael Bombelli writes *Algebra* teatrise and uses imaginary numbers to solve cubic equations.

- 1584 — Zhu Zaiyu calculates equal temperament.

- 1596 — Ludolf van Ceulen computes π to twenty decimal places using inscribed and circumscribed polygons.

17th century

- 1614 — John Napier discusses Napierian logarithms in *Mirifici Logarithmorum Canonis Descriptio*.

- 1617 — Henry Briggs discusses decimal logarithms in *Logarithmorum Chilias Prima*.

- 1618 — John Napier publishes the first references to *e* in a work on logarithms.

- 1619 — René Descartes discovers analytic geometry (Pierre de Fermat claimed that he also discovered it independently).

- 1619 — Johannes Kepler discovers two of the Kepler-Poinsot polyhedra.

- 1629 — Pierre de Fermat develops a rudimentary differential calculus.

- 1634 — Gilles de Roberval shows that the area under a cycloid is three times the area of its generating circle.

- 1636 — Muhammad Baqir Yazdi jointly discovered the pair of amicable numbers 9,363,584 and 9,437,056 along with Descartes (1636).[12]

- 1637 — Pierre de Fermat claims to have proven Fermat's Last Theorem in his copy of Diophantus' *Arithmetica*.

- 1637 — First use of the term imaginary number by René Descartes; it was meant to be derogatory.

- 1654 — Blaise Pascal and Pierre de Fermat create the theory of probability.

- 1655 — John Wallis writes *Arithmetica Infinitorum*.

- 1658 — Christopher Wren shows that the length of a cycloid is four times the diameter of its generating circle.

- 1665 — Isaac Newton works on the fundamental theorem of calculus and develops his version of infinitesimal calculus.

- 1668 — Nicholas Mercator and William Brouncker discover an infinite series for the logarithm while attempting to calculate the area under a hyperbolic segment.

- 1671 — James Gregory develops a series expansion for the inverse-tangent function (originally discovered by Madhava).

- 1671 — James Gregory is the first known person to mention Taylor's Theorem.

- 1673 — Gottfried Leibniz also develops his version of infinitesimal calculus.

- 1675 — Isaac Newton invents an algorithm for the computation of functional roots.

- 1680s – Gottfried Leibniz works on symbolic logic.

- 1691 — Gottfried Leibniz discovers the technique of separation of variables for ordinary differential equations.

- 1693 — Edmund Halley prepares the first mortality tables statistically relating death rate to age.

- 1696 — Guillaume de L'Hôpital states his rule for the computation of certain limits.

- 1696 — Jakob Bernoulli and Johann Bernoulli solve brachistochrone problem, the first result in the calculus of variations.

18th century

- 1706 — John Machin develops a quickly converging inverse-tangent series for π and computes π to 100 decimal places.

- 1708 — Seki Takakazu discovers Bernoulli numbers. Jacob Bernoulli whom the numbers are named after is believed to have independently discovered the numbers shortly after Takakazu.

- 1712 — Brook Taylor develops Taylor series.

- 1722 — Abraham de Moivre states de Moivre's formula connecting trigonometric functions and complex numbers.

- 1722 — Takebe Kenko introduces Richardson extrapolation.

- 1724 — Abraham De Moivre studies mortality statistics and the foundation of the theory of annuities in *Annuities on Lives*.

- 1730 — James Stirling publishes *The Differential Method*.

- 1733 — Giovanni Gerolamo Saccheri studies what geometry would be like if Euclid's fifth postulate were false.

- 1733 — Abraham de Moivre introduces the normal distribution to approximate the binomial distribution in probability.

- 1734 — Leonhard Euler introduces the integrating factor technique for solving first-order ordinary differential equations.

- 1735 — Leonhard Euler solves the Basel problem, relating an infinite series to π.

- 1736 — Leonhard Euler solves the problem of the Seven bridges of Königsberg, in effect creating graph theory.

- 1739 — Leonhard Euler solves the general homogeneous linear ordinary differential equation with constant coefficients.

- 1742 — Christian Goldbach conjectures that every even number greater than two can be expressed as the sum of two primes, now known as Goldbach's conjecture.

- 1748 — Maria Gaetana Agnesi discusses analysis in *Instituzioni Analitiche ad Uso della Gioventu Italiana*.

- 1761 — Thomas Bayes proves Bayes' theorem.

- 1761 — Johann Heinrich Lambert proves that π is irrational.

- 1762 — Joseph Louis Lagrange discovers the divergence theorem.

- 1789 — Jurij Vega improves Machin's formula and computes π to 140 decimal places.

- 1794 — Jurij Vega publishes *Thesaurus Logarithmorum Completus*.

- 1796 — Carl Friedrich Gauss proves that the regular 17-gon can be constructed using only a compass and straight-edge.

- 1796 — Adrien-Marie Legendre conjectures the prime number theorem.

- 1797 — Caspar Wessel associates vectors with complex numbers and studies complex number operations in geometrical terms.

- 1799 — Carl Friedrich Gauss proves the fundamental theorem of algebra (every polynomial equation has a solution among the complex numbers).

- 1799 — Paolo Ruffini partially proves the Abel–Ruffini theorem that quintic or higher equations cannot be solved by a general formula.

19th century

- 1801 — *Disquisitiones Arithmeticae*, Carl Friedrich Gauss's number theory treatise, is published in Latin.

- 1805 — Adrien-Marie Legendre introduces the method of least squares for fitting a curve to a given set of observations.

- 1806 — Louis Poinsot discovers the two remaining Kepler-Poinsot polyhedra.

- 1806 — Jean-Robert Argand publishes proof of the Fundamental theorem of algebra and the Argand diagram.

- 1807 — Joseph Fourier announces his discoveries about the trigonometric decomposition of functions.

- 1811 — Carl Friedrich Gauss discusses the meaning of integrals with complex limits and briefly examines the dependence of such integrals on the chosen path of integration.

- 1815 — Siméon Denis Poisson carries out integrations along paths in the complex plane.

- 1817 — Bernard Bolzano presents the intermediate value theorem—a continuous function that is negative at one point and positive at another point must be zero for at least one point in between.

- 1822 — Augustin-Louis Cauchy presents the Cauchy integral theorem for integration around the boundary of a rectangle in the complex plane.

- 1824 — Niels Henrik Abel partially proves the Abel–Ruffini theorem that the general quintic or higher equations cannot be solved by a general formula involving only arithmetical operations and roots.

- 1825 — Augustin-Louis Cauchy presents the Cauchy integral theorem for general integration paths—he assumes the function being integrated has a continuous derivative, and he introduces the theory of residues in complex analysis.

- 1825 — Peter Gustav Lejeune Dirichlet and Adrien-Marie Legendre prove Fermat's Last Theorem for $n = 5$.

- 1825 — André-Marie Ampère discovers Stokes' theorem.

- 1828 — George Green proves Green's theorem.

- 1829 — János Bolyai, Gauss, and Lobachevsky invent hyperbolic non-Euclidean geometry.

- 1831 — Mikhail Vasilievich Ostrogradsky rediscovers and gives the first proof of the divergence theorem earlier described by Lagrange, Gauss and Green.

- 1832 — Évariste Galois presents a general condition for the solvability of algebraic equations, thereby essentially founding group theory and Galois theory.

- 1832 — Lejeune Dirichlet proves Fermat's Last Theorem for $n = 14$.

- 1835 — Lejeune Dirichlet proves Dirichlet's theorem about prime numbers in arithmetical progressions.

- 1837 — Pierre Wantzel proves that doubling the cube and trisecting the angle are impossible with only a compass and straightedge, as well as the full completion of the problem of constructability of regular polygons.

- 1837 — Peter Gustav Lejeune Dirichlet develops Analytic number theory.

- 1841 — Karl Weierstrass discovers but does not publish the Laurent expansion theorem.

- 1843 — Pierre-Alphonse Laurent discovers and presents the Laurent expansion theorem.

- 1843 — William Hamilton discovers the calculus of quaternions and deduces that they are non-commutative.

- 1847 — George Boole formalizes symbolic logic in *The Mathematical Analysis of Logic*, defining what is now called Boolean algebra.

- 1849 — George Gabriel Stokes shows that solitary waves can arise from a combination of periodic waves.

- 1850 — Victor Alexandre Puiseux distinguishes between poles and branch points and introduces the concept of essential singular points.

- 1850 — George Gabriel Stokes rediscovers and proves Stokes' theorem.

- 1854 — Bernhard Riemann introduces Riemannian geometry.

- 1854 — Arthur Cayley shows that quaternions can be used to represent rotations in four-dimensional space.

- 1858 — August Ferdinand Möbius invents the Möbius strip.

- 1858 — Charles Hermite solves the general quintic equation by means of elliptic and modular functions.

- 1859 — Bernhard Riemann formulates the Riemann hypothesis, which has strong implications about the distribution of prime numbers.

- 1870 — Felix Klein constructs an analytic geometry for Lobachevski's geometry thereby establishing its self-consistency and the logical independence of Euclid's fifth postulate.

- 1872 — Richard Dedekind invents what is now called the Dedekind Cut for defining irrational numbers, and now used for defining surreal numbers.

- 1873 — Charles Hermite proves that e is transcendental.

- 1873 — Georg Frobenius presents his method for finding series solutions to linear differential equations with regular singular points.

- 1874 — Georg Cantor proves that the set of all real numbers is uncountably infinite but the set of all real algebraic numbers is countably infinite. His proof does not use his diagonal argument, which he published in 1891.

- 1882 — Ferdinand von Lindemann proves that π is transcendental and that therefore the circle cannot be squared with a compass and straightedge.

- 1882 — Felix Klein invents the Klein bottle.

- 1895 — Diederik Korteweg and Gustav de Vries derive the Korteweg–de Vries equation to describe the development of long solitary water waves in a canal of rectangular cross section.

- 1895 — Georg Cantor publishes a book about set theory containing the arithmetic of infinite cardinal numbers and the continuum hypothesis.

- 1896 — Jacques Hadamard and Charles Jean de la Vallée-Poussin independently prove the prime number theorem.

- 1896 — Hermann Minkowski presents *Geometry of numbers*.

- 1899 — Georg Cantor discovers a contradiction in his set theory.

- 1899 — David Hilbert presents a set of self-consistent geometric axioms in *Foundations of Geometry*.

- 1900 — David Hilbert states his list of 23 problems, which show where some further mathematical work is needed.

16.3.3 Contemporary

20th century

[13]

- 1900 — David Hilbert publishes Hilbert's problems, a list of unsolved problems

- 1901 — Élie Cartan develops the exterior derivative.

- 1903 — Carle David Tolmé Runge presents a fast Fourier transform algorithm

- 1903 — Edmund Georg Hermann Landau gives considerably simpler proof of the prime number theorem.

- 1908 — Ernst Zermelo axiomizes set theory, thus avoiding Cantor's contradictions.

- 1908 — Josip Plemelj solves the Riemann problem about the existence of a differential equation with a given monodromic group and uses Sokhotsky – Plemelj formulae.

- 1912 — Luitzen Egbertus Jan Brouwer presents the Brouwer fixed-point theorem.

- 1912 — Josip Plemelj publishes simplified proof for the Fermat's Last Theorem for exponent $n = 5$.

- 1919 — Viggo Brun defines Brun's constant B_2 for twin primes.

- 1928 — John von Neumann begins devising the principles of game theory and proves the minimax theorem.

- 1930 — Casimir Kuratowski shows that the three-cottage problem has no solution.

- 1930 — Alonzo Church introduces Lambda calculus.

- 1931 — Kurt Gödel proves his incompleteness theorem, which shows that every axiomatic system for mathematics is either incomplete or inconsistent.

- 1931 — Georges de Rham develops theorems in cohomology and characteristic classes.

- 1933 — Karol Borsuk and Stanislaw Ulam present the Borsuk–Ulam antipodal-point theorem.

- 1933 — Andrey Nikolaevich Kolmogorov publishes his book *Basic notions of the calculus of probability* (*Grundbegriffe der Wahrscheinlichkeitsrechnung*), which contains an axiomatization of probability based on measure theory.

- 1940 — Kurt Gödel shows that neither the continuum hypothesis nor the axiom of choice can be disproven from the standard axioms of set theory.

- 1942 — G.C. Danielson and Cornelius Lanczos develop a fast Fourier transform algorithm.

- 1943 — Kenneth Levenberg proposes a method for nonlinear least squares fitting.

- 1945 — Stephen Cole Kleene introduces realizability.

- 1945 — Saunders Mac Lane and Samuel Eilenberg start category theory.

- 1945 — Norman Steenrod and Samuel Eilenberg give the Eilenberg–Steenrod axioms for (co-)homology.

- 1946 — Jean Leray introduces the Spectral sequence.

- 1948 — John von Neumann mathematically studies self-reproducing machines.

- 1948 — Alan Turing introduces LU decomposition.

- 1949 — John Wrench and L.R. Smith compute π to 2,037 decimal places using ENIAC.

- 1949 — Claude Shannon develops notion of Information Theory.

- 1950 — Stanisław Ulam and John von Neumann present cellular automata dynamical systems.

- 1953 — Nicholas Metropolis introduces the idea of thermodynamic simulated annealing algorithms.

- 1955 — H. S. M. Coxeter et al. publish the complete list of uniform polyhedron.

- 1955 — Enrico Fermi, John Pasta, and Stanisław Ulam numerically study a nonlinear spring model of heat conduction and discover solitary wave type behavior.

- 1956 — Noam Chomsky describes an hierarchy of formal languages.

- 1958 — Alexander Grothendieck's proof of the Grothendieck–Riemann–Roch theorem is published.

- 1960 — C. A. R. Hoare invents the quicksort algorithm.

- 1960 — Irving S. Reed and Gustave Solomon present the Reed–Solomon error-correcting code.

- 1961 — Daniel Shanks and John Wrench compute π to 100,000 decimal places using an inverse-tangent identity and an IBM-7090 computer.

- 1962 — Donald Marquardt proposes the Levenberg–Marquardt nonlinear least squares fitting algorithm.

- 1963 — Paul Cohen uses his technique of forcing to show that neither the continuum hypothesis nor the axiom of choice can be proven from the standard axioms of set theory.

- 1963 — Martin Kruskal and Norman Zabusky analytically study the Fermi–Pasta–Ulam heat conduction problem in the continuum limit and find that the KdV equation governs this system.

- 1963 — meteorologist and mathematician Edward Norton Lorenz published solutions for a simplified mathematical model of atmospheric turbulence – generally known as chaotic behaviour and strange attractors or Lorenz Attractor – also the Butterfly Effect.

- 1965 — Iranian mathematician Lotfi Asker Zadeh founded fuzzy set theory as an extension of the classical notion of set and he founded the field of Fuzzy Mathematics.

- 1965 — Martin Kruskal and Norman Zabusky numerically study colliding solitary waves in plasmas and find that they do not disperse after collisions.

- 1965 — James Cooley and John Tukey present an influential fast Fourier transform algorithm.

- 1966 — E. J. Putzer presents two methods for computing the exponential of a matrix in terms of a polynomial in that matrix.

- 1966 — Abraham Robinson presents non-standard analysis.

- 1967 — Robert Langlands formulates the influential Langlands program of conjectures relating number theory and representation theory.

- 1968 — Michael Atiyah and Isadore Singer prove the Atiyah–Singer index theorem about the index of elliptic operators.

- 1973 — Lotfi Zadeh founded the field of fuzzy logic.

- 1975 — Benoît Mandelbrot publishes *Les objets fractals, forme, hasard et dimension.*

- 1976 — Kenneth Appel and Wolfgang Haken use a computer to prove the Four color theorem.

- 1981 — Richard Feynman gives an influential talk "Simulating Physics with Computers" (in 1980 Yuri Manin proposed the same idea about quantum computations in "Computable and Uncomputable" (in Russian)).

- 1983 — Gerd Faltings proves the Mordell conjecture and thereby shows that there are only finitely many whole number solutions for each exponent of Fermat's Last Theorem.

- 1983 — the classification of finite simple groups, a collaborative work involving some hundred mathematicians and spanning thirty years, is completed.

- 1985 — Louis de Branges de Bourcia proves the Bieberbach conjecture.

- 1987 — Yasumasa Kanada, David Bailey, Jonathan Borwein, and Peter Borwein use iterative modular equation approximations to elliptic integrals and a NEC SX-2 supercomputer to compute π to 134 million decimal places.

- 1991 — Alain Connes and John W. Lott develop non-commutative geometry.

- 1992 — David Deutsch and Richard Jozsa develop the Deutsch–Jozsa algorithm, one of the first examples of a quantum algorithm that is exponentially faster than any possible deterministic classical algorithm.

- 1994 — Andrew Wiles proves part of the Taniyama–Shimura conjecture and thereby proves Fermat's Last Theorem.

- 1994 — Peter Shor formulates Shor's algorithm, a quantum algorithm for integer factorization.

- 1998 — Thomas Callister Hales (almost certainly) proves the Kepler conjecture.

- 1999 — the full Taniyama–Shimura conjecture is proven.

- 2000 — the Clay Mathematics Institute proposes the seven Millennium Prize Problems of unsolved important classic mathematical questions.

21st century

- 2002 — Manindra Agrawal, Nitin Saxena, and Neeraj Kayal of IIT Kanpur present an unconditional deterministic polynomial time algorithm to determine whether a given number is prime (the AKS primality test).

- 2002 — Yasumasa Kanada, Y. Ushiro, Hisayasu Kuroda, Makoto Kudoh and a team of nine more compute π to 1241.1 billion digits using a Hitachi 64-node supercomputer.

- 2002 — Preda Mihăilescu proves Catalan's conjecture.

- 2003 — Grigori Perelman proves the Poincaré conjecture.

- 2007 — a team of researchers throughout North America and Europe used networks of computers to map E_8.[14]

- 2009 — Fundamental lemma (Langlands program) had been proved by Ngô Bảo Châu.[15]

- 2013 — Yitang Zhang proves the first finite bound on gaps between prime numbers.[16]

16.4 See also

- Mathematics portal

- history of mathematical notation

16.5   References

[1] Art Prehistory, Sean Henahan, January 10, 2002.

[2] How Menstruation Created Mathematics, Tacoma Community College, archive link

[3] "OLDEST Mathematical Object is in Swaziland". Retrieved March 15, 2015.

[4] "an old Mathematical Object". Retrieved March 15, 2015.

[5] "Egyptian Mathematical Papyri - Mathematicians of the African Diaspora". Retrieved March 15, 2015.

[6] Carl B. Boyer, *A History of Mathematics*, 2nd Ed.

[7] Corsi, Pietro; Weindling, Paul (1983). *Information sources in the history of science and medicine*. Butterworth Scientific. ISBN 9780408107648. Retrieved July 6, 2014.

[8] Victor J. Katz (1998). *History of Mathematics: An Introduction*, p. 255–259. Addison-Wesley. ISBN 0-321-01618-1.

[9] F. Woepcke (1853). *Extrait du Fakhri, traité d'Algèbre par Abou Bekr Mohammed Ben Alhacan Alkarkhi*. Paris.

[10] O'Connor, John J.; Robertson, Edmund F., "Abu l'Hasan Ali ibn Ahmad Al-Nasawi", *MacTutor History of Mathematics archive*, University of St Andrews.

[11] Arabic mathematics, *MacTutor History of Mathematics archive*, University of St Andrews, Scotland

[12] Various AP Lists and Statistics

[13] Paul Benacerraf and Hilary Putnam, Cambridge University Press, *Philosophy of Mathematics: Selected Readings, ISBN 0-521-29648-X*

[14] Elizabeth A. Thompson, MIT News Office, *Math research team maps E8* Mathematicians Map E8, Harminka, 2007-03-20

[15] Laumon, G.; Ngô, B. C. (2004), *Le lemme fondamental pour les groupes unitaires*, arXiv:math/0404454

[16] "UNH Mathematician's Proof Is Breakthrough Toward Centuries-Old Problem". University of New Hampshire. May 1, 2013. Retrieved May 20, 2013.

   • David Eugene Smith, 1929 and 1959, *A Source Book in Mathematics*, Dover Publications. ISBN 0-486-64690-4.

16.6   External links

   • O'Connor, John J.; Robertson, Edmund F., "A Mathematical Chronology", *MacTutor History of Mathematics archive*, University of St Andrews.

16.7 Text and image sources, contributors, and licenses

16.7.1 Text

- **History of mathematics** *Source:* https://en.wikipedia.org/wiki/History_of_mathematics?oldid=671154105 *Contributors:* AxelBoldt, Ap, Jz-cool, XJaM, William Avery, DavidLevinson, Hannes Hirzel, Hephaestos, Stevertigo, Edward, Michael Hardy, Wshun, Norm, Liftarn, Ixfd64, Dcljr, Looxix~enwiki, William M. Connolley, Djnjwd, H7asan, Rossami, Nikai, Charles Matthews, Timwi, Andrevan, Dino, Reddi, AWhiteC, Jitse Niesen, ThomasStrohmann~enwiki, Jake Nelson, Shizhao, Topbanana, Anupamsr, Donarreiskoffer, Robbot, Fredrik, Matt me, Low-ellian, Gandalf61, Mirv, MathMartin, Henrygb, Seano1, Ancheta Wis, Giftlite, Graeme Bartlett, Wizzy, Harp, Tom harrison, Peruvianllama, Utcursch, Knutux, Slowking Man, Antandrus, Beland, OverlordQ, JoJan, Piotrus, Kaldari, APH, M.e, Almit39, Sam Hocevar, M1ss1ontomars2k4, Barnaby dawson, ELApro, Grstain, PhotoBox, Ta bu shi da yu, Rich Farmbrough, Kristian Ovaska, Dbachmann, Paul August, Stereotek, Ben Standeven, El C, Rgdboer, Haxwell, Jon the Geek, Causa sui, Rockslave, Bobo192, John Vandenberg, Jung dalglish, Nk, Obradovic Goran, Stephen Bain, Jumbuck, Alansohn, ChrisUK, Gorodski, Interiot, Arthena, Sonuwe, Sligocki, Wtmitchell, Welsh spud, Fadereu, Jheald, Kusma, Redvers, Igorpak, Rgrig, Blaxthos, RyanGerbil10, Oleg Alexandrov, Natalya, Alexrudd, Roylee, Simetrical, Woohookitty, Linas, Mindmatrix, Shreevatsa, David Haslam, Merlinme, Ruud Koot, Mpatel, Eilthireach, Gisling, Mandarax, Raguks, Graham87, BD2412, Jshadias, Josh Parris, Sjö, Rjwilmsi, Koavf, MarSch, Salix alba, UriBudnik, Cheesy123456789, Ucucha, Mathbot, Nihiltres, Mark J, Glenn L, Windharp, DVdm, As-dfgl, Siddhant, YurikBot, Wavelength, Mushin, Angus Lepper, RobotE, Deeptrivia, RussBot, Tetzcatlipoca, Bhny, DanMS, Tsch81, Stephenb, Gaius Cornelius, Gustavb, NawlinWiki, Rick Norwood, Wiki alf, Dialectric, Welsh, Trovatore, JFD, Gadget850, Bota47, Igiffin, Wiqi55, Closedmouth, Reyk, Claygate, Scoutersig, JLaTondre, WIN, Katieh5584, Finell, Tom Morris, That Guy, From That Show!, Marquez~enwiki, Sardanaphalus, Attilios, Tadorne, JJL, SmackBot, RDBury, Selfworm, HistoryManiac, KnowledgeOfSelf, Zerida, SaxTeacher, Jagged 85, NickShaforostoff, Miljoshi, Bobzchemist, Canthusus, Edgar181, Flux.books, Srkris, Gilliam, Hmains, Skizzik, Chris the speller, TimBentley, Full Shunyata, Jayanta Sen, Silly rabbit, Akanemoto, DHN-bot~enwiki, Colonies Chris, E946, Can't sleep, clown will eat me, MyNameIsVlad, Jahiegel, Nick Levine, Kelvin Case, Vanished User 0001, Mhym, Khoikhoi, Nakon, TedE, MathStatWoman, Salt Yeung, Kleuske, Drunken Pirate, Mearnhardtfan, Lambiam, Kuru, John, Mmounties, Angrynight, Lazylaces, Minna Sora no Shita, Maziar fayaz, Goodnightmush, JLe-ander, Scetoaux, Aleenf1, Dicklyon, Mets501, JdH, Iridescent, Red 81, Newone, CapitalR, Bharatveer, Gil Gamesh, Courcelles, Shake-speareFan00, JForget, CRGreathouse, Geremia, Eiorgiomugini, CBM, JohnCD, Nczempin, NickW557, Myasuda, Jgtl2, Fair Deal, Travelbird, Flowerpotman, Skittleys, Doug Weller, M a s, Cantoral~enwiki, SteveMcCluskey, Zalgo, ATD~enwiki, Ayzmo, Thijs!bot, Epbr123, Ray-mond Feilner, LeeNapier, Missvain, Sturm55, Nick Number, Albert1ls, AntiVandalBot, Luna Santin, Seaphoto, QuiteUnusual, By Brahma, Neurolinguist, Dylan Lake, Danger, Storkk, P.L.A.R., Hermel, MikeLynch, JAnDbot, Kaobear, MER-C, Dsp13, Jeff560, Thenub314, YK Times, VoABot II, Hasek is the best, JNW, JamesBWatson, Schwarzbichler, Shablog, SwiftBot, Johnbibby, Boffob, Styrofoam1994, Vssun, DerHexer, JaGa, Szczepan1990, Gun Powder Ma, Robin S, MartinBot, Danny6777, Anaxial, R'n'B, CommonsDelinker, Pbroks13, Wiki Raja, J.delanoy, Pharaoh of the Wizards, Trusilver, Bongomatic, Fowler&fowler, Silverxxx, TheSeven, It Is Me Here, Bot-Schafter, Katalaveno, Balt-hazarduju, Krishnachandranvn, Iluvme95, Rosenknospe, SJP, Kansas Bear, Milogardner, DavidCBryant, Vanished user 39948282, Treisijs, Bonadea, PatriciaJH, Anthony.bib, Neo-Vortex, Resfacta, VolkovBot, Dongwenliang, Ali57, NikolaiLobachevsky, Zhou yi777, Philip True-man, Davehi1, Technopat, Anonymous Dissident, Ocolon, Ontoraul, Gigidygla, Martin451, JhsBot, Don4of4, Broadbot, Mohit 29892, AllGlo-ryToTheHypnotoad, Omcnew, LeaveSleaves, Cremepuff222, Mossgiantkiller, Thebof, Geometry guy, Gavin.collins, Kmhkmh, Billinghurst, Monty845, AlleborgoBot, PericlesofAthens, Katzmik, Thony C., Arjun024, SMC89, SieBot, Tresiden, Tiddly Tom, I is a person, Da Joe, Calabraxthis, Xelgen, Flyer22, Oysterguitarist, Guycalledryan, JSpung, Manssour, James07045, Yomomma121, StaticGull, Nic bor, Tel-lyphone, Denisarona, Francvs, Athenean, WikipedianMarlith, Faithlessthewonderboy, ClueBot, Hanine, Zachariel, The Thing That Should Not Be, Smithpith, Jan1nad, Neverquick, Switchcraft, Excirial, Jusdafax, Gtstricky, NuclearWarfare, Cenarium, BOTarate, Al-Andalusi, Djk3, PCHS-NJROTC, Katanada, SoxBot III, XLinkBot, Dthomsen8, DoctorHver, FernoKlump, Avoided, Mitch Ames, Nicolae Coman, Thatguyflint, D.M. from Ukraine, Addbot, Willking1979, Some jerk on the Internet, Jncraton, Fieldday-sunday, L29102004, CanadianLin-uxUser, Leszek Jańczuk, Download, Peter Damian (old), LinkFA-Bot, Ozob, CuteHappyBrute, Tassedethe, LarryJeff, Tide rolls, Teles, Gail, Andhrabhoja, Luckas-bot, Ptbotgourou, Fraggle81, Aasifaalamkhan, CinchBug, Reindra, AnomieBOT, DemocraticLuntz, Ciphers, Jim1138, IRP, Userresuuser, Algorithme, Bluerasberry, Citation bot, Fleaman5000, TettyNullus, The Firewall, Xqbot, Eric Yurken, Corpsegrinderyo, Tacoman108, DSisyphBot, Timmyshin, El Caro, Isheden, BookWormHR, GrouchoBot, Mcoupal, Omnipaedista, RibotBOT, Charvest, Jcmc-clurg, Ringspectrum, Shashank Rajput, 刻意, FrescoBot, Robert pokemon69, KnowledgeAndVision, Cannolis, Citation bot 1, Anthony on Stilts, Tkuvho, Pinethicket, Elockid, Jean de Parthenay, Bejinhan, Wangxiaochun, Sahilsharma0112009, Lzhaobin, Jschnur, Danwhite2010, Space-Flight89, Garald, Tim1357, Pollinosisss, Vrenator, علی ویکی, Sutilcareh, Fastilysock, Adi4094, Diamond912, The Utahraptor, Generalboss3, Kiko4564, EmausBot, John of Reading, WikitanvirBot, Immunize, Super48paul, Anshuman.jrt, Syncategoremata, Tommy2010, Wikipelli, K6ka, Futofma, Saibal11, Daonguyen95, Josve05a, WeijiBaikeBianji, Moorechen, Gz33, Ocaasi, Enragedxltaco, Staszek Lem, MonoAV, Uuրηηιրի, Donner60, Yasekin, Orange Suede Sofa, Matthewrbowker, א28bot, AnthonyMarkes, Petrb, ClueBot NG, Proz, Cwmhiraeth, Wcherowi, MelbourneStar, Satellizer, Kikichugirl, ScottSteiner, Widr, Helpful Pixie Bot, Thisthat2011, Lowercase sigmabot, Northamer-ica1000, Mistory, Razorbliss, R4impulse, Acseitz2, Silvrous, The Determinator, Vilehumanbeing, CitationCleanerBot, Dentalplanlisa, Harizo-toh9, Bobs123456789, Minsbot, Wannabemodel, RudolfRed, Pratyya Ghosh, Mdann52, ChrisGualtieri, Aldosci, Arcandam, Khazar2, Dexbot, Makecat-bot, Siberian Patriot, Jochen Burghardt, Vivek 123456, Weenerlover78, Faizan, Red-eyed demon, Jamesmcmahon0, Jaden uk, Tenti-nator, Gtjw, Babitaarora, NeapleBerlina, IQ125, Sol1, Ugog Nizdast, Finnusertop, Giotach, Ginsuloft, DonaterOfPeace, Bryanrutherford0, A110019459, Anirudhankola, Monkbot, Ozdeger, Iseekyoubarney, Jayakumar RG, Mohammedx2000, ZippyPrune75, Eurodyne, DarkEn-thusiast, Eagleflo, Joyce Waddell, Joshuac3765, For12for11aa, CassandraNicolle, Ronniejuice and Anonymous: 701

- **History of algebra** *Source:* https://en.wikipedia.org/wiki/History_of_algebra?oldid=669378437 *Contributors:* AxelBoldt, The Anome, Heron, Michael Hardy, Reddi, Altenmann, Tonymaric, JoJan, Mike Rosoft, Discospinster, Rich Farmbrough, Bender235, ESkog, Rgdboer, Kwamik-agami, Bobo192, Djlewis, Arthena, Apoc2400, Korrigan, Woohookitty, Mindmatrix, Mandarax, BD2412, Rjwilmsi, Koavf, Bob A, Salix alba, ScottJ, Astrlag, DVdm, Bgwhite, Siddhant, Mukkakukaku, RussBot, Michael Slone, Piet Delport, Stephenb, Rsrikanth05, Anomalocaris, Nawl-inWiki, Dialectric, Grafen, Hv, Cjfsyntropy, NeilN, Bill, SmackBot, Selfworm, KnowledgeOfSelf, Jagged 85, Yamaguchi先生, Gilliam, Hmains, MaxSem, Lambiam, Aleenf1, Ben Moore, JHunterJ, Chris55, CmdrObot, CBM, BeenAroundAWhile, AndrewHowse, Hebrides, Alaibot, Epbr123, Missvain, Dsp13, PhilKnight, LittleOldMe, Bongwarrior, VoABot II, KConWiki, 28421u2232nfenfcenc, JCraw, CommonsDelinker, J.delanoy, Lantonov, DarwinPeacock, Milogardner, DavidCBryant, PatriciaJH, Deor, ABF, Jeff G., Philip Trueman, Kww, Caran Varr, Geometry guy, Falcon8765, Arcfrk, WereSpielChequers, Yintan, S711, Chhandama, JetLover, Weston.pace, JackSchmidt, Nic bor, JL-Bot, Forest Ash, The sunder king, Martarius, De728631, ClueBot, Uncle Milty, Niceguyedc, Excirial, Stonecolblast, Kakofonous, WikiHead, Willking1979, Onmywaybackhome, CanadianLinuxUser, LarryJeff, Gail, Franx97, Thebiggnome, Quantumobserver, Genius101, TheSuave, Yobot, Boulevardier, AnomieBOT, Floquenbeam, Jim1138, Danielt998, LilHelpa, Maddie!, Prunesqualer, Guitarist1897, A.amitkumar, Prari, Personwhoshouldbelistenedto, I dream of horses, Rotorcowboy, Garald, Jujutacular, Tamme babes, JLincoln, King Puffington, WillNess, RjwilmsiBot, Generalboss3, John of Reading, IncognitoErgoSum, GoingBatty, Jsiddall1, Josve05a, D.Lazard, Cymru.lass, Wayne Slam, IGeMiNix, NTox, Rocksajal, ClueBot NG, Wcherowi, Wsws69, Lincoln Josh, Helpful Pixie Bot, KLBot2, Jsiddall12345, BG19bot, Northamerica1000, MusikAnimal, Frze, Davidiad, Rm1271, Oehamilton, Jsiddall123, Killer135, Hmainsbot1, Uche911, Yeahsmart, Lugia2453, Frosty, Eyesnore, Bg9989, Jsiddall987, Jsiddall852, Jianhui67, Crystallizedcarbon, Loraof, Vandan barot, Karthik6596, How0028 and Anonymous: 171

• **History of calculus** Source: https://en.wikipedia.org/wiki/History_of_calculus?oldid=670589683 Contributors: PhilipMW, Michael Hardy, Dougmerritt, Ixfd64, William M. Connolley, Poor Yorick, Jengod, Alex S, Charles Matthews, Stan Lioubomoudrov, Dysprosia, Jitse Niesen, Timc, Kwantus, Matt me, MathMartin, Giftlite, Dbenbenn, Gene Ward Smith, Curps, Waltpohl, Chowbok, Onco p53, Thincat, Pinnerup, PhotoBox, Venu62, Mashford, Rgdboer, Elipongo, JW1805, Alansohn, Arthena, Chiapr, Kurivaim, Mark Williams, Notjim, RyanGerbil10, Oleg Alexandrov, Roylee, Mindmatrix, Shreevatsa, Bonus Onus, JeremyA, Macaddct1984, Mandarax, Magister Mathematicae, Pranathi, Rjwilmsi, Salix alba, Pyb, Mathbot, Srleffler, DVdm, Peter Grey, Mgstone, Deeptrivia, Witan, Gaius Cornelius, Alex Bakharev, Pseudomonas, NawlinWiki, Fabulous Creature, Welsh, Roy Brumback, Zzuuzz, Arthur Rubin, Katieh5584, Runescape rocks, Finell, A bit iffy, SmackBot, RDBury, Selfworm, Rose Garden, Prodego, Jagged 85, Thunderboltz, Srkris, Grokmoo, Famouslongago, RDBrown, Lordkazan, Silly rabbit, Nbarth, Colonies Chris, Leinad-Z, Evlekis, Lambiam, Ulner, Tim bates, Diverman, Jim.belk, RomanSpa, Loadmaster, Zeyad, Sywu, CRGreathouse, Stebulus, Myasuda, Noah Tye, Yaris678, MC10, Tawkerbot4, M a s, JamesAM, Epbr123, CheesemonkeyFrenchperson, Missvain, RobHar, EdJohnston, D.H, Quintote, MER-C, Thenub314, Dougwarrior, Charlesreid1, Adam4445, MartinBot, Bradgib, R'n'B, J.delanoy, Pharaoh of the Wizards, Hippasus, Fowler&fowler, Krishnachandranvn, Chikinsawsage, Belovedfreak, Jamesontai, VolkovBot, Jimmaths, Expitheta, Technopat, Dmcq, Arcfrk, PericlesofAthens, Katzmik, GirasoleDE, StAnselm, Likebox, Paolo.dL, Weston.pace, Dear Reader, Athenean, The Thing That Should Not Be, Smithpith, Mild Bill Hiccup, Niceguyedc, LizardJr8, Oxnard27, Excirial, Sarakoth, Sentriclecub, SchreiberBike, 7, Palnot, Tarheel95, WikiDao, Gingex85, King Pickle, Addbot, Tweeter and the monkey man, Lynx8, Hatashe, Ozob, Tassedethe, דוד שי, Jarble, Andhrabhoja, Freaky, Kilom691, AnomieBOT, Jim1138, Ularevalo98, Citation bot, LovesMacs, Capricorn42, FrescoBot, Sławomir Biały, Citation bot 1, Tkuvho, Pinethicket, 10metreh, Jschnur, Rohitphy, Reach Out to the Truth, Generalboss3, Chenna001, Wikipelli, Slawekb, Érico Júnior Wouters, Jay-Sebastos, Donner60, ClueBot NG, Widr, Rurik the Varangian, Exadrid, Helpful Pixie Bot, Sceptic1954, CityOfSilver, Rm1271, Drift chambers, CitationCleanerBot, Harizotoh9, Lalalaki, Dwade3strahan9, Acc60, The Illusive Man, GoShow, Caleb8clark, Jbeyerl, Gladtobeherenow, Keediethe, Wiki dude2003, KH-1, Starke Hathaway, Jfkdkdfksldk and Anonymous: 183

• **History of combinatorics** Source: https://en.wikipedia.org/wiki/History_of_combinatorics?oldid=663549870 Contributors: Altenmann, Igorpak, Deeptrivia, SmackBot, Colonies Chris, Mhym, Lenoxus, Headbomb, KConWiki, David Eppstein, Indeed123, DarwinPeacock, Kww, Fratrep, J8079s, Addbot, Gtgith, Luckas-bot, Yobot, Joaquin008, Citation bot 1, PigFlu Oink, Trappist the monk, Generalboss3, Aryeolman, Qniemiec, H3llBot, Joel B. Lewis, Helpful Pixie Bot, Brad7777, BattyBot, Monkbot, Loraof and Anonymous: 4

• **History of geometry** Source: https://en.wikipedia.org/wiki/History_of_geometry?oldid=666501204 Contributors: AxelBoldt, LC~enwiki, Brion VIBBER, Zundark, Taw, Manning Bartlett, Eclecticology, Graham Chapman, Youssefsan, XJaM, SimonP, Camembert, Netesq, Quintessent, Michael Hardy, Ixfd64, Fruge~enwiki, TakuyaMurata, Egil, Ellywa, Ahoerstemeier, Fuck You, Glenn, BenKovitz, Andres, Kaihsu, Mxn, Raven in Orbit, Revolver, Charles Matthews, Dino, Joshuabowman, Dysprosia, Phys, Shizhao, Mackensen, Francs2000, Robbot, Altenmann, Dittaeva, Romanm, MathMartin, Lsy098~enwiki, Aetheling, Fuelbottle, Guy Peters, Jleedev, Dina, Tosha, Giftlite, Lethe, Tom harrison, Guanaco, Bobblewik, Gubbubu, Bact, CryptoDerk, Knutux, LucasVB, APH, Cederal, Tomruen, Joyous!, Kate, Grstain, Ta bu shi da yu, Freakofnurture, Discospinster, ElTyrant, Rich Farmbrough, Martinl~enwiki, Spundun, Mani1, Paul August, Brian0918, El C, RoyBoy, Rpresser, Bobo192, Smalljim, BrokenSegue, Maurreen, SpeedyGonsales, Msh210, Gary, Interiot, Arthena, ABCD, Denniss, Angelic Wraith, Gunter, Agutie, MIT Trekkie, Falcorian, Oleg Alexandrov, Dejvid, Stemonitis, Thryduulf, Simetrical, Woohookitty, Linas, Henrik, Mindmatrix, Shreevatsa, Georgia guy, Kokoriko, Kzollman, Jimbryho, Ruud Koot, Jeff3000, Palica, Graham87, Rjwilmsi, Mayumashu, NatusRoma, Linuxbeak, JHMM13, Salix alba, FlaBot, Windchaser, Mathbot, Crazycomputers, RexNL, Sbrools, Lilyth, Sharkface217, Siddhant, YurikBot, Wavelength, RobotE, Deeptrivia, Kafziel, Michael Slone, Aftermath, Retired username, Larry laptop, Nate1481, Happydrifter, Someones life, LifeStar, Googl, Deville, Lt-wiki-bot, Chase me ladies, I'm the Cavalry, Cullinane, Claygate, GraemeL, DVD R W, Marquez~enwiki, Sardanaphalus, SmackBot, Radak, Type O Spud, Hydrogen Iodide, Zerida, Melchoir, Lagalag, Jagged 85, PJM, Frymaster, Gilliam, Brotherbobby, Hmains, Oscarthecat, BirdValiant, Chris the speller, TimBentley, MartinPoulter, Ikiroid, Zinneke, Timothy Clemans, Leinad-Z, Tamfang, Vanished User 0001, Zen611, Lcarscad, Hgilbert, Pilotguy, Dogears, John Reid, ArglebargleIV, Bcasterline, Special-T, LaMenta3, Nehrams2020, Newone, RekishiEJ, Tawkerbot2, Mordan 00, CRGreathouse, Wafulz, Philiprbrenan, GHe, McVities, Lazulilasher, Myasuda, Gregbard, BlueAg09, Scott14, Christian75, JodyB, LeeG, AbcXyz, Curlytop999, Escarbot, AntiVandalBot, Modernist, LibLord, Figma, Gökhan, MER-C, ElkinkF, The Transhumanist, Raanoo, Wisnuops, Bongwarrior, JamesBWatson, SineWave, SwiftBot, KConWiki, Felliax08, DerHexer, Wayne Miller, Geometricarts, Gonzo2008, R'n'B, CommonsDelinker, Nono64, Uncle Dick, Lantonov, Aervanath, Tatrgel, Useight, Dongwenliang, Philip Trueman, Hqb, Olly150, Oxfordwang, Martin451, Sniperz11, Jackfork, LeaveSleaves, Noformation, Enviroboy, PericlesofAthens, Phe-bot, Legion fi, Alexsmail, Yintan, Wilson44691, Gja34, Denisarona, 3rdAlcove, Vmoraru, Athenean, ClueBot, Dustman15, J8079s, Niceguyedc, SamuelTheGhost, DragonBot, Jusdafax, Redroe91, Leonard^Bloom, Glacier Wolf, Rror, Thatguyflint, Mojska, Proofreader77, Non-dropframe, Ka Faraq Gatri, MrOllie, Download, Legobot, Luckas-bot, Yobot, Fraggle81, HairyPerry, BlazerKnight, Citation bot, Maxis ftw, Neurolysis, Xqbot, DSisyphBot, Xashaiar, Omnipaedista, Prunesqualer, HamburgerRadio, Citation bot 1, Tkuvho, SpaceFlight89, Riccardo.fabris, Trappist the monk, Lotje, DARTH SIDIOUS 2, Generalboss3, Slightsmile, Wikipelli, Fæ, Josve05a, TyA, Ego White Tray, ChuispastonBot, ClueBot NG, Gilderien, Shayfalador, O.Koslowski, Helpful Pixie Bot, Thisthat2011, MusikAnimal, Mifter Public, Artmartxx, Harizotoh9, Brad7777, Anbu121, ChrisGualtieri, Hmainsbot1, SFK2, Dicer422, Faizan, Melonkelon, Frozlander and Anonymous: 296

• **History of logic** Source: https://en.wikipedia.org/wiki/History_of_logic?oldid=671167966 Contributors: LC~enwiki, Michael Hardy, Gabbe, Stephen C. Carlson, William M. Connolley, Pratyeka, Denny, Renamed user 4, Reddi, Stone, Markhurd, Tpbradbury, Goethean, Tobias Bergemann, Giftlite, Everyking, 20040302, Siroxo, Junuxx, Kusunose, CSTAR, Alsocal, Hbmartin, D6, Freakofnurture, JTN, Rich Farmbrough, Wclark, Paul August, Bender235, Chalst, Liberatus, Jguk 2, PWilkinson, Snowolf, Velho, Mel Etitis, Linas, Mindmatrix, BD2412, Rjwilmsi, Ucucha, Margosbot~enwiki, Tedder, YurikBot, Deeptrivia, Gaius Cornelius, Marklara, Aftermath, Grafen, MX44, Zagalejo, Tony1, Tomisti, Josh3580, Fram, Amalthea, SmackBot, Fanblade, Jagged 85, Mhss, Clconway, Henning Makholm, Clicketyclack, Ceoil, Ocanter, Bilby,

Beetstra, SandyGeorgia, Iridescent, Mrdthree, CBM, TheTito, Myasuda, Gregbard, Cydebot, Doug Weller, DumbBOT, Malleus Fatuorum, Itsmejudith, Nick Number, AntiVandalBot, Luna Santin, Heysan, Doc Tropics, Itistoday, Hermel, Matthew Fennell, Appraiser, Ling.Nut, Baccyak4H, Seberle, CCS81, R'n'B, CommonsDelinker, Nev1, Cpiral, DarwinPeacock, DadaNeem, KD Tries Again, DorganBot, Steel1943, Tesscass, Ontoraul, Clarince63, The Tetrast, Philogo, Logan, Newbyguesses, SieBot, Kumioko (renamed), Svick, Dabomb87, 3rdAlcove, Francvs, Emptymountains, Classicalecon, Athenean, ClueBot, Nsk92, Singinglemon~enwiki, JustinClarkCasey, Hans Adler, BOTarate, Pich-pich, Algebran, Addbot, Logicist, Ronhjones, Peter Damian (old), Luckas-bot, Yobot, Fraggle81, AnomieBOT, Eumolpo, ChildofMidnight, Why isn't anyone watching the history of the logic?, Anna Frodesiak, FrescoBot, Mfwitten, Citation bot 1, Tkuvho, Pinethicket, U8701, RedBot, Hriber, Generalboss3, Omkargokhale, Dewritech, Syncategoremata, RA0808, Here today, gone tomorrow, Riggr Mortis, HistorianofLogic, Soni Ruchi, Logic Historian, From the other side, Werieth, ZéroBot, H3llBot, Here for a bit, ClueBot NG, Wcherowi, Masssly, Widr, Australopithecus2, Helpful Pixie Bot, Calabe1992, DBigXray, BigEars42, CitationCleanerBot, Harizotoh9, Flosfa, Theconsequentialist, Qetuth, Ighso, Jochen Burghardt, Epicgenius, CsDix, Melody Lavender, Majo statt Senf, NABRASA, Oliveristhingone, Dick P Nuss, Monkbot, KH-1 and Anonymous: 109

- **History of mathematical notation** *Source:* https://en.wikipedia.org/wiki/History_of_mathematical_notation?oldid=670128356 *Contributors:* Michael Hardy, Voidvector, Reddi, Jitse Niesen, HarryHenryGebel, Fredrik, Altenmann, Aetheling, Tobias Bergemann, Giftlite, Beland, Piotrus, Porges, Rgdboer, Kazvorpal, Oleg Alexandrov, Firsfron, Mindmatrix, Ruud Koot, Mpatel, Paul Carpenter, Mandarax, BD2412, Rjwilmsi, Bgwhite, EamonnPKeane, Hairy Dude, RussBot, IanManka, Gaius Cornelius, Dialectric, Trovatore, Tony1, Dv82matt, Tyomitch, SmackBot, F, Reedy, Jagged 85, Hmains, Concerned cynic, Iain.dalton, Colonies Chris, Lambiam, Cesium 133, Agathoclea, Mets501, Dl2000, Nethac DIU, Ken Gallager, Myasuda, Gregbard, SteveMcCluskey, P.L.A.R., Deadbeef, Magioladitis, Faizhaider, Ling.Nut, KConWiki, First Harmonic, David Eppstein, R'n'B, Pomte, Being blunt, FruitMonkey, Plasticup, Sue H. Ping, Agricola44, Mohit 29892, Steve Checkoway, Tomaxer, Svick, Adam Cuerden, Ropata, Rumping, Prohlep, Mild Bill Hiccup, Rhododendrites, Iohannes Animosus, Nettings, D.M. from Ukraine, Bunich, Addbot, Morriswa, Zahd, Tassedethe, Ehrenkater, خالد حسني, Yobot, AnomieBOT, Omnipaedista, FrescoBot, Pillcrow, I dream of horses, Full-date unlinking bot, Lotje, John of Reading, WikitanvirBot, GoingBatty, Zaher kadour, ClueBot NG, Infinitehotel, Helpful Pixie Bot, BG19bot, DPL bot, Eonsko, DavidLeighEllis, Bryanrutherford0, Robevans123, Monkbot, Daytonian Historian and Anonymous: 34

- **Number theory** *Source:* https://en.wikipedia.org/wiki/Number_theory?oldid=669144459 *Contributors:* AxelBoldt, Derek Ross, Calypso, Brion VIBBER, Mav, Zundark, Tarquin, XJaM, Michael Shulman, Christian List, Miguel~enwiki, Twilsonb, Stevertigo, TeunSpaans, Michael Hardy, Booyabazooka, Ixfd64, Dcljr, GTBacchus, Delirium, Minesweeper, Ams80, Ahoerstemeier, Nikai, Rotem Dan, Iorsh, [212], Schneelocke, Hashar, Markb, Revolver, Charles Matthews, Timwi, Dcoetzee, Dysprosia, Jitse Niesen, The Anomebot, Xiaodai~enwiki, Tpbradbury, Sabbut, Jose Ramos, Qianfeng, Finlay McWalter, Bearcat, Robbot, Jaredwf, Fredrik, Romanm, Lowellian, Gandalf61, Fuelbottle, Lupo, PrimeFan, Jleedev, Ancheta Wis, Giftlite, Recentchanges, Pretzelpaws, Lethe, Fropuff, Everyking, Gubbubu, Gadfium, LiDaobing, Antandrus, Beland, Robert Brockway, Bob.v.R, Khaosworks, Karol Langner, APH, Stefan64, Tsemii, Xmlizer, Rich Farmbrough, Paul August, Bender235, ESkog, Ben Standeven, Appleboy, Tompw, El C, Jpgordon, Mysteronald, Maurreen, Hagerman, Pearle, Storm Rider, Msh210, Alansohn, Arthena, Neonumbers, Diego Moya, Velella, SidP, Evil Monkey, CloudNine, Dirac1933, Igorpak, HenryLi, Oleg Alexandrov, Mcsee, Richard Arthur Norton (1958-), Linas, Unixer, Jimbryho, Ruud Koot, Wikiklrsc, GregorB, Dionyziz, Graham87, Dpv, Mayumashu, R.e.b., The wub, DoubleBlue, Sango123, Vuong Ngan Ha, FlaBot, Mathbot, Malhonen, Scythe33, Haonhien, Chobot, Digitalme, YurikBot, Wavelength, Lexi Marie, Lenthe, JabberWok, Stassats, Joth, Welsh, Ino5hiro, Ms2ger, Kompik, Lt-wiki-bot, Arthur Rubin, Willtron, GrinBot~enwiki, That Guy, From That Show!, Marquez~enwiki, Sardanaphalus, SmackBot, Mmernex, Rebollo fr~enwiki, McGeddon, Jagged 85, Rouenpucelle, AustinKnight, Hmains, Anastasios~enwiki, Chris the speller, Bluebot, ChuckHG, PrimeHunter, MalafayaBot, Spellchecker, DHN-bot~enwiki, Colonies Chris, Sct72, Modest Genius, Ianmacm, Lwassink, Bidabadi~enwiki, Andrew Dalby, SashatoBot, Lambiam, ArglebargleIV, UberCryxic, Evildictaitor, Tdudkowski, Jim.belk, Gary13579, Stwalkerster, Childzy, Mets501, Mathsci, Kripkenstein, Joseph Solis in Australia, LDH, Zero sharp, Igoldste, Courcelles, Tawkerbot2, Stifynsemons, CRGreathouse, CmdrObot, Mikeliuk, Sdorrance, Chrisahn, Ken Gallager, Myasuda, Kronecker, Doctormatt, Gogo Dodo, Karl-H, Thijs!bot, Epbr123, Atmd, O, Bhowmickr, Marek69, Woody, RobHar, Sherbrooke, Escarbot, Luna Santin, Guy Macon, Marquess, Qwerty Binary, Normanzhang, JAnDbot, MER-C, Hut 8.5, Magioladitis, JamesBWatson, Wlod, Usien6, Kroposky, Bubba hotep, Systemlover, NJR ZA, Kope, DerHexer, Khalid Mahmood, TheRanger, Vandermude, Glrx, R'n'B, J.delanoy, Trusilver, Numbo3, Maurice Carbonaro, Smeira, Tarotcards, Policron, CompuChip, Milogardner, CombFan, Treisijs, VolkovBot, JohnBlackburne, Philip Trueman, TXiKiBoT, Hotjava, Anna Lincoln, Plclark, Magmi, Pleaseee, Broadbot, Kmhkmh, Blurpeace, Joseph A. Spadaro, Symane, GirasoleDE, SieBot, Calliopejen1, GrooveDog, Iames, KoenDelaere, S2000magician, Amahoney, ClueBot, MeowMeow163, Justin W Smith, Paulsavala, Drmies, Mild Bill Hiccup, DragonBot, PixelBot, Bercant, Cenarium, Arjayay, Jotterbot, H.Marxen, Crowsnest, XLinkBot, Marc van Leeuwen, Killthesteel, Ajcheema, Virginia-American, Addbot, Betterusername, Ronhjones, SpillingBot, Download, Uncia, Ausefi1900, Feketekave, Bluebusy, TeH nOmInAtOr, Luckas-bot, Yobot, MinorProphet, Zhouhaigang, Xylune, AnomieBOT, Rubinbot, Royote, Materialscientist, Citation bot, Maxis ftw, ArthurBot, Gypsydave5, Xqbot, Smk65536, Anne Bauval, Tyrol5, Vaywatch, GrouchoBot, Omnipaedista, Charvest, Raulshc, Aprogrammer, Lexy-lou, Bekus, DanRawsthorne, FrescoBot, Imbalzanog, LucienBOT, Tobby72, BrideOfKripkenstein, Machine Elf 1735, Pinethicket, MarcelB612, Artorio, Garald, Jauhienij, Tim1357, FoxBot, Trappist the monk, Dinamik-bot, Vrenator, Reach Out to the Truth, Korepin, KurtSchwitters, Ccrazymann, EmausBot, Lollipopweare, Fly by Night, EleferenBot, Jmencisom, Wikipelli, Bethnim, ZéroBot, Knight1993, Midas02, D.Lazard, Ian Rastall, Bobdylan1234567, Donner60, Nobrook, Unga Khan, Anita5192, ClueBot NG, Wcherowi, Satellizer, Helpful Pixie Bot, PhnomPencil, JohnChrysostom, AvocatoBot, Xosé Antonio, Brad7777, Weierstrass1, Ducknish, JYBot, Dexbot, Deltahedron, Apdenum, Jamesx12345, Vanamonde93, Prem nath singh, Jodosma, SakeUPenn, Syferion, Programmingcaffeine, ProfessorMoriarty1811, Canto55, SoSivr, J Steed Huang, KasparBot and Anonymous: 285

- **History of statistics** *Source:* https://en.wikipedia.org/wiki/History_of_statistics?oldid=669473992 *Contributors:* Edward, Michael Hardy, Tomi, Alan Liefting, Giftlite, Bkonrad, Dbachmann, Alansohn, Avenue, Woohookitty, Btyner, Rjwilmsi, Jrtayloriv, Bgwhite, Gene.arboit, RussBot, SmackBot, YellowMonkey, Gilliam, Hmains, Betacommand, Nbarth, G716, Rijkbenik, Myasuda, Thijs!bot, Mojo Hand, Headbomb, JustAGal, Jeff560, KConWiki, Nono64, J.delanoy, TheSeven, SJP, DrMicro, Jimmaths, Bc789, Boromir123, Melcombe, The Thing That Should Not Be, Tyler, Qwfp, SF007, Addbot, Ehrenkater, Yobot, Azylber, Ayrton Prost, AnomieBOT, Materialscientist, Citation bot, Quebec99, LilHelpa, Paulkappelle, Samwb123, FrescoBot, Citation bot 1, Pinethicket, Kiefer.Wolfowitz, Stpasha, RedBot, Decstop, RichardFloyd, Dtheobald, BarrySchachter, DARTH SIDIOUS 2, RjwilmsiBot, EmausBot, John of Reading, RMGunton, ZéroBot, Hawkjo, Noodleki, Odysseus1479, ClueBot NG, Jsalvatier, Widr, Helpful Pixie Bot, HMSSolent, BG19bot, Solomon7968, Shaun, Khazar2, Illia Connell, Mogism, Marwa mokhtar, Jianhui67, Monkbot, Virion123, Ricketybridge, Scarlettail, Kreplach123, Hilopmip and Anonymous: 58

- **History of trigonometry** *Source:* https://en.wikipedia.org/wiki/History_of_trigonometry?oldid=669805022 *Contributors:* Heron, Stevertigo, Michael Hardy, Stevenj, Mark Foskey, Charles Matthews, Dcoetzee, Samsara, Robbot, Giftlite, Icairns, Qef, ObsessiveMathsFreak, Dbachmann, Bobo192, Alansohn, Arthena, Shreevatsa, Bkkbrad, Rjwilmsi, Maurog, Salvatore Ingala, Michael Slone, Dialectric, Thiseye, T, Closedmouth, KGasso, SmackBot, Selfworm, Jagged 85, Hmains, Chris the speller, Buermann, Freedom skies, Nbarth, JGXenite, Bill Shillito, Hoof Hearted, Lambiam, Anooshahpour, IronGargoyle, Ckatz, AdultSwim, Iridescent, RekishiEJ, A. Pichler, Conrad.Irwin, Ale jrb, CBM, Myasuda, Yaris678, Xanthoptica, SteveMcCluskey, PamD, Mojo Hand, Bethpage89, Kaobear, Dsp13, YK Times, JNW, Rivertorch, KConWiki, Bruce.Adcock, Szczepan1990, Mdsats, David J Wilson, J.delanoy, Hippasus, Fowler&fowler, Krishnachandranvn, Goingstuckey, Jackfork, Meters, PericlesofAthens, Flyer22, Athenean, ClueBot, Bulluwagger, Shatree, Arakunem, Frank 1729, SamuelTheGhost, Walrasiad, Jusdafax, Dekisugi, Al-Andalusi, Pichpich, D1ma5ad, Addbot, Dgarten5428, SpellingBot, Untitledmind72, Manorupa108, Barjeconiah, Yobot, Andreasmperu, Calle, AnomieBOT, Rjanag, IRP, Flewis, Citation bot, Quebec99, Xqbot, Eagleman205, FrescoBot, Citation bot 1, Tkuvho, Pinethicket, Dude1818, Sanjoydey33, AppuruPan, ZéroBot, GeorgeBarnick, ClueBot NG, Gareth Griffith-Jones, Wcherowi, SusikMkr, Jamison Lofthouse, Baronofbueno, Kasirbot, Widr, Oddbodz, Helpful Pixie Bot, Thisthat2011, Webmail.za, Sphereon, Brad7777, Throwaway1, Cw3ag, Khazar2, Dexbot, Urmin gala, Lugia2453, Frosty, Hair, Lchris314, Faizan, Harlem Baker Hughes, AIGuy1010, Ljkwl99, GeoNerdGuy, AomineDaiki27, Trollwarrior, Piledhighandeep, Retched515, Stink Ninja, Misterugly, For12for11aa and Anonymous: 139

- **Numeral system** *Source:* https://en.wikipedia.org/wiki/Numeral_system?oldid=670657073 *Contributors:* AxelBoldt, Uriyan, Bryan Derksen, Zundark, Hardpack, Eclecticology, XJaM, Tuomas Toivonen, Karl Palmen, Robertolyra, Patrick, Infrogmation, JohnOwens, Michael Hardy, Fred Bauder, Shellreef, SGBailey, Wapcaplet, Qaz, TakuyaMurata, Geoffrey~enwiki, Looxix~enwiki, Cyp, Stevan White, Julesd, Glenn, Bogdangiusca, Cyan, Tristanb, Evercat, Pizza Puzzle, Nikola Smolenski, Timwi, Dcoetzee, Paul Stansifer, Dysprosia, Markhurd, AndrewKepert, Wiwaxia, AnonMoos, Jerzy, Lumos3, Denelson83, Robbot, Jaredwf, Fredrik, Altenmann, Peak, Henrygb, Rorro, Adhemar, 75th Trombone, Mendalus~enwiki, Wereon, J.Rohrer, Jleedev, Pengo, Giftlite, DavidCary, Palapala, Nat Krause, Haeleth, Bfinn, Gus Polly, Rick Block, Eequor, Python eggs, SoWhy, LiDaobing, Sonjaaa, Ran, Noe, Beland, Ctachme, MacGyverMagic, Mzajac, Maximaximax, Icairns, Sam Hocevar, Joyous!, Pinnerup, Chmod007, Lacrimosus, EugeneZelenko, Bornintheguz, Noisy, Discospinster, 4pq1injbok, Rich Farmbrough, Dbachmann, Paul August, Goochelaar, Sunborn, DanP, Fgrosshans, JoanQ, Theinfo, Kwamikagami, Mark R Johnson, Spoon!, Rajah, Shorne, Celada, Raj2004, JYolkowski, Eric Kvaalen, Atlant, Mr Adequate, Jesset77, Andrewpmk, Sligocki, Stephan Leeds, Suruena, Gpvos, DV8 2XL, Kazvorpal, Oleg Alexandrov, Velho, Woohookitty, Blumpkin, Shreevatsa, Ruud Koot, Kmg90, Cbdorsett, Andrea.gf, Eras-mus, Daniel Lawrence, Paxsimius, Graham87, BD2412, Jshadias, Phoenix-forgotten, Koavf, Mo-Al, AySz88, Nihiltres, Garyvdm, SDaniel, Intgr, Viznut, Le Anh-Huy, Glenn L, Chobot, DVdm, EamonnPKeane, YurikBot, Deeptrivia, Jimp, RussBot, Lucinos~enwiki, Wiki alf, Grafen, R.e.s., Nascigl, Dogcow, Davilla~enwiki, Jpbowen, Bobak, Arthur Rubin, Mad Cat, SmackBot, Adam majewski, Incnis Mrsi, KnowledgeOfSelf, McGeddon, Unyoyega, Blue520, Trojo~enwiki, Jagged 85, Eskimbot, Gilliam, Buck Mulligan, Julian Diamond, Srulikbd~enwiki, Bduke, EncMstr, Octahedron80, A. B., Zachorious, Nick Levine, Ioscius, SundarBot, Krich, Jiddisch~enwiki, Dreadstar, Ruwanraj, Maelnuneb, Henning Makholm, Mlpkr, The undertow, EMU CPA, Easytoremember, SilkTork, Ezra Katz, JorisvS, Ckatz, The Man in Question, 16@r, Loadmaster, JHunterJ, Pjrm, Yugyug, Geekygator, Lenoxus, Blehfu, OMGsplosion, Tim Andrews, FilipeS, Cydebot, Grahamec, Vanished user vjhsduheuiui4t5hjri, Carifio24, M a s, Epbr123, Najro, SomeHuman, Escarbot, LachlanA, AntiVandalBot, WinBot, Vanjagenije, Myanw, JAnDbot, Britcom, Greensburger, Twanderson, Bongwarrior, VoABot II, Efini, Mbarbier, Twsx, Abednigo, David Eppstein, Patstuart, Billcito, IL-Kuma, Cowenby, Mm1972, STBot, Jim.henderson, Rettetast, Trusilver, Silverxxx, Milogardner, DavidCBryant, Nvram20, Pedalist, Trumpet7~enwiki, Gogobera, Deor, JohnBlackburne, LokiClock, Philip Trueman, Gavroche42, JhsBot, Geometry guy, Mightyeldude, Skarz, Ceranthor, Date20070309, Logan, EmxBot, SieBot, Diego Grez, Anchor Link Bot, ClueBot, The Thing That Should Not Be, Krajcsi, Mild Bill Hiccup, Turbojet, Quanstro, Dan 9111, LusciniaMegarhynchos, Aitias, Anoopan, DerBorg, Qwfp, DavidRabahy, XLinkBot, Avoided, Addbot, Rashaani, MrOllie, Favonian, Ehrenkater, Lightbot, Fryed-peach, Luckas-bot, OrgasGirl, Ptbotgourou, Fraggle81, TaBOTzerem, AnomieBOT, Galoubet, Materialscientist, LilHelpa, JimVC3, Prunesqualer, Mathonius, Aaron Kauppi, A.amitkumar, Thehelpfulbot, Machine Elf 1735, Pinethicket, Sfax.tn, Pianoplonkers, Beao, Lemmiwinks2, Idunius, Dinamik-bot, Felipito1.966, EmausBot, MrFawwaz, RenamedUser01302013, Miss Manzana, Werieth, ZéroBot, D.Lazard, Lilbry25, Donner60, Chewings72, Jokg, Rmashhadi, ClueBot NG, Gareth Griffith-Jones, Vacation9, FinFihlman, O.Koslowski, Widr, MerlIwBot, OAnimosity, HMSSolent, Hummidy Humphrey, Félix Wolf, Anbu121, BattyBot, 2jokster, YFdyh-bot, Perić Željko, Vishnu.makam, Kitiiy, Mathdata, Technical math, Aegyptopithecus, Nezamieh, Chiragbapat, Cr.marthi, Slapsal, StevenD99, Monkbot, Valgodeskom, Gmalaven1008, Rezena, KasparBot, Yo Yo Harp and Anonymous: 328

- **Kenneth O. May Prize** *Source:* https://en.wikipedia.org/wiki/Kenneth_O._May_Prize?oldid=643804578 *Contributors:* Ruud Koot, Mhym, WOSlinker, Addbot, Kaktus Kid, Bjornsm, Omnipaedista, Tkuvho, Plucas58, ZéroBot, Flying Fische, Solomon7968, پارسی باپور شا and Anonymous: 1

- **List of important publications in mathematics** *Source:* https://en.wikipedia.org/wiki/List_of_important_publications_in_mathematics?oldid=659846398 *Contributors:* Gareth Owen, Edward, Michael Hardy, Dcljr, TakuyaMurata, Angela, Smack, Charles Matthews, Dfeuer, Jitse Niesen, WhisperToMe, Markhurd, Altenmann, Gandalf61, MathMartin, Wile E. Heresiarch, Tobias Bergemann, Giftlite, Thorne, Fropuff, Duncharris, Alensha, Jason Quinn, Gzornenplatz, CryptoDerk, Kaldari, APH, Sam Hocevar, Kevyn, Spiffy sperry, Rich Farmbrough, Nparikh, Gadykozma, Mateo SA, Gauge, Rgdboer, Nickj, Kanzure, Maurreen, Ardric47, JohnyDog, Proteus71, Theaterfreak64, Sl, Oleg Alexandrov, Simetrical, Woohookitty, Linas, Shreevatsa, LOL, Nefertum17, Mpatel, Poetsoutback, Isnow, Gisling, Blisco, Adking80, Rjwilmsi, Salix alba, R.e.b., Bharathrangarajan, Mathbot, Who, Adandrews, Sodin, Masnevets, Bgwhite, JamesLee, Deeptrivia, RMcGuigan, PaulGarner, Paki.tv, Dtrebbien, Trovatore, Hv, Tony1, Pegship, Closedmouth, Kungfuadam, Sardanaphalus, SmackBot, Haza-w, Reedy, Zerida, SaxTeacher, Jagged 85, Thunderboltz, XudongGuan~enwiki, Hmains, Chris the speller, Bluebot, Bduke, Silly rabbit, Taxipom, Clconway, Nbarth, Berland, Leland McInnes, Jon Awbrey, Wybot, Henning Makholm, Tesseran, Lambiam, Xdamr, Syrcatbot, DVanDyck, Mets501, JdH, Colonel Warden, Madmath789, LadyofShalott, Yendor1958, CRGreathouse, Gyopi, TheTito, Myasuda, Gregbard, Equendil, M a s, Headbomb, RobHar, Nick Number, Marokwitz, Mathisreallycool, JenLouise, Sangwinc, Turgidson, Dsp13, Dream Focus, JamesBWatson, A3nm, David Eppstein, Grantsky, Nono64, McSly, C quest000, Krishnachandranvn, Burkhard.Plache, PMajer, Marcosaedro, PDFbot, Milowent, Howard Cleeves, Sethkills, GirasoleDE, Cmdr Clarke, Taemyr, Svick, Anchor Link Bot, Melcombe, Randy Kryn, Niceguyedc, StephanNaro, Mathieu ottawa, Addbot, DOI bot, Upeksharuvani, Humbugde, Uncia, Ozob, Yobot, Dizzyjosh, AnomieBOT, WebsterRiver, Citation bot, Bci2, Geregen2, LilHelpa, Xqbot, RJGray, Gilo1969, Omnipaedista, Tobby72, Citation bot 1, Kiefer.Wolfowitz, RobinK, Tcnuk, Shay Falador, DolphinL, Fly by Night, Slawekb, Pro translator, WeijiBaikeBianji, H3llBot, Quondum, RockMagnetist, ClueBot NG, Wcherowi, Snotbot, Moritz37, MerlIwBot, Helpful Pixie Bot, Curb Chain, BG19bot, Lockwoods, Northamerica1000, Alexjbest, Letsbefiends, Ybidzian, AustinBuchanan, K9re11, Monkbot, Federico Leva (BEIC), Lúcia D. Coelho and Anonymous: 67

- **Lists of mathematicians** *Source:* https://en.wikipedia.org/wiki/Lists_of_mathematicians?oldid=668002610 *Contributors:* AxelBoldt, Zundark, Amillar, Eclecticology, Kowloonese, Danny, XJaM, JeLuF, Toby Bartels, Deb, DavidLevinson, David spector, N8chz, Olivier, Chris Q, Renata, Tomo, Lir, Infrogmation, DrewT2, Michael Hardy, Oliver Pereira, Isomorphic, Dominus, Nixdorf, Liftarn, Lquilter, Sannse, TakuyaMurata, GTBacchus, Karada, Minesweeper, Looix~enwiki, Mbogelund, Nanshu, Docu, Muriel Gottrop~enwiki, Snoyes, Den fjättrade ankan~enwiki, Nanobug, Andrewa, Bogdangiusca, LouI, Cimon Avaro, EdH, Harry Potter, Mxn, Wfeidt, Epo~enwiki, Hike395, Loren Rosen, Bjcairns, Karl Schalike, EL Willy, Charles Matthews, Guaka, Tantalate, Timwi, Viz, RickK, NilsB, Dysprosia, Jitse Niesen, Maximus Rex, Imc, Fibonacci, Jose Ramos, Topbanana, Patrick McDonough, Secretlondon, Jerzy, Qertis, Carbuncle, Jni, Aleph4, Robbot, Paul Klenk, Dzhuo, Jaredwf, Fredrik, Psychonaut, Romanm, Big Jim Fae Scotland, Gandalf61, MathMartin, Rholton, Hemanshu, SchmuckyTheCat, Cautious, Seth Ilys, Cutler, Carnildo, Filemon, Agendum, Weialawaga~enwiki, Kevin Saff, Giftlite, Dbenbenn, Mikez, Pretzelpaws, Ævar Arnfjörð Bjarmason, Banach~enwiki, Lupin, MathKnight, Herbee, Dissident, Marcika, Fropuff, Ds13, Everyking, Curps, Waltpohl, Duncharris, DO'Neil, Steve-o~enwiki, Finn-Zoltan, Cambyses, JillandJack, Gzornenplatz, Dan Gardner, Pne, Bobblewik, Christopherlin, Gadfium, Woggly, Jonel, Alberto da Calvairate~enwiki, HorsePunchKid, MarkSweep, DNewhall, Gene s, Gauss, Pmanderson, Icairns, Almit39, Wyllium, Sam nead, Udzu, Flex, Xmlizer, Mormegil, Jayjg, Monkeyman, Mikolt, Noisy, ArnoldReinhold, MeltBanana, Crescent Moon, Paul August, Mwng, BACbKA, Bennylin, EmilJ, CeeGee, Hurricane111, Billymac00, Cje~enwiki, Hesperian, Mdd, Marc van Woerkom, Jumbuck, Msh210, Siim, Orzetto, Avenue, Bucephalus, Helixblue, Esprungo, Emvee~enwiki, Harej, Dan100, AndrejBauer, Oleg Alexandrov, Mindmatrix, Gruepig, Ruud Koot, Pixeltoo, Mpatel, MFH, GregorB, Eruionnyron, Btyner, Rachack, Mekong Bluesman, Avia, Hochnebel, Daderot, Mishuletz, Mathbot, Vince Vatter, Cpcheung, Paxwel, Roboto de Ajvol, Borgx, Encyclops, Gene.arboit, Lenthe, Trovatore, Anser~enwiki, Starze, CLW, Stefan Udrea, FF2010, VodkaJazz, Curpsbot-unicodify, Jonathan.s.kt, Ahonc, FocalPoint, Josh Pak, Melchoir, Londoneye, Vald, Jagged 85, Stifle, Geneb1955, Guido Kanschat, AbdullahOmar, Constanz, Aleksandar Šušnjar, Ewjw, Mhym, Calbaer, Cybercobra, G716, Yakovkronrod, TurboCat, Michael Bednarek, Green Giant, Highpriority, Syrcatbot, Mets501, JdH, A. Pichler, Dpinotsis, CmdrObot, Leujohn, Xaman, Ntsimp, Jennifer seberry, Omicronpersei8, Riojajar~enwiki, LaGrange, LeBofSportif, Egel, AntiVandalBot, Lamentation, Cripet, Tbernste, JAnDbot, Turgidson, Magioladitis, Apoc107, David Eppstein, Afil, Alro, Cafeturque, KylieTastic, Bogdan~enwiki, SieBot, Sanya3, ClueBot, Excirial, Jusdafax, Cenarium, Addbot, Uncia, LinkFA-Bot, David0811, TaBOT-zerem, DemocraticLuntz, Jim1138, RandomAct, Clark89, Xqbot, Melmann, Omnipaedista, Mathonius, Dougofborg, Mmm-kkk, GWST11, Calmer Waters, Foobarnix, DixonDBot, Oracleofottawa, Innotata, EmausBot, Solarra, EWikist, ClueBot NG, MerlIwBot, Hallows AG, South19, Klilidiplomus, Illia Connell, Dexbot, Kovácsjózsi, SoSivr, D-4597-aR, Amberlast12345 and Anonymous: 187

- **List of mathematics history topics** *Source:* https://en.wikipedia.org/wiki/List_of_mathematics_history_topics?oldid=642178601 *Contributors:* Michael Hardy, Charles Matthews, Gandalf61, Emax, D6, ZeroOne, Rgdboer, Dirac1933, Mathbot, Fplay, Syrcatbot, Yendor1958, PKT, Cenarium, Ozob, Tkuvho and Anonymous: 2

- **Timeline of mathematics** *Source:* https://en.wikipedia.org/wiki/Timeline_of_mathematics?oldid=669118522 *Contributors:* AxelBoldt, Derek Ross, Bryan Derksen, Zundark, The Anome, Gareth Owen, Andre Engels, XJaM, DavidLevinson, FvdP, Heron, Hephaestos, Mrwojo, Edward, Michael Hardy, Oliver Pereira, SGBailey, Minesweeper, Looix~enwiki, Ams80, Snoyes, Den fjättrade ankan~enwiki, Tim Retout, Nikai, Cimon Avaro, Schneelocke, EL Willy, Charles Matthews, Timwi, Dino, Reddi, Jitse Niesen, Dandrake, Sanxiyn, Greenrd, Chuunen Baka, Smallweed, Gandalf61, MathMartin, Aetheling, Seano1, Ancheta Wis, Takanoha, Giftlite, Tom harrison, Herbee, Joe Kress, Python eggs, Gadfium, Sigfpe, Rajasekaran Deepak, Beland, Salasks, Icairns, Arcturus, Todd Kloos, PhotoBox, D6, Discospinster, Qutezuce, Paul August, Ben Standeven, RoyBoy, Jung dalglish, Hesperian, Cburnett, Zereshk, Falcorian, Oleg Alexandrov, Roylee, Mindmatrix, Stolee, Nefertum17, Ruud Koot, Mpatel, Tabletop, Kelisi, Waldir, Mandarax, Graham87, Jshadias, Rjwilmsi, Susiesushi, Salix alba, Mike s, Krash, Maurog, FlaBot, KarlFrei, Ewlyahoocom, Shawn@garbett.org, Quuxplusone, DVdm, Gdrbot, Algebraist, Phantomsteve, Jpetry, Yamara, Pseudomonas, Anomalocaris, Arichnad, Trovatore, Apokryltaros, IvanDurak, Tony1, Redgolpe, JRawle, That Guy, From That Show!, Aforencich, SmackBot, Zerida, SaxTeacher, Pokipsy76, Jagged 85, Sloman, Hmains, Saros136, Schwallex, E946, Cícero, Brutha~enwiki, Jbergquist, Lambiam, Leoberacai, JoshuaZ, Onionmon, JeffW, Lottamiata, Newone, Klovelace, CRGreathouse, Suto, CBM, Myasuda, M a s, AntiVandalBot, Steelpillow, Ekabhishek, Kaobear, Kurrgo master of planet x, Matthew Komorowski, Hroðulf, VoABot II, David Eppstein, Felliax08, DerHexer, MartinBot, Danny6777, Dashnick, Krishnachandranvn, Jgoizueta, Anthony.bib, Sternkampf, Funandtrvl, Lights, Philip Trueman, Clarince63, Wloveral, Williamanthony, AlleborgoBot, Katzmik, GirasoleDE, SieBot, Macgyver89, Oxymoron83, Smaug123, Hobartimus, Chillum, Smithpith, Mild Bill Hiccup, J8079s, Niceguyedc, He7d3r, Abrech, Sun Creator, SchreiberBike, Addbot, Hgladney, Ozob, Ace45954, Pomimo~enwiki, Yobot, AnomieBOT, IRP, AdjustShift, Xqbot, RJGray, SassoBot, Pythagoras0, PigFlu Oink, Rushbugled13, Jujutacular, Trappist the monk, Lotje, Catinator, Tinman44, EmausBot, Energy Dome, John of Reading, Akerans, 1234r00t, Chewings72, Petrb, ClueBot NG, Proz, Catlemur, Helpful Pixie Bot, BG19bot, Northamerica1000, Odytso2, Brad7777, DanielHendrycks, Jojobill, Backendgaming, Singaporemathematician, Bryanrutherford0, K9re11, Paintherface and Anonymous: 104

16.7.2 Images

- **File:005-a-Ruby-kindles-in-the-vine-810x1146.jpg** *Source:* https://upload.wikimedia.org/wikipedia/commons/0/08/005-a-Ruby-kindles-in-the-vine-810x1146.jpg *License:* Public domain *Contributors:* The Rubaiyat of Omar Khayyam (1905, 1912)[1] *Original artist:* Adelaide Hanscom

- **File:10_DM_Serie4_Vorderseite.jpg** *Source:* https://upload.wikimedia.org/wikipedia/commons/0/0d/10_DM_Serie4_Vorderseite.jpg *License:* Public domain *Contributors:* http://www.bundesbank.de/Redaktion/DE/Standardartikel/Kerngeschaeftsfelder/Bargeld/dm_banknoten.html#doc18118bodyText2 *Original artist:* Deutsche Bundesbank, Frankfurt am Main, Germany

- **File:Academia_mosaic.jpg** *Source:* https://upload.wikimedia.org/wikipedia/commons/6/66/Academia_mosaic.jpg *License:* Public domain *Contributors:* http://www.stoa.org/diotima/icons/academia_mosaic.jpg *Original artist:* anonimous

- **File:AlfredTarski1968.jpeg** *Source:* https://upload.wikimedia.org/wikipedia/commons/7/71/AlfredTarski1968.jpeg *License:* GFDL *Contributors:* The Oberwolfach photo collection, http://owpdb.mfo.de/detail?photo_id=6091 *Original artist:* George M. Bergman

- **File:Ambox_globe_content.svg** *Source:* https://upload.wikimedia.org/wikipedia/commons/b/bd/Ambox_globe_content.svg *License:* Public domain *Contributors:* Own work, using File:Information icon3.svg and File:Earth clip art.svg *Original artist:* penubag

- **File:Anil_Kumar_Gain.png** *Source:* https://upload.wikimedia.org/wikipedia/en/d/d4/Anil_Kumar_Gain.png *License:* Public domain *Contributors:*

16.7.3   Content license

www.ingramcontent.com/pod-product-compliance
Lightning Source LLC
Chambersburg PA
CBHW080759180526
45168CB00006B/2265